SCIENTIFIC PRINCIPLES AND TECHNIQUE OF OPTICAL FABRICATION PROCESSES

SCIENTIFIC PRINCIPLES AND TECHNIQUE OF OPTICAL FABRICATION PROCESSES

Edited by:

Maria Emilova Velinova

ARCLER
PRESS

www.arclerpress.com

Scientific Principles and Technique of Optical Fabrication Processes

Maria Emilova Velinova

Arcler Press
2010 Winston Park Drive,
2nd Floor
Oakville, ON L6H 5R7
Canada
www.arclerpress.com
Tel: 001-289-291-7705
001-905-616-2116
Fax: 001-289-291-7601
Email: orders@arclereducation.com

ISBN: 978-1-77407-200-4 (Hardcover)

DECLARATION

Some content or chapters in this book are open access copyright free published research work, which is published under Creative Commons License and are indicated with the citation. We are thankful to the publishers and authors of the content and chapters as without them this book wouldn't have been possible.

ABOUT THE EDITOR

Maria Velinova is Ph.D. holder in Quantum chemistry at the University of Sofia since April 2012. Her major research experience is in the field of Computational Chemistry, especially in statistical mechanics methods applied to different sorts of biomolecules. Member of the Laboratory of Quantum and Computational Chemistry at the University of Sofia.

TABLE OF CONTENTS

LIST OF CONTRIBUTORS

Yan Zhang
Fitzpatrick Institute for Photonics, Duke University, Durham, NC 27708, USA
Department of Biomedical Engineering, Duke University, Durham, NC 27708, USA

Anuj Dhawan
Fitzpatrick Institute for Photonics, Duke University, Durham, NC 27708, USA
Department of Biomedical Engineering, Duke University, Durham, NC 27708, USA

Tuan Vo-Dinh
Fitzpatrick Institute for Photonics, Duke University, Durham, NC 27708, USA
Department of Biomedical Engineering, Duke University, Durham, NC 27708, USA
Department of Chemistry, Duke University, Durham, NC 27708, USA

John Ballato
The Center for Optical Materials Science and Engineering Technologies (COMSET) and the Department of Materials Science and Engineering, Clemson University, Clemson, SC 29634, USA

Peter Dragic
Department of Electrical and Computer Engineering, University of Illinois at Urbana-Champaign, Urbana, IL 61801, USA

Rita S. Rodrigues Ribeiro
INESC TEC, Rua do Campo Alegre, 687, Porto, Portugal and Departamento de Física e Astronomia, Faculdade de Ciências, Universidade do Porto, Rua Campo Alegre, 687, Porto, Portugal
Masdar Institute of Science and Technology, Abu Dhabi, United Arab Emirates

Pabitra Dahal
Masdar Institute of Science and Technology, Abu Dhabi, United Arab Emirates

Ariel Guerreiro
INESC TEC, Rua do Campo Alegre, 687, Porto, Portugal and Departamento de Física e Astronomia, Faculdade de Ciências, Universidade do Porto, Rua Campo Alegre, 687, Porto, Portugal

Pedro A. S. Jorge
INESC TEC, Rua do Campo Alegre, 687, Porto, Portugal and Departamento de Física e Astronomia, Faculdade de Ciências, Universidade do Porto, Rua Campo Alegre, 687, Porto, Portugal

Jaime Viegas
Masdar Institute of Science and Technology, Abu Dhabi, United Arab Emirates

Marcel Roeder
Hahn-Schickard, Allmandring 9b, 70569 Stuttgart, Germany
Institute for Micro Integration (IFM), University of Stuttgart, Allmandring 9 b, 70569 Stuttgart, Germany

Thomas Guenther
Institute for Micro Integration (IFM), University of Stuttgart, Allmandring 9 b, 70569 Stuttgart, Germany

André Zimmermann
Hahn-Schickard, Allmandring 9b, 70569 Stuttgart, Germany
Institute for Micro Integration (IFM), University of Stuttgart, Allmandring 9 b, 70569 Stuttgart, Germany

Fang Jiang
State Key Laboratory of High Performance Ceramics and Superfine Microstructure, Shanghai Institute of Ceramics, Chinese Academy of Sciences, Shanghai 200050, China
Graduate University of the Chinese Academy of Sciences, Beijing 100049, China

Yan Liu
State Key Laboratory of High Performance Ceramics and Superfine Microstructure, Shanghai Institute of Ceramics, Chinese Academy of Sciences, Shanghai 200050, China

Yong Yang
State Key Laboratory of High Performance Ceramics and Superfine Microstructure, Shanghai Institute of Ceramics, Chinese Academy of Sciences, Shanghai 200050, China

Zheng-Ren Huang
State Key Laboratory of High Performance Ceramics and Superfine Microstructure, Shanghai Institute of Ceramics, Chinese Academy of Sciences, Shanghai 200050, China

Dan Li
State Key Laboratory of High Performance Ceramics and Superfine Microstructure, Shanghai Institute of Ceramics, Chinese Academy of Sciences, Shanghai 200050, China
Graduate University of the Chinese Academy of Sciences, Beijing 100049, China

Gui-ling Liu
State Key Laboratory of High Performance Ceramics and Superfine Microstructure, Shanghai Institute of Ceramics, Chinese Academy of Sciences, Shanghai 200050, China

Xue-Jian Liu
State Key Laboratory of High Performance Ceramics and Superfine Microstructure, Shanghai Institute of Ceramics, Chinese Academy of Sciences, Shanghai 200050, China

Yuan Liu
School of Astronautics, Harbin Institute of Technology, Harbin 150001, China

La Han
Center for Precision Engineering, Harbin Institute of Technology, Harbin 150001, China

Haiying Liu
Xiaguang Optical Electron Co., Ltd, Yangzhou 225127, China

Yikai Shi
Science and Technology on Integrated Logistics Support Laboratory, National University of Defense Technology, Changsha 410073, China

Junjie Zhang
Center for Precision Engineering, Harbin Institute of Technology, Harbin 150001, China

Jiang Guo
Key Laboratory for Precision and Non-Traditional Machining Technology of Ministry of Education, Dalian University of Technology, Dalian 116024, China

Hirofumi Suzuki
Department of Mechanical Engineering, Chubu University, 1200 Matsumoto-cho, Kasugai, Aichi 487-8501, Japan

Jingfei Yin
Key Laboratory for Precision and Nontraditional Machining of Ministry of Education, Dalian University of Technology, Dalian 116024, China

Qian Bai
Key Laboratory for Precision and Nontraditional Machining of Ministry of Education, Dalian University of Technology, Dalian 116024, China

Bi Zhang
Key Laboratory for Precision and Nontraditional Machining of Ministry of Education, Dalian University of Technology, Dalian 116024, China

Depart ment of Mechanical and Energy Engineering, South University of Science and Technology, Shenzhen 518055, China

Georg Schnurbusch
Laboratory for Precision Machining (LFM), University of Bremen, Badgasteiner Str. 2, 28359 Bremen, Germany

Ekkard Brinksmeier
Foundation Institute of Materials Science, Badgasteiner Str. 3, 28359 Bremen, Germany

Oltmann Riemer
Laboratory for Precision Machining (LFM), University of Bremen, Badgasteiner Str. 2, 28359 Bremen, Germany

Yan Gu
School of Mechatronic Engineering, Changchun University of Technology, Changchun 130012, China

Wenhui Zhu
School of Mechatronic Engineering, Changchun University of Technology, Changchun 130012, China

Jieqiong Lin
School of Mechatronic Engineering, Changchun University of Technology, Changchun 130012, China

Mingming Lu
School of Mechatronic Engineering, Changchun University of Technology, Changchun 130012, China

Mingshuo Kang
School of Mechatronic Engineering, Changchun University of Technology, Changchun 130012, China

Beatriz De Agustina
Department of Manufacturing Engineering, Industrial Engineering School, National University of Distance Education (UNED), C/Juan del Rosal, 12, E28040-Madrid, Spain

Marta María Marín
Department of Manufacturing Engineering, Industrial Engineering School, National University of Distance Education (UNED), C/Juan del Rosal, 12, E28040-Madrid, Spain

Roberto Teti
Department of Materials and Production Engineering, University of Naples Federico II Piazzale Tecchio, 80, Naples 80125, Italy

Eva María Rubio
Department of Manufacturing Engineering, Industrial Engineering School, National University of Distance Education (UNED), C/Juan del Rosal, 12, E28040-Madrid, Spain

Hirokazu Yoshino
Loughborough University, Leicestershire LE11 3TU, UK
Taylor Hobson Ltd., Leicestershire LE4 9JD, UK

John Michael Walls
Loughborough University, Leicestershire LE11 3TU, UK

Roger Smith
Loughborough University, Leicestershire LE11 3TU, UK

Jessica M. Werrell
School of Physics and Astronomy, Cardiff University, Cardiff, UK

Soumen Mandal
School of Physics and Astronomy, Cardiff University, Cardiff, UK

Evan L. H. Thomas
School of Physics and Astronomy, Cardiff University, Cardiff, UK

Emmanuel B. Brousseau
Cardiff School of Engineering, Cardiff University, Cardiff, UK

Ryan Lewis
School of Biosciences, Cardiff University, Cardiff, UK

Paola Borri
School of Biosciences, Cardiff University, Cardiff, UK

Philip R. Davies
School of Chemistry, Cardiff University, Cardiff, UK

Oliver A. Williams
School of Physics and Astronomy, Cardiff University, Cardiff, UK

Lei Pan
MIIT Key Laboratory of Critical Materials Technology for New Energy Conversion and Storage, School of Chemistry and Chemical Engineering, Harbin Institute of Technology, Harbin, 150001, China

Hongbo Xu
MIIT Key Laboratory of Critical Materials Technology for New Energy Conversion and Storage, School of Chemistry and Chemical Engineering, Harbin Institute of Technology, Harbin, 150001, China

Ruizhen Lv
Center for Composite Materials and Structure, Harbin Institute of Technology, Harbin, 150001, China

Jun Qiu
School of Energy Science and Engineering, Harbin Institute of Technology, Harbin, 150001, China

Jiupeng Zhao
MIIT Key Laboratory of Critical Materials Technology for New Energy Conversion and Storage, School of Chemistry and Chemical Engineering, Harbin Institute of Technology, Harbin, 150001, China

Yao Li
Center for Composite Materials and Structure, Harbin Institute of Technology, Harbin, 150001, China

Yuan Li
Department of Physics, University of Science and Technology Beijing, Beijing, 100083, China
Laser Fusion Research Center, China Academy of Engineering Physics, Mianyang, 621900, China

Hongwei Yan
Laser Fusion Research Center, China Academy of Engineering Physics, Mianyang, 621900, China

Ke Yang
Laser Fusion Research Center, China Academy of Engineering Physics, Mianyang, 621900, China

Caizhen Yao
Laser Fusion Research Center, China Academy of Engineering Physics, Mianyang, 621900, China

Zhiqiang Wang
Laser Fusion Research Center, China Academy of Engineering Physics, Mianyang, 621900, China

Xinshu Zou
Laser Fusion Research Center, China Academy of Engineering Physics, Mianyang, 621900, China

Chunyan Yan
Department of Physics, University of Science and Technology Beijing, Beijing, 100083, China

Xiaodong Yuan
Laser Fusion Research Center, China Academy of Engineering Physics, Mianyang, 621900, China

Xin Ju
Department of Physics, University of Science and Technology Beijing, Beijing, 100083, China

Liming Yang
Fine Optical Engineering Research Center, Chengdu, 610041, China

Christoph Gerhard
Faculty of Engineering and Natural Sciences, Technical University of Applied Sciences Wildau, Hochschulring 1, 15745 Wildau, Germany

Marco Stappenbeck
Faculty of Natural Sciences and Technology, University of Applied Sciences and Arts, Von-Ossietzky-Straße 99, 37085 Göttingen, Germany

LIST OF ABBREVIATIONS

AE	Acoustic emission
AFM	Atomic force microscope
BPT	Benzo[a]pyrene tetrol
BIT	Bonded-interface technique
CMP	Chemical mechanical polishing
CVD	Chemical vapour deposition
CSI	Coherence scanning interferometry
CT	Computed tomography
CLSM	Confocal laser scanning microscopy
CW	Continuous wave
CPCM	Cross-polarization confocal microscopy
COC	Cyclic olefin copolymer
COP	Cyclic olefin polymer
DND	Detonation nanodiamond
DBD	Dielectric barrier discharge
DFT	Digital Fast Fourier
DOP	Dioctylphthalate
DF	Disordered films
DOE	Diffractive optical elements
DLS	Dynamic light scattering
ECD	Eddy current detection
EDM	Electric discharge machining
EFI	Electric field intensity
ECM	Electrochemical machining
ELID	Electrolytic in-process dressing
FFT	Fast Fourier Transform
FTS	Fast tool servo

FEM	Finite element model
FIB	Focused ion beam
FWM	Four-Wave Mixing
FPP	Fresnel phase plate
FZP	Fresnel zone plate
FWHM	Full width at half maximum
FOM	Fundamental optical mode
GPD	Gross domestic product
HCF	Helical complex field
HOMI	Higher order mode instabilities
HOMs	Higher order modes
HF	Hydrofluoric acid
IC	Integrated circuit
ISR	Interfacial surface roughness
IRHD	International rubber hardness degree
IBE	Ion beam etching
ISRNC	ISR with noise compensation
LMA	Large mode area
LASIK	Laser-Assisted in-situ Keratomileusis
LDW	Laser direct writing
LIDT	Laser induced damage threshold
LMJ	Laser Megajoule
LIGA	Lithographie, Galvanik and Abformung
MBN	Magnetic Barkhausen noise
MRF	Magnetorheological finishing
MRF	Magneto rheological Finishing
MSFC	Marshall Space Flight Center
MRR	Material removal rate
MCP	Mechanical chemical polishing
MEMS	Micro-electro-mechanical systems
MI	Modal instability'
MFD	Mode field diameter

MCVD	Modified chemical vapor deposition
MD	Molecular dynamics
NCD	Nanocrystalline diamond
NIL	Nanoimprint lithography
NIF	National Ignition Facility
NSOM	Near-field scanning optical microscope
NAs	Numerical apertures
OCT	Optical coherent tomography
OTs	Optical tweezers
OVD	Outside vapor deposition
PV	Peak-to-valley
PC	Personal computer
PBGs	Photonic band gaps
PCFs	Photonic crystal fibers
PCVD	Plasma-assisted chemical vapor deposition
PIAD	Plasma-ion-assisted deposition
PC	Polycarbonate
PDI	Polydispersity index
PAHs	Polynuclear aromatic hydrocarbons
POM	Polyoxymethylene
PS	Polystyrene
QWOT	Quarter-wavelength optical thickness
RF	Radio frequency
RB	Reaction-bonded
RAP	Robot Assisted Polishing
RMS	Root mean square
RPM	Rounds per minute
SAM	Scanning acoustic microscopy
SEM	Scanning electron microscope
SPM	Self-Phase Modulation
SiC	Silicon carbide
SCD	Single crystal diamond

SMF	Single mode optical fibres
STS	Slow Tool Servo
SBS	Stimulated Brillouin scattering
SRS	Stimulated Raman scattering
SSD	Subsurface damage
SAW	Surface acoustic wave
SSD	Surface and subsurface damage
SR	Surface roughness
SRCT	Synchrotron radiation computed tomography
3D-PCs	Three-dimensional photonic crystals
TIS	Total integral scattering
TIRM	Total internal reflection microscopy
TEM	Transmission electron microscopy
WC	Tungsten carbide
UPM	Ultra-Precision Machining
VAD	Vapor axial deposition
WLI	White light interferometer
XRD	X-ray diffraction
XPS	X-ray photoelectron spectroscopy
YAG	Yttrium aluminum garnet

PREFACE

Optical fabrication processes have obviously been developed to respond to the growing new requirements, such as high surface qualities, in terms of low roughness level, minimized subsurface damage and high form accuracies. Among these processes, polishing is the most common technology used, being an essential step in optics manufacturing and in-mold finishing operations. This book aims to support decision making when selecting the most suitable optical fabrication technology by providing from a materials-science perspective an overview of available technologies.

Section 1 of *Scientific Principles and Technique of Optical Fabrication Processes* book introduces the optical fabrication processes by showing several kinds of applications. In particular, it treats of design and fabrication of fiber optic nanoprobes for optical sensing, of Fresnel plates on optical fibers, of glass optical waveguides, of optical mould inserts. In the end, Section 1 presents a pulsed IR-laser assisted additive manufacturing process to print freeform optics.

Section 2 focuses on the processes of surface polishing. In detail, it presents the optical surface processes and the recent developments of Silicon Carbide (SiC) substrate, and it discusses an analytical and experimental optimization of annular polishing parameters of reaction-bonded SiC. Moreover, it treats the effects of process parameters, aiming at clarifying qualitatively the interrelations between material removal mechanism and the polishing process. Lastly, it deals with the influence of different polishing materials in the material removal of steel samples.

Section 3 treats the subsurface damages ranging between the methods for their detection to the influence processes.

Section 4 deals with surface roughness. In particular, it presents the analysis of acoustic emission (AE) signals and the coherence scanning interferometry using noise compensation, for evaluating the surface roughness. Moreover, it discusses the effect of slurry composition on the chemical mechanical polishing of thin diamond films and the importance of optimum slurry injection for obtaining a more cost-effective and environmentally benign chemical mechanical planarization process.

Finally, the last Section 5 reviews laser damage resistance during a manufacturing process by taking into account the material and its surface molecular structure defects, the polishing suspension concentration, and the presence of scratches in the surface. In the end, it discusses the ways of controlling and improving laser damage threshold during the optical fabrication processes.

SECTION I: INTRODUCTION TO OPTICAL FABRICATION PROCESSES

Chapter 1

Design and Fabrication of Fiber-Optic Nanoprobes for Optical Sensing

Yan Zhang[1,2], **Anuj Dhawan**[1,2], **Tuan Vo-Dinh**[1,2,3]

[1]Fitzpatrick Institute for Photonics, Duke University, Durham, NC 27708, USA

[2]Department of Biomedical Engineering, Duke University, Durham, NC 27708, USA

[3]Department of Chemistry, Duke University, Durham, NC 27708, USA

ABSTRACT

This paper describes the design and fabrication of fiber-optic nanoprobes developed for optical detection in single living cells. It is critical to fabricate probes with well-controlled nanoapertures for optimized spatial resolution and optical transmission. The detection sensitivity of fiber-optic nanoprobe depends mainly on the extremely small excitation volume that is determined by the aperture sizes and penetration depths. We investigate the angle dependence of the aperture in shadow evaporation of the metal coating onto

Citation: Zhang, Yan et al. "Design and Fabrication of Fiber-Optic Nanoprobes for Optical Sensing." Nanoscale research letters vol. 6,1 (2010): 18. https://dx.doi.org/10.1007%2Fs11671-010-9744-5

the tip wall. It was found that nanoaperture diameters of approximately 50 nm can be achieved using a 25° tilt angle. On the other hand, the aperture size is sensitive to the subtle change of the metal evaporation angle and could be blocked by irregular metal grains. Through focused ion beam (FIB) milling, optical nanoprobes with well-defined aperture size as small as 200 nm can be obtained. Finally, we illustrate the use of the nanoprobes by detecting a fluorescent species, benzo[a]pyrene tetrol (BPT), in single living cells. A quantitative estimation of the numbers of BPT molecules detected using fiber-optic nanoprobes for BPT solutions shows that the limit of detection was approximately 100 molecules.

Keywords: Fiber-optic nanoprobes, Focused ion beam, Shadow evaporation, Optical sensing

INTRODUCTION

The emergence of nanotechnology opens new horizons for nanosensors and nanoprobes that are suitable for intracellular measurements. Nanosensors provide critical information for monitoring biomolecular processes within a single living cell, thus could provide great advances in biomedical research and clinical applications. Fiber-optic nanosensors with nanoscale dimensions are capable of sensing intracellular/intercellular physiological and biological parameters in submicron environments. Tapered fibers with distal diameters between 20 and 500 nm have been demonstrated for near-field scanning optical microscopy (NSOM) [1, 2]. Chemical nanosensors were developed for monitoring calcium and nitric oxide, among other physico-chemicals in single cells [3, 4]. Vo-Dinh and coworkers have developed nanobiosensors to detect biochemical targets inside living single cells [5, 6, 7, 8, 9, 10, 11, 12]. Fiber-optic nanoprobe promises to be an area of growing research that could potentially provide an imaging tool for monitoring individual cells and even biological molecules. Single-molecule detection and imaging schemes using nanofibers could open new possibilities in the investigation of the complex biochemical reactions and pathways in biological and cellular systems leading to important applications in medicine and health effect studies.

Optical nanotips were first developed as scanning probes in near-field optical microscopes [2]. Such nanoprobes can achieve resolution as high as $\lambda/50$, where λ is the wavelength of light [1]. It is important to control aperture size, taper shape, and metal coating to achieve a better performance

[13]. The fiber-optic probes were fabricated by laser-heated pulling or chemical etching [14, 15, 16]. Laser-pulled fiber tips can achieve diameters smaller than 50 nm with small cone angles [14]. Chemical etching tips have larger cone angles and similar apex sizes [15]. However, it is often difficult to control the etching process. The side of the fiber was further coated with silver, aluminum, or gold films to confine the light [9, 14, 17]. Traditional manufacturing processes still limit the quality of metal-coated fiber probes. Optical throughput of pulled nanoprobes is limited by the sharp taper angle. Chemical-etched tips have higher throughput; however, they do not have a flat distal end as laser-pulled ones which are difficult to form well-defined nanoapertures in shadow evaporation. Moreover, shadow evaporation often leads to either complete or irregular coated tip. Grainy structures of metal thin film increase the distance between the aperture and the sample, which reduce the resolution and intensity. It is also easy to form pin holes at the tapered region that could cause light-leaking. The aperture deviates from ideal circular shape because of grains. A quantitative analysis of probe transmission efficiency becomes difficult. Focused ion beam (FIB) fabrication of nanostructures has been applied on optical fibers for chemical sensing [18]. FIB milling for nanostructure formation allows precise control of size and shape in nanometer accuracy. This paper deals specifically with the metal coating on the formation of nanoaperture at the tip end. Coating materials and angles greatly affect the quality of the nanoprobe. By combining with focused ion beam milling, nanoprobes with well-defined aperture as small as 200 nm have been obtained. We investigate the capacity of the nanoprobes by detect benzo[a]pyrene tetrol in living cells.

EXPERIMENTAL PROCEDURES

Nanoprobe Fabrication

Optical nanoprobes were fabricated through laser pulling method, which consists of local heating of an optical fiber (Polymicro Technologies FVP400440480) using a laser and subsequently pulling the fiber apart. Fabrication of nanosensors requires techniques capable of making reproducible optical fibers with submicron-size-diameter core. Figure 1 illustrates the experimental setup for the fabrication of nanofibers using the micropipette puller (Sutter Instruments P-2000) [9]. As the laser pulling process is a time-dependent heating effect, laser power, timing of pulling, velocity setting, and pulling force all contribute to the taper shape and tip size. Since transmission efficiency is highly dependent on the taper shape, it is crucial to control the tip shape in the fabrication of high-quality nanoprobes.

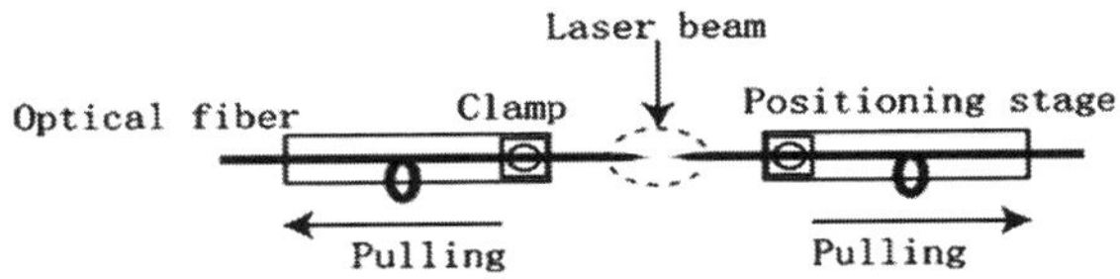

Figure 1: Fabrication of nanofibers by laser pulling.

The sidewall of the tapered end was then coated with a thin layer of metal, such as silver, aluminum, or gold to prevent light leakage of the excitation light on the tapered side of the fiber. An array of fiber probes was attached on a rotating motor inside a thermal evaporation chamber (Quorum Technologies E6700). The rotation rate was controlled by a microcontroller board (Parallax). While the probes were rotating, the metal was allowed to evaporate onto the tapered side of the fiber tip to form a thin coating. The nanoaperture was formed through shadowed evaporation as the fibers were tilting away from the source. The nanoprobes were characterized with scanning electron microscopy (FEI XL30).

In order to fabricate well-defined fiber-optic nanoprobe tips, we employed focused ion beam (FIB) milling of nanoapertures in the metallic films deposited on tapered tips of optical fibers. Before carrying out FIB milling, the optical fibers were coated with metallic films (aluminum, silver or gold) using electron beam evaporation (CHA Industries Solution E-Beam). During the evaporation process, the fiber-optic tips faced the metal source to ensure that the fiber side walls and the tips were completely covered with a thin metallic layer (100–150 nm). The sample mount was rotated to improve uniformity and the thickness of the metallic film was monitored by a quartz crystal monitor. The deposition rate was varied between 0.05 and 0.2 nm s^{-1} at a chamber pressure of ~3×10^{-6} Torr for the electron beam evaporated films.

A Hitachi FB2100 focused ion beam milling machine with a gallium ion source was used to fabricate the nanoapertures on the fiber tips. Beam currents and accelerating voltages of 0.01 nA and 40 keV energy were typically used. The desired nanostructures were milled by rastering the ion beam and employing a beam blanker. The beam blanker shuts on and off according to a 8-bit grayscale, 512 by 512 pixel image file. Tapered optical fiber tips with nanoapertures were fabricated by employing FIB milling at magnifications varying between 3000× and 18000× depending on the desired minimum aperture size. To form metallic nanostructures on the tips of optical fibers, a special fiber holder that could fit in the FIB stage was fabricated.

Optical Measurement

The optical measurement system used for nanoprobe is schematically illustrated in Figure 2. For nanoprobe measurements, the 325 nm line of a HeCd laser (CVI Melles Griot, 15 mW laser power) was focused onto a 400 μm delivery fiber. A tapered fiber was coupled to the delivery fiber through a capillary tubing and was secured to the micromanipulators (Narishige MLW-3) on a Zeiss Axiovert 200M microscope (Zeiss). The fluorescence emitted from the region beyond the aperture was collected by the microscope objective and passed through a bandpass filter (386 nm) and then focused onto a photomultiplier tube (PMT, Hamamatsu, HC125-2) for detection. The output from the PMT was recorded on a universal counter (Agilent 53131A), and a personal computer (PC) was used for further data treatment.

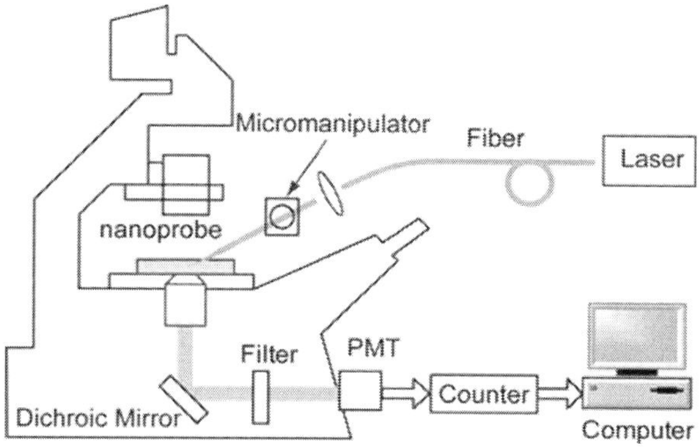

Figure 2: Instrumental system for fluorescence measurements using nanoprobes.

Nanoprobes were also used to investigate BPT in single cells. PC3 human prostate cancer cells were incubated with 1 μM BPT in PBS for 2 h. Control cells are treated with PBS only. All dishes were rinsed with PBS prior to measurement. Nanoprobes were controlled by the micromanipulator to puncture the cell and keep inside while taking the measurement.

RESULTS AND DISCUSSION

Effect of the Metal Evaporation Angle

A scanning election microscopy (SEM) photograph of one of the nanofibers fabricated by the laser pulling method is shown in Figure 3a. The distal end of the nanofiber is approximately 40 nm. The fiber was pointing away from the evaporation source with an angle of approximately 25°. The tapered end was coated with ~75–100 nm of metal in the thermal evaporator. With the metal coating, the size of the probe tip is approximately 200–250 nm (Figure 3b).

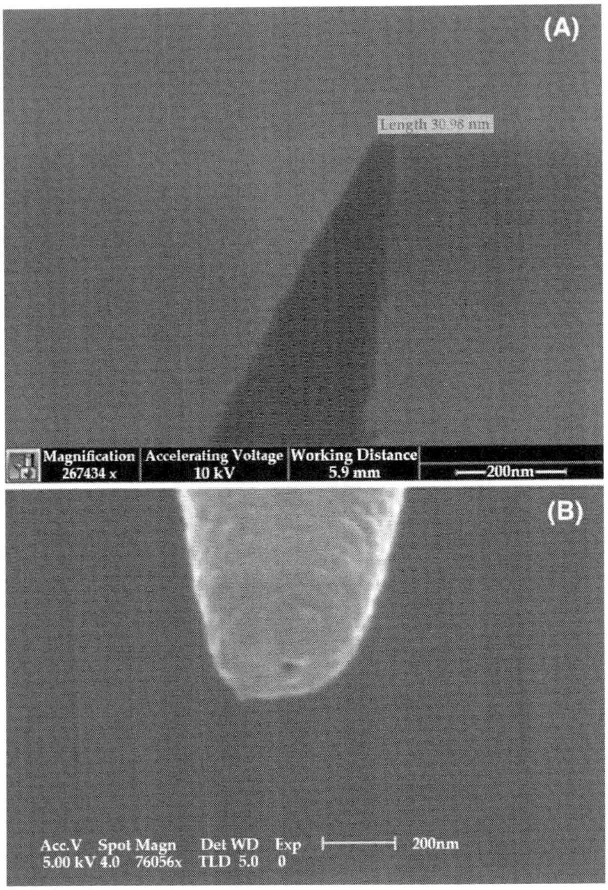

Figure 3: SEM images of a an uncoated nanofiber and b a gold-coated nanofiber.

Due to the inclination angle, the tip ends are shadowed from evaporation when the fiber tips are tilted away from the source. The effect of shadow evaporation angle is illustrated in Figure 4. A nanoaperture was formed on the tip end for optical excitation. The size of the nanoaperture is related to the angle between fiber axis and evaporation direction. For example, if the angle is less than 20°, most of the fibers are fully covered with metal and no aperture is visible using SEM. On the contrary, if the angle is higher than 30°, a larger area of the distal end of the fiber tip will be exposed (Figure 5). The optimal angle of inclination can be determined through characterization of nanoapertures under SEM. The SEM can determine the actual size of the tip aperture. However, having a nanometer-sized aperture does not guarantee a good near-field probe. The tip aperture, even though it may appear small on the SEM, could be a result of aluminum over-coating, and hence not be a functional light aperture for actual measurements. Therefore, a functional scan is necessary to reveal the near-field effect from the probe. Near-field scanning optical microscope (NSOM) enables functional analysis of the nanosensor probe by performing a scan on a standard sample, e.g., a Fischer pattern. The standard sample usually consists of patterns with size less than the diffraction limits (0.5 λ) that can be determined by an atomic force microscope. The nanosensor probe was attached to an NSOM system working as an NSOM probe. Typically, the aperture of the probe roughly determines the resolution of the image. In other word, the image quality thus represents the quality of the probe.

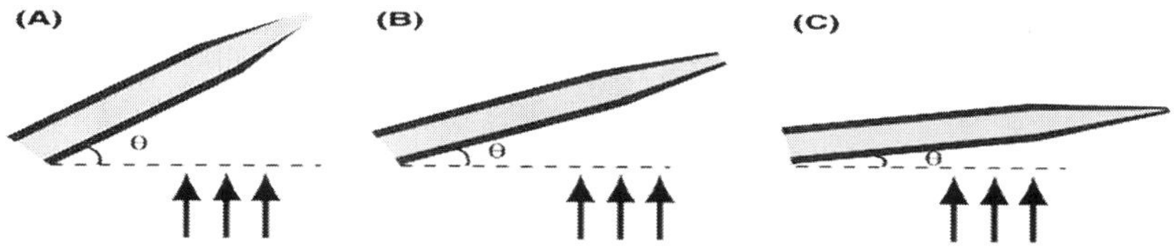

Figure 4: Nanoaperture formation by shadow evaporation with a high angle (>25°), b medium angle (~25°), and c low angle (<25°).

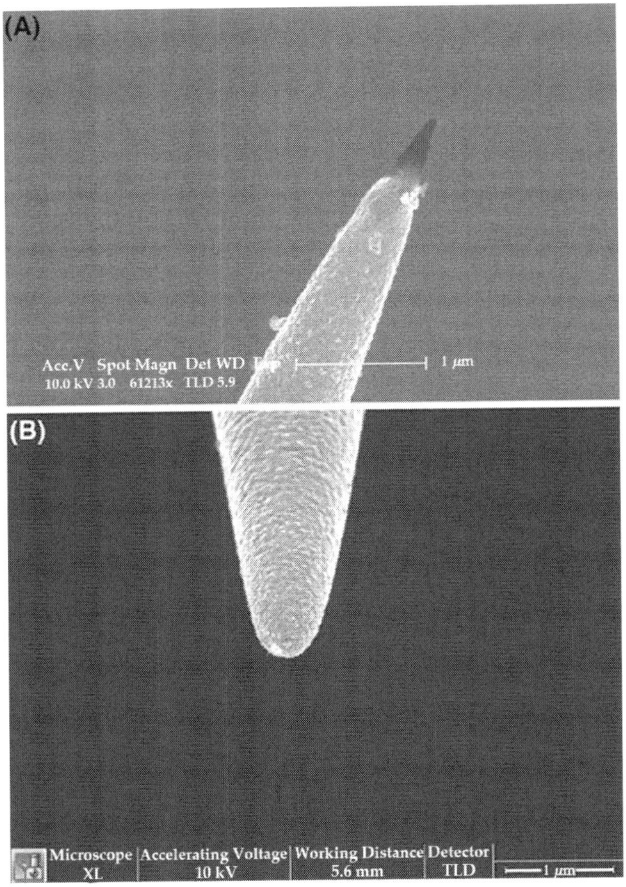

Figure 5: SEM images of silver-coated nanofibers under different angles: a 40° and b 20°.

Effect of the Different Metal Coatings on the Nanoprobes

The metallic coating process is a critical step in nanoprobe fabrication. A thin film of an optically opaque metal such as aluminum, silver, or gold is coated along the outside walls of the tapered optical fiber tip to form an optical light pipe free of defects, which would permit photons to escape from the tapered sides of the optical fiber. An optical aperture to allow evanescent wave excitation is formed at the tip's apex by angled evaporation. Silver has been used in nanoprobe fabrication [9]. It has high reflectivity in the visible and IR range and very stable in aqueous solutions as long as oxidizing agents or complexing agents are not present. But a silver layer will oxidize rapidly under ordinary atmospheric conditions and will not exhibit a high reflectance below 400 nm. Therefore, it is desirable to use the nanoprobe right after metal evaporation. Otherwise, the light shielding will deteriorate or even the coating will peel off after a few days in air.

Gold thin film was demonstrated to be a very stable coating under environmental conditions although it does not have high reflectance in visible range. An interface layer such as Cr is required to increase adhesion between gold and the fiber silica surface. Gold has a high melting temperature (660°C for Al, 960°C for Ag, and 1,060°C for Au) and a good thermal resistance. The thermal stress generated during metallic film deposition damages the aperture due to very different thermal expansion coefficients of metal and quartz. Gold coating has the lowest thermal expansion coefficient (23.1×10^{-6}/°C for Al, 18.9×10^{-6}/°C for Ag, 14.2×10^{-6}/°C for Au, 0.55×10^{-6}/°C for SiO_2), which will reduce the thermal destruction of the fiber tip.

Aluminum is a desirable material to use because it has the highest extinction coefficient of all metals. Aluminum adheres to fibers more firmly than silver or gold so that no interface layer is required and general cleaning does not affect the coating. Figure 6 compares the nanofibers after argon plasma cleaning (Emitech K-1050X, 100 W, 5 min). Silver coatings are easily peeled off while aluminum coatings exhibit no changes under the same condition. Aluminum is inert toward corrosive agents since a protective oxide layer is formed readily upon contact to the air. However, it is difficult to evaporate aluminum as a thin film while maintaining smooth films with small grain sizes [19]. Grainy films contribute to the high background in near-field sensing. The grain diameter is highly and sensitively dependent on the deposition pressure. Below 5×10^{-6} torr, the size of the individual grains is smaller than 100 nm. There was a relationship between rate of metallic deposition and subsequent surface roughness, and studies revealed that higher coating rate (>10 nm/s) resulted in better smoothness and the film opacity required for our intended sensor applications.

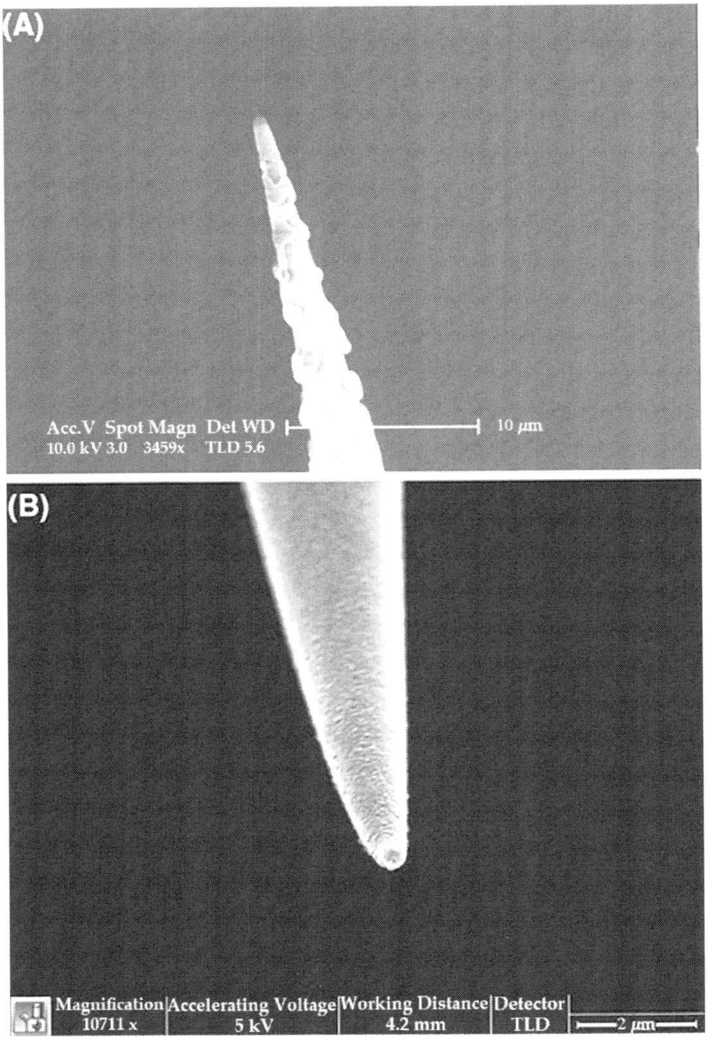

Figure 6: SEM images of a silver- and b aluminum-coated nanofibers after plasma cleaning.

Nanoprobe Fabrication using Focus Ion Beam (FIB) Milling

The first FIB milling process involved placing the metal-coated optical fiber tips horizontally, i.e., orthogonal to the focused ion beam and then cutting the tips (both the tapered silica fiber and the metal over-coating) such that an aperture could be developed at the tip. Milling of the nanoapertures using this process has an advantage that it is not time-consuming as several tips placed adjacent to each other can be cut with the same beam raster, it gives reliable nanoprobes with well-defined nanoapertures of circular geometry, and the length of the optical fiber nanoprobes can be longer, which can make coupling of light into the optical fibers easier. The second process involved positioning the fiber tips such that they faced the focused ion beam and then carrying out the milling of the nanoapertures at the tip. Although this process enables fabrication of nanoapertures of different geometries and sizes in a very controllable manner, it limits the length of the fiber-optic probe as only a certain length of the optical fiber probe can be placed vertically in the Hitachi FB2100 focused ion beam milling machine. By milling with a focused ion beam, an aperture with controllable shape and diameter as small as 200 nm was achieved (Figure 7). The angle of evaporation is not necessary in FIB, therefore reducing the chance of pin-hole formation. A clean aperture free from grains also facilitates the subsequent functionalization of bioreceptor molecules on the fiber distal end for biosensing applications. FIB processing is a promising technique in nanoprobe fabrication in addition to laser pulling and chemical etching.

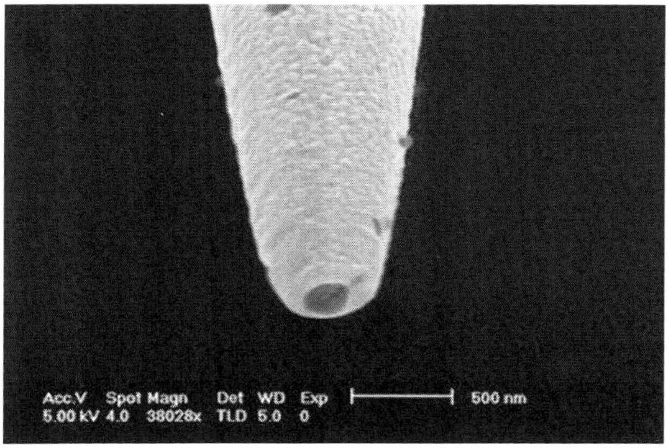

Figure 7: FIB-etched nanoprobe with aperture diameter of 200 nm.

Fluorescence Measurement of Benzo[a]Pyrene Tetrol (BPT) Using Nanoprobes

Chemical analysis of polynuclear aromatic hydrocarbons (PAHs) is of great environmental and toxicological interest because many of them have been shown to be mutagens and/or potent carcinogens in laboratory animal assays. Benzo[a]pyrene (BaP), which has been extensively investigated, is one of the more potent carcinogens among PAHs and is a fundamental indicator of exposure and carcinogenic activity of all PAHs. In order to facilitate the study of intracellular dynamics of benzo[a]pyrene tetrol (BPT), the related biomarker under BaP exposure, a quantitative estimation of the numbers BPT molecules detected using fiber-optic nanoprobes for solutions containing different BPT concentrations was performed and shown in Figure 8. The limit of detection that corresponds to the amount of analyte emitting a signal 3 times the standard deviation of the noise was determined to be 1 μM for BPT. Under 1 μM, the dark count noise from PMT was stronger than the signal. The detection volume can be estimated as 17 aL for 200 nm aperture probe based on Bethe–Bouwkamp theory. Therefore, the limit of detection was approximately 100 BPT molecules. Figure 9 shows the intracellular measurement of BPT in PC3 human prostate cancer cells. The cells were incubated with 1 μM BPT in PBS for 2 h at 37°C. Control cells are PC3 cells treated with PBS only. All dishes were rinsed with PBS prior to measurement. It is apparent that the treated cells exhibited higher fluorescence reading than the control group. Although in our preliminary experiments the living cells were directly incubated with BPT, the results illustrate that the nanoprobes can be employed to detect very low concentrations of fluorescent species such as BPT molecules that are important biomarkers of exposure and carcinogenic activity of related PAHs, inside living cells.

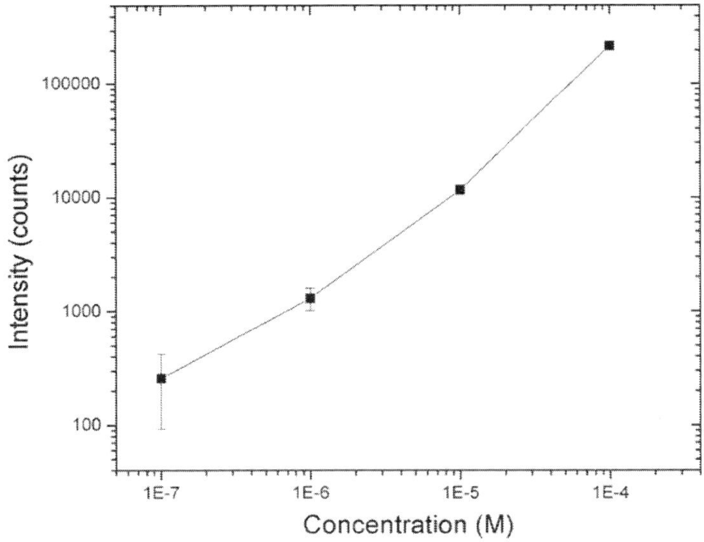

Figure 8: Fluorescence intensities of benzo[a]pyrene tetrol (BPT) measured with a fiber-optic nanoprobe.

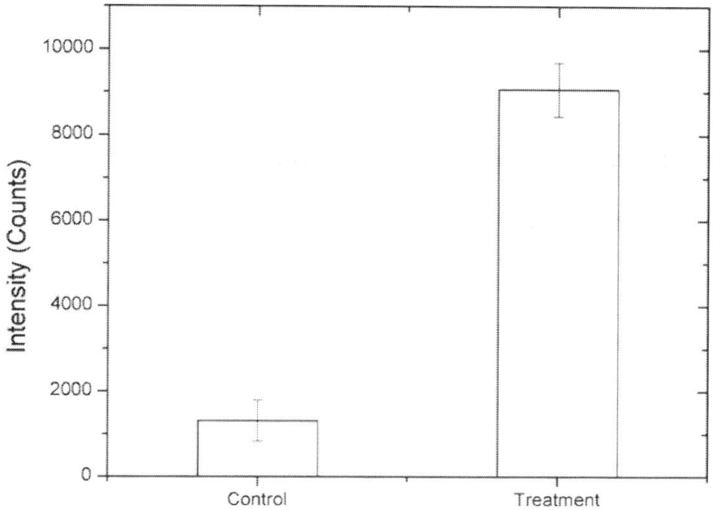

Figure 9: Intracellular measurement of benzo[a]pyrene tetrol (BPT) with nanoprobes. PC3 human prostate cancer cells were incubated with 1 μM BPT in PBS for 2 h. Control cells are treated with PBS only. All *dishes* were rinsed with PBS prior to measurement.

CONCLUSION

Fiber-optic nanoprobes have opened up new applications in molecular biology and medical diagnostics. Due to their small sizes, nanosensor provides important tools for minimal invasive analysis at single cellular or sub-cellular level. Because transmission efficiency is highly related to the aperture size, control the nanoaperture size is essential in the fabrication of high-quality nanoprobes. Subtle changes in the tilt angle during metal evaporation can greatly affect the size or even the existence of the aperture. A much more rational fabrication process would involve a nanofabrication technique such as FIB, in which aperture size could be independently controlled from evaporation. The detection sensitivity of fiber-optic nanoprobes depends mainly on the extremely small excitation or detection volume set by the aperture sizes and penetration depths. This effectively reduces background fluorescence, thereby enhance detection sensitivity. Nanofabrication would also greatly improve the reproducibility of aperture shapes and hence the optical performance of near-field probes.

ACKNOWLEDGEMENTS

The author acknowledges the contribution of G.D. Griffin, J.P. Alarie, B.M. Cullum, and P. Kasili. This research is sponsored by the National Institutes of Health (1R01ES014774 and R01-EB006201) and US Army Medical Research and Material Command (W81XWH-09-1-0064).

REFERENCES

1. Betzig E, Chichester RJ: *Science*. 1993, 262: 1422–1425. 10.1126/science.262.5138.1422
2. Pohl DW: *Advances in Optical and Electron Microscopy. Volume 12*. Academic London; 1991:243–312.
3. Tan WH, Shi ZY, Kopelman R: *Anal Chem*. 1992, 64: 2985–2990. 10.1021/ac00047a019
4. Tan WH, Shi ZY, Smith S, Birnbaum D, Kopelman R: *Science*. 1992, 258: 778–781. 10.1126/science.1439785
5. Cullum BM, Griffin GD, Miller GH, Vo-Dinh T: *Anal Biochem*. 2000, 277: 25–32. 10.1006/abio.1999.4341
6. Kasili PM, Song JM, Vo-Dinh T, Am J: *Chem Soc*. 2004, 126: 2799–2806. 10.1021/ja037388t
7. Song JM, Kasili PM, Griffin GD, Vo-Dinh T: *Anal Chem*. 2004, 76: 2591–2594. 10.1021/ac0352878
8. Vo-Dinh T: *Spectrochim Acta, Part B*. 2008, 63: 95–103. 10.1016/j.sab.2007.11.027
9. Vo-Dinh T, Alarie JP, Cullum BM, Griffin GD: *Nat Biotechnol*. 2000, 18: 764–767. 10.1038/77337
10. Vo-Dinh T, Griffin GD, Alarie JP, Cullum BM, Sumpter B, Noid DJ: *J Nanopart Res*. 2000, 2: 17–27. 10.1023/A:1010005908586
11. Vo-Dinh T, Kasili P: *Anal Bioanal Chem*. 2005, 382: 918–925. 10.1007/s00216-005-3256-7
12. Vo-Dinh T, Kasili P, Wabuyele M: *Nanomedicine*. 2006, 2: 22–30.
13. Essaidi N, Chen Y, Kottler V, Cambril E, Mayeux C, Ronarch N, Vieu C: *Appl Opt*. 1998, 37: 609–615. 10.1364/AO.37.000609
14. Valaskovic GA, Holton M, Morrison GH: *Appl Opt*. 1995, 34: 1215–1228. 10.1364/AO.34.001215
15. Stockle R, Fokas C, Deckert V, Zenobi R, Sick B, Hecht B, Wild UP: *Appl Phys Lett*. 1999, 75: 160–162. 10.1063/1.124305
16. Lambelet P, Sayah A, Pfeffer M, Philipona C, Marquis-Weible F: *Appl Opt*. 1998, 37: 7289–7292. 10.1364/AO.37.007289
17. Vo-dinh T, Zhang Y: Optical Nanosensors for Medicine and Health Effect Studies. Handbook of nanophysics: nanomedicine and nanorobotics, in press.

18. Dhawan A, Muth JF, Leonard DN, Gerhold MD, Gleeson J, Vo-Dinh T, Russell PE: *J Vac Sci Technol, B: Microelectron Nanometer Struct.* 2008, 26: 2168–2173. 10.1116/1.3013329
19. Holland L: *Vacuum Deposition of Thin Films.* Wiley, New York; 1956.

Chapter 2

Materials Development for Next Generation Optical Fiber

John Ballato [1] **and Peter Dragic** [2]

[1]The Center for Optical Materials Science and Engineering Technologies (COMSET) and the Department of Materials Science and Engineering, Clemson University, Clemson, SC 29634, USA

[2]Department of Electrical and Computer Engineering, University of Illinois at Urbana-Champaign, Urbana, IL 61801, USA

ABSTRACT

Optical fibers, the enablers of the Internet, are being used in an ever more diverse array of applications. Many of the rapidly growing deployments of fibers are in high-power and, particularly, high power-per-unit-bandwidth systems where well-known optical nonlinearities have historically not been especially consequential in limiting overall performance. Today, however, nominally weak effects, most notably stimulated Brillouin scattering (SBS) and stimulated Raman scattering (SRS) are among the principal phenomena restricting continued scaling to higher optical power levels. In order to address

Citation: Ballato, John; Dragic, Peter. 2014. "Materials Development for Next Generation Optical Fiber." Materials 7, no. 6: 4411-4430. https://doi.org/10.3390/ma7064411

these limitations, the optical fiber community has focused dominantly on geometry-related solutions such as large mode area (LMA) designs. Since such scattering, and all other linear and nonlinear optical phenomena including higher order mode instability (HOMI), are fundamentally materials-based in origin, this paper unapologetically advocates material solutions to present and future performance limitations. As such, this paper represents a 'call to arms' for material scientists and engineers to engage in this opportunity to drive the future development of optical fibers that address many of the grand engineering challenges of our day.

Keywords: optical fiber, high energy lasers, stimulated Brillouin scattering, stimulated Raman scattering

INTRODUCTION

Light today enables an annual global economic impact of about $7.5 trillion (USD) [1], which is approximately half of the gross domestic product (GPD) of the United States; and light's economic value is growing. The three largest market sectors relying on light-based products are those associated with telecommunications, transportation, and biotechnologies [2]. The telecommunications sector constitutes over 50% of the economic value. Core technologies enabled by light include e-commerce, Internet music and movies, and information routing and storage. The transportation sector includes laser-based manufacturing (e.g., precision cutting, drilling, and welding) of automobiles and airplanes. The biotechnological sector includes laser and optics-based diagnostic and surgical systems as well as Laser-Assisted in-situ Keratomileusis (LASIK) eye surgery, blood testing, and gene sequencing [2]. Smaller by market value but critical nonetheless are laser-based defense, security, and sensor systems, which will continue to drive technology development especially at "specialty" (high value and performance, low volume) levels.

Another way to view this cumulative effect on global goods and services is that light-based solutions exist (or could reasonably exist with further advancements) for many of the technology platforms identified as being most central to society over the next decade [3]. Despite this centrality of light to the many modern conveniences there has been decades-long erosion in the appreciation given to the materials as enablers of such technologies [4]. Modern and prototypic future systems that are prescribed as solutions to present performance limitations are strongly driven by 'structural' (geometric) as opposed to material considerations. Perhaps the best exemplars of these are photonic crystal fibers (PCFs) which are transversely periodic arrays of air holes surrounded by glass that run longitudinally down an optical fiber. Such designs range from the fairly straight-forward (note that the word "simple" was not used) to the intricately absurd; at least with an eye towards scalability and industrial, albeit low volume, manufacturing. Yet while PCFs can force light to do its bidding in a wide variety of useful ways, many phenomena, such as reduced nonlinearities, can be accomplished more simply through materials; or so this paper plans to promote and prove.

In reality and with practicality in mind, the true solution is both. It takes little more than opening a basic physics book to remember that properties are dependent upon both shape and substance; e.g., the capacitance of a capacitor—amongst the most elementary electronic devices—is dependent upon its geometry (area and thickness) and the material (specifically its permittivity) from which it is made. However, we, the materials community, are at a cross-road. A recent (albeit US-centric) study by Corning Incorporated [5] concludes by noting that "... less than one-quarter of students at US universities who are doing research in glass science are studying systems that would make them well prepared for a future career in the glass industry. In our experience ... students with expertise in glass families that are industrially relevant ... are more likely to be hired into a position in industry and also require less on the job technical training after being hired." The authors of that work go on to identify a very rich range of problems that materials scientists and engineers can work on (and get employed to solve!) that would have significant societal benefits.

Accordingly, this paper seeks to amplify further the recommendations of two recent works [4,5] by providing a review of materials-related solutions to power-scaling in high energy optical fiber laser-based systems. Given that the lowest threshold nonlinearities that presently limit performance are stimulated Brillouin scattering (SBS) and stimulated Raman scattering (SRS), these phenomena will be used as primary exemplars. Also to be discussed are higher order mode instabilities (HOMI) and management of the nonlinear refractive index. It is the authors' hope that this work, along with [4,5], will help to reinvigorate the global materials community to rejoin the development of next generation optical fibers.

RESULTS AND DISCUSSION

The intensity of light in both high-power continuous wave (CW) and high power-per-unit-bandwidth pulsed laser systems is sufficient to induce nonlinearities in the material response to the propagating electromagnetic field. The result is a number of nonlinear and parametric photonic processes that are widely used or alternatively considered to be parasitic to system performance [6], including Raman scattering (or stimulated Raman scattering, SRS), Brillouin scattering (or stimulated Brillouin scattering, SBS), Self-Phase Modulation (SPM), and Four-Wave Mixing (FWM). Each of these is described in greater detail in Section 2.1 as are their materials' origins.

Since these parasitic effects are intensity-dependent, one approach to their control is to spread the optical power out over a larger cross-sectional area thereby reducing the effective power-per-unit area. In practice, this has led to so-called "large mode area" (LMA) fibers where the dimension of the core in which the light is propagating is expanded relative to more conventional designs. However, such large core sizes, coupled with typical fiber numerical apertures (NAs) which together generally render such fibers 'few-moded' almost always promote multimode guidance such that the beam quality of the resultant laser light is deteriorated. Several methods to overcome this obstacle have been proposed and implemented. A few of these are outlined next.

In principle, restricting the light launch conditions such that only the fundamental optical mode (FOM) is excited is one way to achieve the highest beam quality [7] available from such fibers. However, mode coupling is brought about through fiber bends, packaging, and environment, leading to degradation in the quality (or M^2) of the beam in most practical situations. Therefore, an early approach to overcoming the inevitable multimode nature of the beam in these LMA fibers was the discovery that via careful bending of the fiber, higher order modes (HOMs) could be 'peeled away' from the core [8] leaving what effectively amounts to single mode operation. The technique calls for coiling the fiber into a preselected diameter such that the differential loss between the fundamental mode and HOMs is large. In this way, the fundamental mode dominates the laser characteristics and captures essentially all of the available gain. It too is possible to design a microstructured fiber for 'endlessly single mode operation' [9] with extremely low effective NA, but the intolerance of such fibers to bending has largely confined them to the rod-like regime, with the resultant lasers more closely resembling a cross-over of sorts between bulk and fiber lasers. In the restricted mode launch regime, it was subsequently found that excitation of a HOM in a few-moded fiber offered increased robustness of the mode propagation which was coupled with the added benefit that the HOMs have larger mode areas than the FOM [10].

Other "effectively" single mode LMA fibers were also subsequently developed to force what otherwise is a multimode design into behaving like a single mode fiber. As with the simple fiber coiling method described above, the goal is to provide some sort of differential loss mechanism such that the HOMs experience vastly greater propagation losses than the FOM. Hence the FOM will experience the highest gain and dominate the modal distribution at the laser output. A few recent fiber designs include the chirally-coupled core fiber (HOMs are coupled to and shed from one or more satellite waveguides that form a helix around the fiber core) [11], photonic bandgap Bragg fiber (where only the FOM propagates with low loss) [12], solid photonic bandgap fibers (where bending loss is exacerbated on HOMs relative to the FOM) [13], leakage channel fibers (wherein all modes are lossy, but the loss is minimized for the FOM) [14], and selectively-doped triple clad fibers (where the FOM has the highest overlap with the active region in the fiber) [15]. While there is clearly no shortage of methods utilized to improve the beam quality from an LMA fiber, such fibers generally are quite complex in their cross-sectional geometries. Furthermore, while effectively single mode operation does result, there is a power threshold above which single mode operation is unstable [16,17]. This 'modal instability' (MI) or 'higher order mode instability' (HOMI) yields a time varying output beam profile that negates many of the benefits of large mode areas fibers.

The authors have recently articulated a different approach to mitigating and, in some cases, completely removing from consideration, these parasitic effects based on the judicious choice of materials rather than on fiber geometry [4]. Specific examples are provided in Section 2.2 below. However, it is necessary first to describe the interaction of light with the optical material (fiber) through which they propagate and how such nonlinear and/or parasitic phenomena arise. Only then can materials-related solutions be more fully envisioned and appreciated.

The Materials Physics of Linear and Nonlinear Optical Phenomena

Depending on the application, nonlinear phenomena in optical fiber can be useful or deleterious. From the perspective of achieving high powers from active optical fiber systems, these processes are generally parasitic in nature. They give rise to significant limitations to power scaling and tremendous effort has been made in suppressing them (or in other words, elevating the threshold power at which these processes become significant). Since these processes are intensity-dependent, a larger mode diameter in an optical fiber usually implies a higher 'turn-on' threshold for these interactions. As described above, this has led to a number of fiber designs aimed at increasing the mode field diameter (MFD) in an optical fiber while maintaining robust single mode operation. While nonlinear effects can be controlled to some extent with proper waveguide design, each of these processes can be described mathematically via a combination of materials properties that cooperate to define a 'gain' or 'gain coefficient'. With careful selection of precursor materials, it is possible to tailor or design the composition such that these gain coefficients are reduced to acceptable levels for a given application.

The first nonlinear process investigated is stimulated Brillouin scattering (SBS). SBS is an interaction between hypersonic (thermally excited) acoustic waves and the optical signal in a fiber. Brillouin scattering begins as a spontaneous process with optical back-scattering most efficiently taking place from acoustic waves that are Bragg-matched to the optical wave. Simply put, the acoustic wave is a periodic longitudinal pressure, and therefore also density, variation. The spatially modulated density of the material then corresponds to a spatially modulated refractive index. Via electrostriction, the interference between the forward-propagating signal and back-scattered light feeds the acoustic (pressure) wave. This 'positive feedback' process increases in efficiency as the power is increased until 'threshold' is reached wherein the acoustic wave becomes a highly-efficient reflector to the optical signal. At this point Brillouin scattering is in the stimulated state. In general, SBS limits the amount of light per unit bandwidth that can be transmitted down or generated in an optical fiber. As such, it typically has the lowest threshold of all the nonlinear processes in narrow linewidth systems.

The Brillouin gain coefficient for a given optical wavelength (λ) is given to be $g_B = 2\pi n^7 p^2_{12}/c\lambda^2 \rho V_a \Delta\nu_B$ ([18], which has units of m/W), where n is the refractive index, p_{12} is a Pockels photoelastic constant, c is the speed of light, ρ is the mass density, V_a is the acoustic velocity, and $\Delta\nu_B$ is the Brillouin spectral width (which is inversely proportional to the acoustic damping time). Given the dependencies of the Brillouin gain on these physical parameters, beneficial materials properties for the suppression of SBS include: (1) a large acoustic velocity; (2) relatively low index; (3) relatively large mass density; (4) large Brillouin spectral width (i.e., large acoustic attenuation coefficient); and (5) low p_{12}. Unfortunately, a large density typically suggests a relatively large refractive index, so the density and index tend to mitigate one another with respect to g_B. Regardless, a clear guideline with selection criteria can be identified for choosing materials for reduced Brillouin gain.

To model the system, an additive approach based on the work of Schott and Winkelmann [19,20] is utilized to calculate each of the terms in this equation for a mixed multicomponent glass (such as a germanosilicate glass). Specifically, these quantities are

(a) the acoustic velocity

$V_a = \left(\sum_{i=1}^{N}(m_i - m_{i-1})/V_{a,i}\right)^{-1}$;

(b) the mass density

$\rho = \sum_{i=1}^{N}(m_i - m_{i-1})\rho_i$;

(c) the refractive index

$n = \sum_{i=1}^{N}(m_i - m_{i-1})n_i$; and

(d) the Brillouin spectral width ($\Delta\nu_B$),

$\Delta\nu_B = (V_a/\pi)\sum_{i=1}^{N}(m_i - m_{i-1})(\nu_a/\nu_{ref})^2\alpha_i(\nu_{ref})$

which is proportional to the acoustic attenuation coefficient α (units of m^{-1}). The scaling term $(\nu_a/\nu_{ref})^2$ accounts for the usual frequency-squared dependence of the intrinsic linewidth, while α is always specified at a reference acoustic frequency ν_{ref}. It is important to note that it is possible (but seemingly rare) for the scaling term to be a sub-square function [21], depending on the material and thermal history of the glass. For each calculation $(m_i - m_{i-1})$ is the volume of constituent i in the fiber. The Pockels' elasto-optic coefficient p_{12} is calculated utilizing a similar additive approach, but it is lumped in with the refractive index rather than calculated directly. For details, the reader is asked to refer to Reference [22]. The key point with regard to p_{12} is that for a material this value can be either positive or negative. Therefore a combination of materials with photoelastic constants of the former and the latter types can give rise to a composition wherein p_{12} is very close to zero. As a result, the possibility of a material with zero Brillouin gain exists. Such negation of certain glass coefficients in order to suppress or eliminate a nonlinear effect can also be applied to HOMI in high power optical fibers. It is well-known that restricted mode launch and careful modal control can give rise to effective single-mode operation in a few-moded fiber utilized in a fiber laser. However, at some sharp threshold power a dynamic randomization of the mode distribution at the laser output is observed (referred to as higher order mode instability, or HOMI), and is believed to be the result of a thermally-induced refractive index grating. The process is believed to be driven in part by dn/dT [16,17]; the grating forms via a longitudinally-periodic thermal variation in the fiber. It is interesting to point out that HOMI is not a process that can be reduced with a wider MFD, as it becomes more significant of a problem as the core diameter is increased [16,17].

While rendering a gain coefficient zero may be possible for some interactions, this may not be the case for others. An example is stimulated Raman scattering (SRS). Raman scattering is an interaction between an optical wave and optical phonons [18]. In the case of silica-dominated materials systems, the Raman spectrum has a peak gain positioned roughly 13.2 THz from the signal frequency. It can be considered to be a parasitic effect in high-peak-power fiber laser systems where wavelength control is mandatory [23]. The excitation of SRS from spontaneous scattering can lead to wavelength shifts and power instabilities that degrade laser performance, thus presenting the need for its suppression. However, the dependence of the Raman gain is on material properties (Raman cross section, refractive index) that do not appear to necessarily facilitate a zeroing of this value [23]. Instead, materials can be selected such that one or more of the following conditions are met [24]: (1) the material is highly disordered in order to broaden the Raman gain spectrum thereby reducing the peak value; (2) high concentrations of materials with low Raman gain are utilized; and (3) materials are utilized that have Raman spectra with minimal overlap. For (3), a mixture of two materials with similar-strength but not overlapping Raman gain spectra could cut the peak Raman gain coefficient in half.

Processes such as Self-Phase Modulation (SPM) and Four-Wave Mixing (FWM) are results of the dependence of the refractive index on the optical intensity [18]. SPM can be observed in pulsed (time-varying) systems, and it results from an intensity-induced time-varying refractive index (due to the time-varying nature of the signal itself). In multi-wavelength systems, FWM is the transfer of an amount of energy to a new wavelength by the amount that is (plus or minus) the difference in photon energy between two extant optical frequencies. In the case of non-degenerate FWM, if the existing optical frequencies are ν_1 and ν_2 (and $\nu_1 \neq \nu_2$) then the new frequencies will be at $2\nu_1 - \nu_2$ and $2\nu_2 - \nu_1$. Similarly, degenerate FWM can also occur (where $\nu_1 = \nu_2$) with new frequency components appearing both to the blue and red side of ν_1. The general effect of these processes is to broaden and modify the optical spectrum, and depending on the application this may be undesirable in high peak power and multi-wavelength optical amplifier systems. Since the dependence of the refractive index is given by $n(I) = n_0+n_2I$ where n_0 is the linear refractive index, I is the intensity of the optical signal, and n_2 is the nonlinear refractive index, suppression of these parametric phenomena will require a minimization of n_2.

From the above descriptions, a materials solution to these performance-limiting parasitic phenomena rely on developing optical fibers using materials with reduced (if possible, zero) photoelasticity (p_{12}), peak Raman gain, nonlinear refractive index, and thermo-optic coefficients. Though one composition that exhibits all of these effects would be ideal, it generally is not necessary since each phenomena has a different threshold condition and so often is application specific. The authors therefore suggest a more thoughtful consideration of the fiber's compositions based on the specific application and anticipated optical power levels.

Materials Solutions

As noted above, stimulated Brillouin scattering (SBS) and stimulated Raman scattering (SRS) are two of the principal nonlinearities that plague continued power scaling-up in fiber-based laser systems. The sections below provide an overview of selected optical fiber material solutions to mitigate and, in the case of SBS potentially negate entirely, these parasitic effects.

Intrinsically Low Brillouin Gain Glasses and Optical Fibers

The fundamental question at play is less so "what materials can be used to lower the Brillouin gain" but, rather, "do the materials that reduce Brillouin gain (when added into silica) form a sufficiently stable glass to draw into optical fiber?" While this question may seem straight-forward to answer, it is not always the case. Firstly, glass stability is a kinetic consideration and so the method of fabrication needs to be determined. For the purposes of this work, the conventional method of manufacturing will be one of the chemical vapor deposition methods employed in the commercial production of optical fiber; i.e., outside vapor deposition (OVD), modified chemical vapor deposition (MCVD), vapor axial deposition (VAD), or plasma-assisted chemical vapor deposition (PCVD). In each case, the combination of temperature cycling between soot deposition and translation of the torch along the forming preform followed by subsequent densification and consolidation (and eventual fiber drawing) leads to a fairly extreme set of time and temperature conditions. Second, for practical reasons, the glass from which the fibers are made must be intrinsically low loss, mechanically robust, and compatible with the other fiber components used to make the fiber-based laser system. These conditions conspire to yield silica and high silica-content glasses as the standard family from which practical optical fibers are made.

This, of course, is not an impediment as silica is, and has been for millennia, a marvel when it comes to glass stability and properties. However, in returning to the more fundamental question of which glass systems could be used as the core phase in high performance optical fibers, the answer turns out to be a fairly small range of compositions. From a materials perspective, the principal additions to silica in commercial optical fibers are GeO_2, which raises the refractive index and can enhance photosensitivity, Al_2O_3, raises the refractive index and can enhance the solubility of active rare-earth dopants, P_2O_5, which raises the refractive index and reduces glass viscosity, B_2O_3, which reduces the refractive index and increases the expansion coefficient, and F, which reduces the refractive index and glass viscosity. Importantly, there are limits: for GeO_2 [25] and P_2O_5 [26], there is a complete solid solution with the base SiO_2 such that liquid-liquid or solid-solid immiscibility is not problematic and the optimum dopant concentration is mainly determined by its impact on refractive index, thermal expansion, and viscosity. For a surprisingly wide range of other additives into silica, such immiscibilities exist and greatly restrict the compositions from which optical fibers can be made. This material limitation has the added detriment of restricting the range of useful properties that the glass can exhibit. Accordingly, 'mixing the unmixable' can only be accomplished through novel manufacturing methods that obviate the combinations of time and temperature that conspire to facilitate the phase instabilities [27]. In the sections that follow, the fibers discussed are made using the "molten core" approach, which is described briefly in Section 3 below and references therein. For the purposes of this Review, suffice it to say that the molten core approach is a super-liquidus (or super-consulate point) process where the fiber is drawn at a temperature that exceeds the melting point (or immiscibility limit) of the core phase so that it is a homogeneous liquid. The cladding glass is selected such that it possesses the appropriate viscosity to draw into high quality fiber at this same temperature. Since the core is molten, and generally reasonably fluid, there is expected to be some dissolution of the cladding glass by the core melt during the fiber drawing process. As such, the core of the resultant fiber, which quenches into a glass as the fiber cools, is an 'interaction product' of the initial precursor composition and that of the cladding glass. Given that pure SiO_2 arguably is the most refractory of the conventional glasses, not to mention naturally compatible with commercial optical fiber components, it tends to be the cladding material of choice. Accordingly, the dissolution of some silica into the core melt during molten core processing has the (generally) added benefit of enhancing glass-forming stability of the resultant core.

In specific regard to suppressing stimulated Brillouin scattering (SBS), the natural consequence of the discussion above is whether or not a material solution exists rather than address the issue by spatially distributing the propagating mode over a larger cross-section of the core, i.e., a large mode area approach. As noted above, the Brillouin gain at a given wavelength is dependent upon the refractive index, n, Pockels photoelastic constant, p_{12}, mass density, ρ, acoustic velocity, V_a, and Brillouin spectral width, Δv_B, of the material through which the light it transmitted. In terms of glass modifiers to silica, one desires compounds that increase the refractive index (for optical waveguide formation), density, acoustic velocity, and Brillouin line-width while reducing the photoelastic coefficient. Divalent metal oxides, such as the alkaline earths (MgO, CaO, SrO, and BaO) and a range of trivalent metal oxides, such as Al_2O_3, Y_2O_3, and the lanthanide oxides (Ln_2O_3 where Ln are the rare-earth elements) are known to impart the aforementioned features into silica glasses. However, in each case, there are quite limited compositional ranges over which miscibility exists in the silicate melt. A classic example of this, relevant to optical fiber systems, is the Al_2O_3-SiO_2 system, which exhibits a well-known two-phase, liquid-liquid immiscibility region for binary aluminosilicates [28]. Accordingly, the concentration of Al_2O_3 that can be homogeneously mixed into SiO_2given the processing conditions of conventional optical fiber preforms is about 12 weight percent. Unfortunately, the binary systems Y_2O_3-SiO_2 [29], MgO-SiO_2 [30], CaO-SiO_2, SrO-SiO_2 and BaO-SiO_2 [31], and ternary systems Y_2O_3-Al_2O_3-SiO_2 [32] and MgO-Al_2O_3-SiO_2 [30] also suffer from similar limitations though the exact immiscibility range differs for each system.

As introduced briefly above, the molten core method offers a powerful route to the direct fabrication of optical fibers using materials that otherwise would be very difficult, if not impossible to form into conventional glass preforms and draw. Given the inevitable dissolution of some cladding glass into the core melt, it is often beneficial to begin with a core precursor composition that possesses more than the amount of a given compound desired in the final fiber. Further, and for completeness, it is worth noting that said core precursor phase can be amorphous or crystalline (single or polycrystalline) since it melts during the fiber draw process and therefore has no "memory" of its starting structure. Accordingly, for fiber possessing high concentrations of the aforementioned alkaline earth oxides and sesquioxides the pure end member was employed. As an example, for high Al_2O_3-content fibers, pure sapphire was employed [33]. Similarly for high Y_2O_3-Al_2O_3, BaO- and MgO-Al_2O_3-content fibers, YAG ($Y_3Al_5O_{12}$), BaO and spinel ($MgAl_2O_4$) were employed [34,35,36]. Since such compounds, at the concentrations realized, typically phase-separate when mixed with silica, these fibers exhibited a range of useful properties that had not previously been observed including a record low Brillouin scattering [33], an athermal Brillouin frequency condition [33], and the identification of an atensic (zero stress-dependent) Brillouin frequency composition [35].

In specific regard to Brillouin scattering, one particularly interesting opportunity that presents itself by using these novel glass compositions enabled by the molten core method is the ability to realize optical fibers possessing zero, or near-zero photoelasticity. As noted above, the Brillouin gain is dependent on several factors, including a square-dependence on the p_{12} photoelastic coefficient. While silica glass possesses a p_{12} value that is positive, many of the materials noted above possess also about negative p_{12} values. The result is a glass that can balance the positive and negative photoelastic contributions of its constituent compounds to yield a fiber with very low Brillouin scattering. As a case in point, sapphire (Al_2O_3) possesses a slightly negative p_{12} value (−0.03). When the sapphire melts during the molten core process and some of the silica (p_{12} = 0.271) from the cladding dissolves into it, the resulting aluminosilicate core exhibited an SBS gain that was reduced by nearly 20 dB in comparison to conventional glasses [33]. So 'mixing the unmixable' via the molten core approach can be very useful for obtaining glasses not previously fabricated yet ones with quite extraordinary properties; particularly as they relate to Brillouin scattering. Table 1 provides the derived values for the BGC parameters from a series of novel material fibers made recently. As can also be seen from Table 1, there is a wide range of tailorability in the parameters influencing the Brillouin gain that judicious choice of materials can affect.

Table 1: Brillouin-Gain-Related Material Parameters from Selected Crystal-Derived All-Glass Optical Fibers

Parameter	Unit	Silica (SiO_2)	YAG-Derived Fiber	Al_2O_3-Derived Fiber	MgO-Derived Fiber	BaO-Derived Fiber	Yb_2O_3-Derived Fiber	La_2O_3-Derived Fiber
V_a	m/s	5970	7649	9790	8731	3131	4110	3979
ρ	kg/m³	2200	3848	3350	3322	4688	8102	5676
$\Delta\nu_B$	MHz	17	253	274	†	178	1375	181
n	–	1.444	1.868	1.653	1.810	1.792	1.881	1.877
p_{12}	–	0.271	0.022	−0.03	†	−0.33	−0.123	−0.027
Reference	–	22	34	22,33	36	35	37	49

Intrinsically Low Raman Gain Glasses and Optical Fibers

Unlike Brillouin gain, Raman gain is not linked to material properties that could be potentially driven to zero by choice of composition. However, by analogy to LMA fibers where the propagating optical mode is spread out over the fiber cross-section to reduce its intensity at a given spatial location, the distribution of the glasses' bond energies can be broadened thereby reducing the effective Raman gain at a given wavelength. As a case in point, all-glass optical fibers derived from yttrium aluminum garnet (YAG) were fabricated [24]. It was found that the Raman peak at 440 cm^{-1}, which is associated with the Si–O–Si stretching mode, did not change position with core glass composition but was much broader spectrally likely due to a larger distribution of Si–O–Si bond angles in these more highly modified glasses. The defect line peaks at 490 cm^{-1} (D_1) and 600 cm^{-1} (D_2), which are attributable to the 4- and 3-member ring 'breathing' modes respectively, diminished with the combined (Y_2O_3 + Al_2O_3) content in the resultant glass though their spectral position did not change suggesting that yttria and alumina may lessen the number of ring structures in the glass. These same trends were also later observed for the sapphire-derived (high alumina content) fiber. Raman spectra for the YAG-derived fibers can be found in [24] while Figure 1 shows the set of Raman spectra obtained for the sapphire-derived fiber described above. As mentioned, the features of the spectra trend in ways similar to the YAG-derived fiber. In fact, the Raman gain spectra for the YAG- and sapphire-derived fibers are nearly identical, except that the sapphire-derived fiber appears to be missing the line observed at 950 cm^{-1} (and attributed to the oxygen hole center [24]) from the YAG-derived fibers [37].

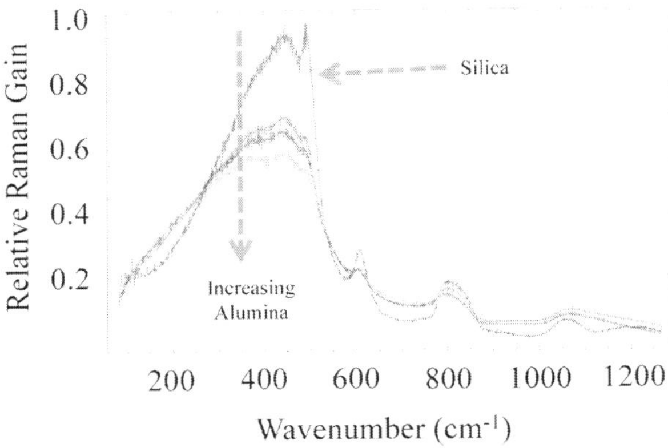

Figure 1: Raman gain spectra (normalized to silica) measured from the sapphire derived fiber for three different alumina concentrations. The spectrum appears to broaden and get weaker relative to the cladding (silica, black line).

The spectra obtained suggest that the core material possesses much more disorder than the pure silica cladding glass. As a consequence of these modifications to the glass structure, the spontaneous Raman scattering intensity from the yttrium-aluminosilicate fibers was found to increase linearly with increasing silica content [24] with a reduction of about 3 dB measured for a silica content of about 67.5 mol%. This trend is similar for the sapphire-derived fiber, with the concentrations of alumina being 26.9, 30.8, and 41.2 mol% for the three aluminosilicate fibers and a maximum reduction in Raman gain of about 2.5 dB for the highest alumina content (see Figure 1). While the reduction in Raman gain can partly be attributable to the replacement of silica with materials of relatively lower Raman gain, it is interesting to note that other common fiber dopant materials (e.g., GeO_2, P_2O_5, and B_2O_3) have Raman gain coefficients larger than that of silica. Thus, the YAG-derived fiber results highlighted here suggest that alumina and yttria impart intrinsically lower Raman gain into the glass. An additional benefit, therefore, of these low-silica content yttrium aluminosilicate glasses for optical fibers is a reduced Raman (and Brillouin) gain and therefore increased SRS (and SBS) threshold.

It is important to note that even with the addition of 'high-Raman-gain' materials to silica it is still possible to achieve some level of reduction to the Raman gain. As an example, the Raman spectra for the baria-derived fiber [35] are shown in Figure 2. The baria concentrations in the three fibers are 10.4, 12.3, and 18.4 mol%. Relative to the cladding spectrum (pure silica), many of the conclusions about the shape of the spectra obtained from the

YAG- and sapphire-derived fiber also seem to hold for the bariosilicate fibers. However, it is also clear that several new lines appear, including a prominent one near 1071 cm^{-1} likely attributable to the Ba-O bond. From the spectra and given the concentrations of baria in the fibers, it can be deduced that the strength of the aforementioned Raman line is relatively large in the glass. In fact, one can deduce by extrapolation that at a baria concentration in the vicinity of 25–30 mol%, the Raman gain will be larger in the baria-derived fiber than that in pure silica. However, at lower BaO concentrations, the contribution to the Raman gain spectrum by baria is at a different phonon frequency, not overlapping with a strong silica phonon line. Thus, as long as the concentration of baria is kept low enough such that its contribution to the Raman spectrum is weaker than silica, the absolute maximum Raman gain coefficient can be reduced. The bottom line is that, in the case of the bario-silicate glasses, the net Raman gain can still be reduced by up to 30% relative to silica.

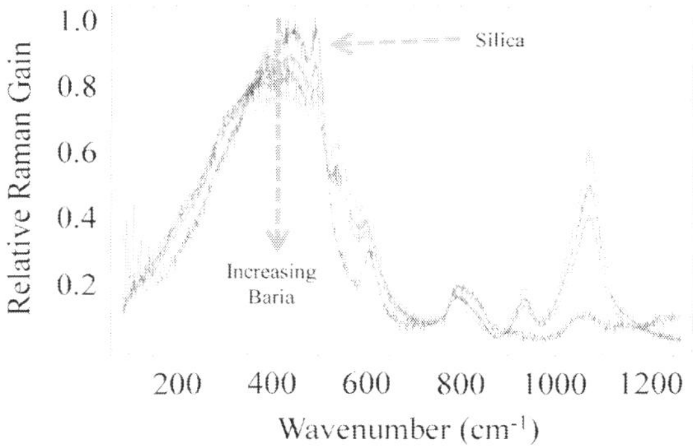

Figure 2: Raman gain spectra (normalized to silica) measured from the barium oxide (BaO) derived fiber for three different baria concentrations. Several new lines appear, likely attributable to the Ba-O bond, including a strong one near 1071 cm^{-1}. The silica spectrum (black line) was obtained from the fiber cladding of one of the fibers.

Optical Fibers with Enhanced Thresholds for Higher Order Mode Instabilities (HOMI)

As noted in the Introduction, there exists in active LMA fibers a power threshold where dynamic randomization of the mode distribution at the laser output is observed, known as "higher order mode instability" (HOMI), and is believed to result from a thermally-induced refractive index grating. Since the process is believed to be driven in part by the thermo-optic coefficient (dn/dT) the natural question is whether marked increases in the HOMI threshold can be obtained through judicious tailoring of the core material's thermo-optic coefficient?

While it has already been shown that YAG-derived fibers possess larger thermal conductivities than their silica counterparts [38] (as another potential way to reduce HOMI), the material dn/dT can also be tailored. Simply stated, if dn/dT = 0, such modal instabilities should be improved, if not completely removed. Much like the aforementioned work on intrinsically-low-Brillouin-gain glasses, combining materials with thermo-optic coefficients of opposite sign can give rise to a significant reduction in dn/dT, and possibly even its taking on a value of zero. Materials such as SiO_2, GeO_2 (dn/dT larger than silica), and Al_2O_3 (dn/dT similar to silica) have positive dn/dT, but this value can also be negative such as in P_2O_5 and B_2O_3 [39]. As an illustrative example, and based on experimental measurements on B_2O_3-doped optical fibers, the thermo-optic coefficient for a borosilicate glass is shown in Figure 3 including a composition where dn/dT = 0 is identified, along with points where the dn/dT is 75% and 50% that of silica. The additive materials model was utilized to generate the curve in Figure 3. The data point where dn/dT reached 50% that of silica is the lowest that the authors have observed to date.

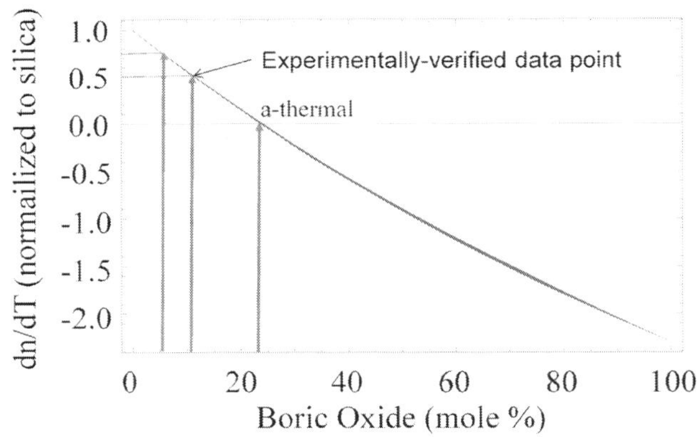

Figure 3: Thermo-optic coefficient of borosilicate glass with increasing boria content. A measured data point on a fabricated fiber is provided. The compositions where dn/dT reduces by 25% and becomes zero (athermal) are identified on the plot.

Indeed if the HOMI threshold is proportional to dn/dT, and if dn/dT can be zeroed through careful materials selection, then the HOMI problem can, in principle, be negated. However, for the example above, this requires the use of large quantities of boria, which may not be compatible with conventional commercial methods of fiber fabrication. Therefore, targeting more modest increases in the HOMI threshold (25% or 50% for example) enables the use of these conventional fabrication methods, but while still taking advantage of significant improvements in performance. It is worth noting that it is not anticipated nonconventional fiber fabrication methods such as the molten core approach will represent limitations to fiber commercializability in the future. In the particular case of the molten core method, conventional draw processes are employed lessening the obstacles to technology transfer.

Nonlinear Refractive Index and Parasitics Depending Thereon

As noted in the Introduction, processes such as self-phase modulation (SPM) and four-wave mixing (FWM) result from the dependence of the refractive index on the optical intensity (the optical Kerr Effect). As with all other nonlinear optical processes, these may be utilized or be considered parastitics, depending on the requirements of an optical system. FWM and SPM may be useful, for example, for wavelength shifters or for generating of optical solitons, which could be useful for high bit-rate optical communications or short pulse lasers, respectively. However, their broadening and modifying of the optical spectrum may be undesirable in high peak power and multi-wavelength optical amplifier systems. The strength of these interactions is related to the nonlinear refractive index n_2, i.e., to the material properties of the glass from which the fibers are made. To first order, the nonlinear index likely carries through the additive model via the refractive index (since, as mentioned above, $n(I) = n_0 + n_2 I$, where n_0 is the linear refractive index, I is the intensity of the optical signal, and n_2 is the nonlinear refractive index) much like the Pockels coefficients do through the strain and stress optic coefficients [22]. To date, however, this has not yet been verified.

There are numerous publications dedicated to the modeling of the nonlinear refractive index [40,41,42]. It is strongly dependent on the linear (nominal) refractive index of the material, and therefore lower refractive index materials tend to have lower n_2 values. An expression for n_2 in terms of widely used glass engineering coefficients, e.g., the Abbe number v_d and the refractive index at the Fraunhofer d-line, is [40]

$$n_2(esu) = \frac{K(n_d - 1)(n_d^2 + 2)^2}{v_d\sqrt{1.517 + (n_d^2 + 2)(n_d + 1)\frac{v_d}{6n_d}}} \quad (1)$$

where K is a material constant that is dependent upon the effective oscillator strength (or strength of the induced dipole moment) and shape of the potential well of the oscillator [40]. Upon inspection it is seen that materials with large Abbe numbers and low n_d are desirable for achieving low nonlinear refractive index values. As a result, classes of fluoride glasses, for example, have relatively low n_2 [42]. Unfortunately, the ranges of Abbe numbers and refractive indices available for most common glass-forming materials are limited (perhaps 50–100 for the former and 1.4 to 1.6 for the latter), and the effectiveness of reducing n_2 in this way is therefore also limited. Of perhaps greater interest is that heretofore a more 'chemical' approach may be taken in the design of materials with predetermined n_2. The nonlinear refractive index depends strongly on the polarizability of the material and, therefore, on the nature of the chemical bond (which strongly influence the parameter K). It has been found that in some covalent bonding systems, in contrast to ionic ones, the value of n_2 (via K [40]) might even take on negative values [43]. Therefore, if an abundance of covalent bonds can be engineered into the material, for example the Al-P bond in the $AlPO_4$ system [44], significant reductions in n_2 may be possible. This work has only just begun but offers exciting possibilities to control n_2-based nonlinearities.

Limitations

Knowledge of the appropriate phase behavior does not always guarantee success. As one example of this, consider the system: Yb_2O_3-SiO_2, which could be generalized to the lanthanide oxides. High rare-earth content glasses, particularly silicate glasses for their potential compatibility with silica-based optical fibers, are of interest for their magneto-optic properties. Specifically, high lanthanide oxide silicate glasses could be useful for Faraday isolators in high energy laser systems.

The Yb_2O_3-SiO_2 phase diagram exhibits a two-liquid immiscibility over a compositional range from about 75–98 mole percent SiO_2 for temperatures above about 1973K to the upper consolute temperature of 2473 K [45]. In this system a eutectic [$L \rightarrow Yb_2O_3 + Yb_2Si_2O_7$] exists at a composition of about 38 mole percent SiO_2 and a temperature of 2125K (1852 °C). The liquidus temperature for this eutectic composition is below the conventional draw temperature for a silica-clad fiber using the molten core approach and so seemed an ideal way to obtain a high Yb-content silicate fiber. Accordingly, homogeneous powders of composition 38 mol% SiO_2/62 mol% Yb_2O_3 were synthesized, pressed into pellets, sleeved into a pure silica tube, and drawn at about 2000 °C.

Figure 4 provides a scanning electron microscope image of the resultant fiber; clearly observed is phase separation. Upon compositional analysis, the average composition as measured to be about 86 mol% SiO_2 (~50 weight percent), this SiO_2 dissolving into the core melt from the pure silica cladding glass. Said composition is right in the center of the aforementioned liquid-liquid immiscibility and the subsequent phase separation was quenched into a solid form as the molten-core fiber cooled.

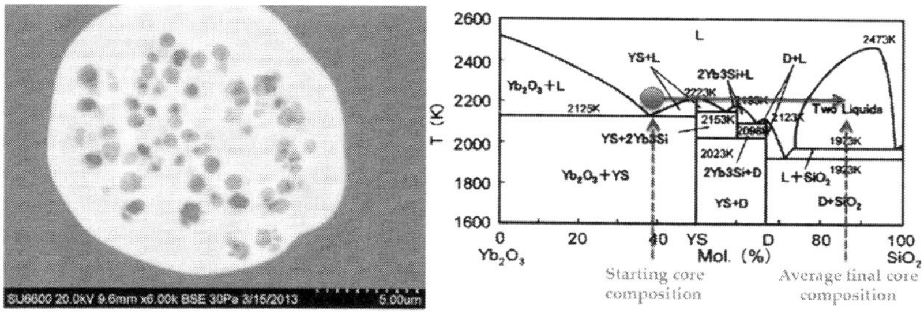

(a) (b)

Figure 4: Scanning electron micrograph of the cross-section of a pure-silica clad, Yb_2O_3-SiO_2 core fiber clearly showing (a) phase separation and (b) the phase diagram for the Yb_2O_3-SiO_2 system (reproduced with permission from [45], Copyright 2005 the Journal of the Ceramic Society of Japan). The phase diagram also depicts the starting core composition and the final average core composition following molten core fiber processing whereby additional SiO_2 dissolves into the core from the cladding and shifts the final core composition into the two liquid immiscibility region.

A Note on Costs

There is a natural tendency to assume that anything "specialty" is necessarily expensive due, in part, to the lack of an economy of scale associated with something that is not commodity. Here are two specific counter-examples to this assumption.

Described above was an optical fiber derived from sapphire. While sapphire is most commonly thought of in the form of a gem stone, industrial sapphire is quite common and inexpensive. Exemplifying this was the sapphire used in said work [33], which measured 4.2 mm in diameter by 100 mm in length (constituting a weight of 27.5 carats) and cost $175 (USD). The telecommunication-grade silica cladding tube costs about $2,000 (USD), bringing the total materials cost to less than $3,000 (USD; only in optics could silica cost 1100% more than alumina). For comparison, an equivalently-sized (27 carat) sapphire gemstone sold at auction for well over $15,000 (USD); the difference between sapphire as gemstone and sapphire as transparent industrial alumina is clear (pun intended). Obviously one needs a fiber draw tower (but not a lathe) to make the fiber but it was, in the grand scheme of achieving a record 19 dB suppression in SBS, very inexpensive. Photonic crystal fibers, which have not achieved this level of SBS suppression and are far more labor-intensive to make, and cost significantly more in material and labor costs.

A second example was the recent report of the first glass optical fiber to propagate light via transverse Anderson localization [46,47] as opposed to the conventional total internal reflection. The use of Anderson localization as a propagation modality opens the door to the use of optical fibers for spatially multiplexed high data capacity information links. The glass used in that work was drawn from a porous artisan glass known as "satin quartz", which costs $75 (USD) for a meter-length rod. As with the sapphire, one still needs a draw tower (but no lathe) and the appropriate know-how and expertise but those are constant no matter what is being drawn.

From the perspective of material and labor costs, these specialty fibers can be remarkably inexpensive. The difference is not cost but understanding of the enabling materials and glass science and chutzpah to try something new.

EXPERIMENTAL SECTION

The specific optical fibers whose properties are described herein have been reported on individually. For detailed information on the material and fiber fabrication and characterization, the reader is referred to the following References: YAG-derived fibers [24,34], sapphire-derived fiber [33], spinel-derived fiber [36], baria-derived fiber [35], ytterbia-derived fiber [48], and lanthana-derived fiber [49].

More generally, though, optical fibers conventionally are prepared via one of a series of chemical vapor deposition methods [50,51,52]. In such cases, the core and clad glasses possess reasonably similar thermal (particularly glass transition temperatures, T_g) and thermomechanical properties such that they can be co-drawn into a waveguiding fiber that possesses high strength and appropriate optical properties. This thermal and thermomechanical compatibility tends to imply that the compositions of said core and clad glasses are not overly different.

However, as noted above, most of the fibers treated in this work possess compositions that suffer from phase instabilities, such as liquid-liquid immiscibility, and so cannot be processed using the aforementioned conventional measures. As such, a new method had to be developed that yielded high quality optical fibers while still being industrially relevant. Based on the original work of Reference [53], the molten core approach was employed. References [4,54] provides a more detailed review of the molten core method and material solutions primarily for SBS and SRS mitigation but does not address issues associated with SPM, FWM, and HOMI (which is discussed in this work). Briefly, instead of relying on core/clad material compatibility, the molten core approach requires that the core phase melt at a temperature below where the cladding phase draws into fiber; in other words, during the fiber fabrication, the core is above its liquidus (or upper consulate point if an immiscibility exists). Silica is a preferred cladding glass for the molten core approach not only because it facilitates compatibility of the molten core-derived fibers with conventional silica-based fibers, but also it is a highly refractory glass such that many core phases exist that melt at temperatures below the draw temperature for pure silica (ranging from about 1925 to 2050 °C, depending on size of silica preform).

While advantageous in many cases, the cladding glass is attacked by the core melt and is somewhat dissolved. This makes the final post-drawn fiber an interaction product of the initial core phase and the cladding glass, depending upon the draw time and temperature as well as preform and fiber dimensions. In the specific case of SBS suppression, where most of the core phases chosen possess negative photoelastic constants, the dissolution of the silica (possessing a positive photoelastic constant) can give rise to a resultant multi-component glass with near-zero photoelasticity.

CONCLUSIONS

Provided herein has been a review of the present state of next generation optical fibers that meet the needs of present and future high energy laser systems from the specific perspective of SBS and SRS mitigation through a purely materials approach. Simple materials systems, including those in the Al_2O_3-SiO_2, BaO-SiO_2, Al_2O_3-Y_2O_3-SiO_2, and Al_2O_3-MgO-SiO_2 families, have been shown to yield optical fibers of classic core/clad geometries that exhibit reductions in SBS by nearly 20 dB and SRS by about 3 dB, though much work remains. Also described were fibers with marked increases in the threshold for higher order mode instabilities (HOMI) though judicious tailoring of the core materials' thermo-optic coefficient, dn/dT.

Though most of the focus here has been placed on crystal-derived all-glass optical fibers fabricated using the cost-effective and industrially compatible molten core approach, the more general message is that materials clearly offer solutions to present and future problems plaguing the high energy fiber-based laser sector. Accordingly, significant opportunities exist for the optical materials community in terms of scholarly topics for study, funding to do so, and job prospects.

ACKNOWLEDGMENTS

Many have contributed to the work described herein—both experimentally as well as conceptually. The authors are especially indebted to Bob Rice (Dreamcatchers Consulting), Wade Hawkins (Clemson University), Courtney Kucera (Clemson University), Roger Stolen (Clemson University), Anna Peacock (University of Southampton), Mark Dubinskiy (US Army Research Lab), Josh Rothenberg (Northrop Grumman Corporation), and thought-leaders as relates to this opportunity for materials-related optical technology enablers, particularly John Mauro, Claudio Mazzali, and Mike Pambianchi, all of Corning Incorporated. Numerous grants and contracts financially supported various parts of this work including the US Department of Defense Joint Technology Office (JTO) through their High Energy Laser Multidisciplinary Research Initiative (HEL-MRI) program contracts W911NF-05-1-0517, FA9550-07-1-0566, and W911NF-12-1-0602, US Army Defense University Research Instrumentation Program (DURIP) award W911NF-07-1-0325, and the Northrop Grumman Corporation.

AUTHOR CONTRIBUTION

John Ballato analyzed data and wrote some of the manuscript. Peter Dragic conducted experimental work, analyzed data and wrote some of the manuscript. John Ballato and Peter Dragic jointly conceived of the work.

REFERENCES

1. National Research Council. Optics and Photonics: Essential Technologies for Our Nation; The National Academies Press: Washington, DC, USA, 2013.
2. Baer, T.; Schlachter, F. Report to the OSTP—Lasers in Science and Technology 2010; Stanford University: Stanford, CA, USA, 2013.
3. Mazzali, C. The Timing for Specialty Optical Fibers. In Proceedings of the Workshop on Specialty Optical Fibers and their Applications, Optical Society of America Topical Digest, paper T3.3, Sigtuna, Sweden, 28–30 August 2013.
4. Ballato, J.; Dragic, P. Rethinking Optical Fiber: New Demands, Old Glasses. J. Am. Ceram. Soc. 2013, 96, 2675–2692. [CrossRef]
5. Mauro, J.; Philip, C.; Vaughn, D.; Pambianchi, M. Glass Science in the United States: Current Status and Future Directions. Int. J. Appl. Glass Sci. 2014, 4, 1–14.
6. Stolen, R. The early years of fiber nonlinear optics. J. Light. Technol. 2008, 26, 1021–1031.
7. Fermann, M. Single-mode excitation of multimode fibers with ultrashort pulses. Opt. Lett. 1998, 23, 52–54.
8. Koplow, J.; Kliner, D.; Goldberg, L. Single-mode operation of a coiled multimode fiber amplifier. Opt. Lett.2000, 25, 442–444.
9. Limpert, J.; Schmidt, O.; Rothhardt, J.; Röser, F.; Schreiber, T.; Tünnermann, A.; Ermeneux, S.; Yvernault, P.; Salin, F. Extended single-mode photonic crystal fiber lasers. Opt. Express 2006, 14, 2715–2720.
10. Ramachandran, S.; Nicholson, J.; Ghalmi, S.; Yan, M.; Wisk, P.; Monberg, E.; Dimarcello, F. Light propagation with ultralarge modal areas in optical fibers. Opt. Lett. 2006, 31, 1797–1799.
11. Sosnowski, T.; Kuznetsov, A.; Maynard, R.; Ma, X.; Zhu, C.; Hu, I.; Galvanauskas, A.; Koponen, J.; McComb, T. 3C Yb-doped Fiber Based High Energy and Power Pulsed Fiber Lasers. In Proceedings of the SPIE 8601, Fiber Lasers X: Technology, Systems, and Applications, Photonics West, San Fancisco, CA, USA, 2 February 2013.
12. Gaponov, D.; Février, S.; Devautour, M.; Roy, P.; Likhachev, M.; Aleshkina, S.; Salganskii, M.; Yashkov, M.; Guryanov, A. Management of the high-order mode content in large (40 μm) core photonic bandgap Bragg fiber laser. Opt. Lett. 2010, 35, 2233–2235.

13. Kashiwagi, M.; Saitoh, K.; Takenaga, K.; Tanigawa, S.; Matsuo, S.; Fujimaki, M. Effectively single-mode all-solid photonic bandgap fiber with large effective area and low bending loss for compact high-power all-fiber lasers. Opt. Express 2012, 20, 15061–15070.
14. Dong, L.; Wu, T.; McKay, H.; Fu, L.; Li, J.; Winful, H. All-Glass Large-Core Leakage Channel Fibers. IEEE J. Sel. Top. Quantum Electron. 2009, 15, 47–53.
15. Laperle, P.; Paré, C.; Zheng, H.; Croteau, A. Yb-Doped LMA Triple-Clad Fiber for Power Amplifiers. In Proceedings of the SPIE 6453, Fiber Lasers IV: Technology, Systems, and Applications, San Jose, CA, USA, 20 January 2007.
16. Otto, H.; Stutzki, F.; Jansen, F.; Eidam, T.; Jauregui, C.; Limpert, J.; Tünnermann, A. Temporal dynamics of mode instabilities in high-power fiber lasers and amplifiers. Opt. Express 2012, 20, 15710–15722.
17. Jauregui, C.; Eidam, T.; Otto, H.; Stutzki, F.; Jansen, F.; Limpert, J.; Tünnermann, A. Physical origin of mode instabilities in high-power fiber laser systems. Opt. Express 2012, 20, 12912–12925.
18. Agrawal, G.P. Nonlinear Fiber Optics, 2nd ed.; Academic Press: San Diego, CA, USA, 1995.
19. Winkelmann, A. Ueber die specifischen Wärmen verschieden zusammengesetzter Gläser [On the specific heats of different composite glasses]. Ann. Phys. Chem. 1893, 49, 401–420. (In German)
20. Winkelmann, A.; Schott, O. Über die Elastizität und über die Zugund Druckfestigkeit verschiedener neuer Gläser in ihrer Abhängigkeit von der chemischen Zusammensetzung [On the elasticity and the tensile and compressive strength of several new glasses in their dependence on the chemical composition]. Ann. Phys. Chem. 1894, 51, 697–730. (In German)
21. Law, P.; Dragic, P. Wavelength dependence of the Brillouin spectral width of boron doped germanosilicate optical fibers. Opt. Express 2010, 18, 18852–18865.
22. Dragic, P.; Ballato, J.; Morris, S.; Hawkins, T. Pockels' coefficients of alumina in aluminosilicate optical fiber. J. Opt. Soc. Am. B 2013, 30, 244–250.
23. Stolen, R.; Ippen, E. Raman gain in glass optical waveguides. Appl. Phys. Lett. 1973, 22, 276–278.
24. Dragic, P.; Ballato, J. Characterisation of Raman gain spectra in

Yb:YAG-derived optical fibres. Electron. Lett.2013, 49, 895–897.

25. Baret, G.; Madar, R.; Bernard, C. Silica-Based Oxide Systems I. Experimental and Calculated Phase Equilibria in Silicon, Boron, Phosphorus, Germanium, and Arsenic Oxide Mixtures. J. Electrochem. Soc.1991, 138, 2830–2835.
26. Tien, T.; Hummel, F. The System SiO_2–P_2O_5. J. Am. Ceram. Soc. 1965, 48, 75–80.
27. Dragic, P.; Ballato, J. 120 Years of Optical Glass Property Calculations: From the Law of Mixtures and the Birth of Glass Science to Mixing the Unmixable. Opt. Photonics News 2014, in press.
28. MacDowell, J.; Beall, G. Immiscibility and Crystallization in Al_2O_3-SiO_2 Glasses. J. Am. Ceram. Soc. 1969, 52, 17–25.
29. Hyatt, M.; Day, D. Glass Properties in the Yttria-Alumina-Silica System. J. Am. Ceram. Soc. 1987, 70, C283–C287.
30. Jung, I.; Decterov, S.; Pelton, A. Critical Thermodynamic Evaluation and Optimization of the MgO-Al_2O_3, CaO-MgO-Al_2O_3, and MgO-Al_2O_3-SiO_2 Systems. J. Phase Equilibria Diffus. 2004, 25, 329–345.
31. Frantz, J.; Mysen, B. Raman Spectra and Structure of BaO-SiO_2, SrO-SiO_2, and CaO-SiO_2 melts at 1600 °C. Chem. Geol. 1995, 121, 155–176.
32. Zhang, Y.; Navrotsky, A. Thermochemistry of Glasses in the Y_2O_3-Al_2O_3-SiO_2 System. J. Am. Ceram. Soc.2003, 86, 1727–1732.
33. Dragic, P.; Hawkins, T.; Morris, S.; Ballato, J. Sapphire-derived all-glass optical fibers. Nat. Photon. 2012, 6, 629–635.
34. Dragic, P.; Law, P.; Ballato, J.; Hawkins, T.; Foy, P. Brillouin Spectroscopy of YAG-Derived Optical Fibers. Opt. Express 2010, 18, 10055–10067.
35. Dragic, P.; Kucera, C.; Furtick, J.; Guerrier, J.; Hawkins, T.; Ballato, J. Brillouin Spectroscopy of a Novel Baria-doped Silica Glass Optical Fiber. Opt. Express 2013, 21, 10924–10941.
36. Mangognia, A.; Kucera, C.; Guerrier, J.; Furtick, J.; Dragic, P.; Ballato, J. Spinel-Derived Single Mode Optical Fiber. Opt. Mater. Express 2013, 3, 511–518.
37. Liu, Y.; Galvin, T.; Hawkins, T.; Ballato, J.; Dong, L.; Foy, P.; Eden, J.; Dragic, P. Linkage of Oxygen Deficiency Defects and Rare Earth Concentrations in Glass Optical Fiber Probed by Ultraviolet Absorption and Laser Excitation Spectroscopy. Opt. Express 2012, 20,

14494–14507.

38. Ballato, J.; Hawkins, T.; Foy, P.; Kokuoz, B.; Stolen, R.; McMillen, C.; Daw, M.; Su, Z.; Tritt, T.; Dubinskii, M.; et al. On the Fabrication of All-Glass Optical Fibers from Crystals. J. Appl. Phys. 2009, 105.
39. Shima, K.; Himeno, K.; Sakai, T.; Okude, S.; Wada, A.; Yamauchi, R. A novel temperature-insensitive long-period fiber grating using a boron-doped-germanosilicated-core fiber. In Proceedings of the Conference on Optical Fiber Communication, Dallas, TX, USA, 16–21 February 1997.
40. Boling, N.; Glass, A.; Owyoung, A. Empirical Relationships for Predicting Nonlinear Refractive Index Changes in Optical Solids. IEEE J. Quantum Electron. 1978, QE-14, 601–608.
41. Fournier, J.; Snitzer, E. The nonlinear refractive index of glass. IEEE J. Quantum Electron 1974, QE-10, 473–475.
42. Töpfer, T.; Hein, J.; Philipps, J.; Ehrt, D.; Sauerbrey, E. Tailoring the nonlinear refractive index of fluoride-phosphate glasses for laser applications. Appl. Phys. B 2000, 71, 203–206.
43. Agrawal, G.P.; Flytzanis, C. Delocalization and superalternation effects in the nonlinear susceptibilities of one-dimensional systems. Chem. Phys. Lett. 1976, 44, 366–370.
44. DiGiovanni, D.J.; MacChesney, J.B.; Kometani, T.Y. Structure and properties of silica containing aluminum and phosphorus near the AlPO4 join. J. NonCryst. Solids 1989, 113, 58–64.
45. Yamamoto, H.; Akiyama, K.; Hirata, T.; Murakami, Y. Dependence of Yb_2O_3/SiO_2 Molar Ratio on High temperature Characteristics of Gas Pressure Sintered Si_3N_4. J. Ceram. Soc. Jpn. 2005, 113, 325–329.
46. Karbasi, S.; Hawkins, T.; Ballato, J.; Koch, K.; Mafi, A. Transverse Anderson localization in a disordered glass optical fiber. Opt. Mater. Express 2012, 2, 1496–1503.
47. Karbasi, S.; Frazier, R.; Koch, K.; Hawkins, T.; Ballato, J.; Mafi, A. Image Transport through a Disordered Optical Fiber Mediated by Transverse Anderson Localization. Nat. Commun. 2014, 5.
48. Dragic, P.; Ballato, J.; Morris, S.; Hawkins, T. The Brillouin gain coefficient of Yb-doped aluminosilicate glass optical fibers. Opt. Mater. 2013, 35, 1627–1632.
49. Dragic, P.; Litzkendorf, D.; Kucera, C.; Ballato, J.; Schuster, K. Brillouin Scattering Properties of Lanthano-Aluminosilicate-Core

Optical Fiber. Appl. Opt. 2014. submitted.

50. MacChesney, J.; DiGiovanni, D. Materials Development of Optical Fiber. J. Am. Ceram. Soc. 1990, 73, 3537–3556.
51. Nagel, S.; MacChesney, J.; Walker, K. An Overview of the Modified Chemical Vapor Deposition (MCVD) Process and Performance. IEEE Trans. Microw. Theory Tech. 1982, 30, 305–322.
52. Izawa, T. Early days of VAD process. IEEE J. Sel. Top. Quantum Electron. 2000, 6, 1220–1227.
53. Ballato, J.; Snitzer, E. Fabrication of Fibers with High Rare-Earth Concentrations for Faraday Isolator Applications. Appl. Opt. 1995, 34, 6848–6854.
54. Morris, S.; Ballato, J. Molten Core Fabrication of Novel Optical Fibers. Bull. Am. Ceram. Soc. 2013, 92, 24–29.

Chapter 3

Fabrication of Fresnel Plates on Optical Fibres By Fib Milling For Optical Trapping, Manipulation and Detection of Single Cells

Rita S. Rodrigues Ribeiro [1,2], Pabitra Dahal[2], Ariel Guerreiro[1], Pedro A. S. Jorge[1] & Jaime Viegas[2]

[1]INESC TEC, Rua do Campo Alegre, 687, Porto, Portugal and Departamento de Física e Astronomia, Faculdade de Ciências, Universidade do Porto, Rua Campo Alegre, 687, Porto, Portugal

[2]Masdar Institute of Science and Technology, Abu Dhabi, United Arab Emirates

ABSTRACT

The development of economical optical devices with a reduced footprint foreseeing manipulation, sorting and detection of single cells and other

Citation: Rita S. Rodrigues Ribeiro, Pabitra Dahal, Ariel Guerreiro, Pedro A. S. Jorge & Jaime Viegas "Fabrication of Fresnel plates on optical fibres by FIB milling for optical trapping, manipulation and detection of single cells" Scientific Reports volume 7, Article number: 4485 (2017). https://doi.org/10.1038/s41598-017-04490-2

micro particles have been encouraged by cellular biology requirements. Nonetheless, researchers are still ambitious for advances in this field. This paper presents Fresnel zone and phase plates fabricated on mode expanded optical fibres for optical trapping. The diffractive structures were fabricated using focused ion beam milling. The zone plates presented in this work have focal distance of ~5 µm, while the focal distance of the phase plates is ~10 µm. The phase plates are implemented in an optical trapping configuration, and 2D manipulation and detection of 8 µm PMMA beads and yeast cells is reported. This enables new applications for optical trapping setups based on diffractive optical elements on optical fibre tips, where feedback systems can be integrated to automatically detect, manipulate and sort cells.

INTRODUCTION

Light driven tools, such as, optical tweezers (OTs), are one of the main breakthroughs of the last decades. Optical manipulation was first demonstrated by A. Ashkin, in 1970, where a micro particle was trapped by two counter propagating laser beams due to radiation pressure[1].

Optical tweezers are regularly used in the immobilization and manipulation of a wide range of particles, while precisely measuring positions and applied forces. In microrheology, OTs play an important role in the measurement of viscosity of fluids using trapped beads as test probes[2]. In particle physics, the generation of cold atoms is attained due to magneto-optical traps[3,4], enabling further advances in quantum technology[5]. At the same time, biomedicine is equally benefiting from recent advances on OTs, with a diversity of applications being reported, such as research on red blood cell deformation and aggregation[6,7,8], dynamics of molecular motors[9], such as myosin and kinesin[10,11], or evaluation of forces generated by the transcription of enzymes in DNA strands[12]. Simultaneously, cell separation and sorting using optical trapping systems are also tools frequently used in biology with growing applicability[13].

The platforms employed in the analysis of the particle targets normally rely on modified microscopes with high numerical aperture objectives, usually adapted to accommodate other parts, such as, spatial light modulators or quadrant photo detectors, for the generation of spatial traps and particle displacement monitoring[14], respectively. Nonetheless, the development of OTs is growing towards optofluidic platforms, envisioning lab-on-a-chip tools. The lenses responsible for the optical trapping are one of the key elements that may be improved. With this in mind, this paper presents a set of Fresnel zone and phase plates, fabricated on common optical fibres using focused ion beam milling for trapping purposes. The literature on trapping using single optical fibres has seen advances in the design and capabilities of the lenses[15]. Early on, optical fibre probes were limited to tapered fibres[16], fabricated by chemical etching[17], thermal processes[18], among others[19]. For instance, Baojun Li has reported the use of tapered fibres for trapping, and manipulation of single or multiple targets[20,21]. The vast range of applications of these fibre probes show their flexibility[22,23,24]. However, progress in microfabrication technology has enabled better-quality trapping probes, envisaging higher trapping control[25,26]. FIB milling is a fabrication method that allows to control, with optical subwavelength resolution, the features of the designed structures, contrarily to the methods mentioned above. At the same time, besides trapping and manipulation, the structures presented in this work are also used for size-based detection of particles.

In this work, the choice of Fresnel diffractive structures is linked with three main reasons: Fresnel lenses are an alternative to conventional objectives, since their design allows tailoring features like focusing distance and numerical aperture, matching the desired values necessary in OTs; the planar design offers significant chances to be implemented in microfluidic channels[27,28], incorporated in optofluidic devices; and finally, the possibility to replicate these structures using nanoimprinting lithography, to fabricate a large number of fibre probes[29]. There is some literature reporting the fabrication of Fresnel plates on optical fiber tips using femtosecond laser micromachining and FIB milling. The first method was employed in the fabrication of structures mainly focused on light coupling devices[30,31,32]. Nevertheless, the second process was used to fabricate a structure designed for optical trapping of subwavelength particles[33], on the contrary, the present work reports trapping, manipulation and detection of micron-sized particles and cells, foreseeing applications to larger target sizes, such as mammalian cells.

FABRICATION AND OPTICAL CHARACTERIZATION

Design

A Fresnel zone plate (FZP), is a diffractive optical element, composed by a sequence of concentric alternating opaque and transparent zones, with axial symmetry34. The radii of the successive rings are given by:

$$r_n = \sqrt{n\lambda f + \frac{n^2\lambda^2}{4}}, \tag{1}$$

where n is an integer, λ is the wavelength and f is the focal length. For $\lambda \ll f$, Eq. (1) reduces to:

$$r_n = \sqrt{n\lambda f}. \tag{2}$$

The intensity of the focal point is proportional to the number of zones. These structures have chromatic aberration, such as convex lenses, since f is inversely proportional to λ. In contrast, Fresnel plates have multiple foci: composed by the main focal point, f, and higher order focal points at f/3, f/5, …, with decreasing brightness35. From this, a Fresnel zone plate is basically an amplitude device. Nevertheless, an alternative plate may be achieved if the opaque zones are replaced by transparent π-phase steps. In this case the resulting focal spot will be brighter, increasing the optical conversion efficiency of the plate36, since the light is no longer blocked. This is the so-called Fresnel phase plate (FPP). The depth (d) of the phase zones is given by

$$d = \frac{\lambda}{2(n_{plate} - n_{medium})}, \tag{3}$$

where n_{plate} and n_{medium} are the refractive index of the plate and the surrounding medium, respectively. The size of FZP and FPP on single mode optical fibres (SMF) is limited by their mode field diameter, which is normally a few micrometres wide. Since the number of zones directly affects the intensity and width of the central peak of the diffraction pattern, it makes sense that mode-expanded optical fibres are used30. A scheme illustrating the optical fibre tips is visible in Fig. 1(a). The single mode fibre is spliced to a multimode one, allowing the Gaussian mode that propagates in the SMF to expand in the multimode segment. This results in a broader effective working area on the top of the tip. The cross section that is covered by the beam depends of the length of the MMF, and can be calculated through the numerical aperture relation.

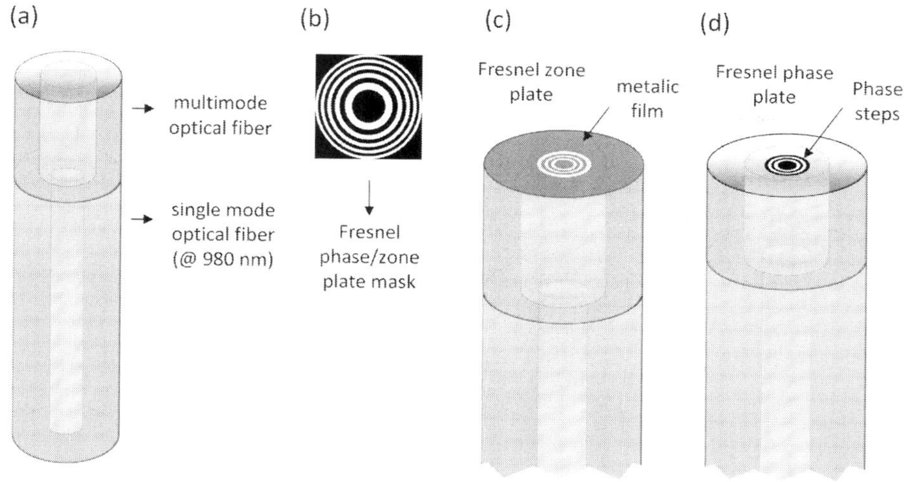

Figure 1: (a) optical fibre probe design composed by the SM fibre and the MM segment; (b) Fresnel phase/zone plate characteristic ring structure; (c) Fresnel zone plate on the tip of the optical fibre (coated with 20 nm Pt film); (d) Fresnel phase plate on the tip of the optical fibre (no film).

Fabrication

The production of diffractive structures, such as Fresnel plates (see Fig. 1(b)), requires high resolution fabrication methods. To address this challenge, several authors have proposed the use of distinct processes, such as: femtosecond pulsed laser micromachining30, 31, e-beam lithography27, focused ion beam milling37,38,39, among others28, 40. In this work, the fabrication of the Fresnel zone and phase plates is carried out using focused ion beam (FIB) milling41, 42.

The single and multi-mode fibres described in this paper are SM 980 (Thorlabs) and AFS 50/125Y (Thorlabs), respectively. The procedure starts with the splice of a SM with a MM fibre segment. After this, the MMF is cleaved at an optimal length. A USB camera with a magnifying lens is placed above the cleaving machine allowing the observer to control the process. The fibre is moved, relatively to the blade, using a micrometric positioner placed before the machine. The imaging setup allows a continuous magnification range from 50x to 500x, and the micro stage has a resolution of 0.5 μm thus enabling to cut the MMF with enough accuracy. The length of the MM is important because it determines the dimensions of the fibre cross section covered by the beam. In this case, the extent of the MMF is optimized so that the beam covers the diameter of the plate that will be milled on the fibre tip. Once cleaved and cleaned, the fibres are finally mounted in a 45° stub, that is suitable to be used in the FIB.

The FIB is a nano resolution fabrication process, where Gallium ions are accelerated towards the sample43. Depending on the acceleration current, it can either be used in the removal of atoms from the sample's surface, or to image it. In this case the FIB is integrated in a dual-beam system with a scanning electron microscope (Quanta 3D, FEI Company), providing high resolution imaging.

In the fabrication of the FZP, initially a layer of 20 nm of Platinum (Pt) is deposited over the optical fibres, using e-beam evaporation. The metallic film is necessary for two reasons: to avoid charging effects on the dielectric fibre surface during the FIB milling and to block the light, by the odd zones of the plate (Fig. 1(c)). In contrast, for the fabrication of FPP, the optical fibres are coated with a thin layer, 5 nm, of Gold/Palladium (Au/Pd) (see Fig. 1(d)). In this case, the goal of this layer is merely to avoid the charging effects during the milling and will be removed after the fabrication of the phase plates. In this paper, the fabrication of four different types of plates is described: two FZP (FZP-1 and FZP-2) and two FPP (FPP-1 and FPP-2).

FZP-1, composed by 10 zones, is projected to have a focal distance of 3 μm and a radius (r_{10}) of 5.42 μm. The amplitude mask used during the fabrication process is shown in Fig. 2(a). During the fabrication, the exposure dose is controlled by the dwell time, corresponding to a bitmap pixel value (Fig. 2(a)). In this fashion, for a white pixel (level 255) the dwell time was set to 2 μs, while for a black pixel (level 0) the beam is blanked. The ion beam current was set to 0.3 nA (beam diameter 31 nm), corresponding to a milling time of approximately 41 seconds. An image of the optical fibre with the FZP can be seen in Fig. 2(b), corresponding zoom in Fig. 2(c). Analysing this image, a radius of 5.57 μm is measured. This value is slightly larger than the projected one, having a relative error of 2.73%.

The second amplitude plate, FZP-2 is composed by 22 zones, and is dimensioned to have a focal distance of 6 μm, corresponding to a radius (r_{22}) of 11.37 μm. The amplitude mask correspondent to the profile imprinted on the top of the fibre is presented in Fig. 2(d). The beam current and dwell time were 0.3 nA and 1 μs, respectively. The overall milling time was approximately 4 minutes. The resulting zone plate can be seen in Fig. 2(e,f). From the images, a radius (r_{22}) of 12.18 μm was estimated, with a relative error of 7.05%. Please note that the structures visible in the boarder or the fibre are just for control of the fabrication conditions, the performance of the central plate is not influenced by those.

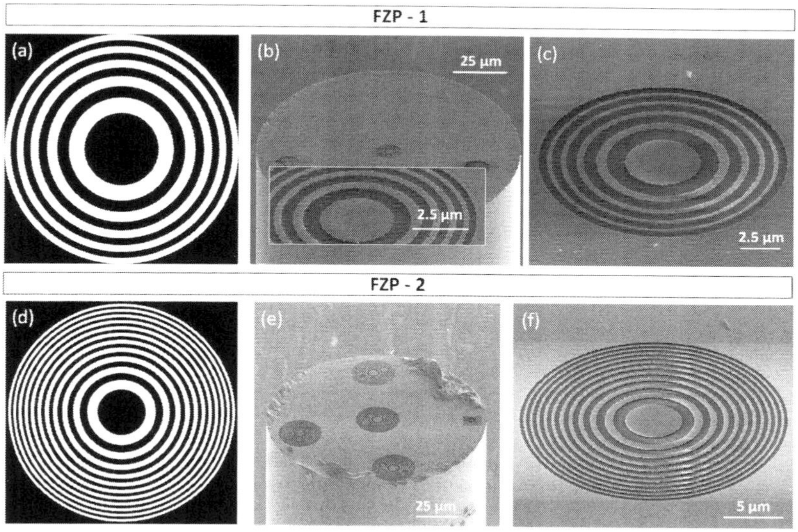

Figure 2: Fresnel zone plates: (a) amplitude mask used to fabricate FZP-1; (b) image of the optical fibre with FZP-1, (inset) zoom of a FZP fabricated with a metallic film of Au/Pd; (c) zoom of FZP-1; (d) amplitude mask used to fabricate FZP-2; (e) image of the optical fibre with FZP-2; (f) zoom of FZP-2.

Previous to the fabrication of FZP-1 and 2, some tests were carried out, to check the final conditions of the plates surfaces, regarding the smoothness. When using Au/Pd to cover the optical fibres, after the milling, the surface was quite rough. This is due to the different sputtering rates for each type of metal. An example of a zone plate with an irregular surface is visible in the inset of Fig. 2(b). In this regard, the fibres were covered by Pt.

FPP-1 is composed by 4 zones, of alternated depths. It was adjusted to have a focal distance of 10 µm, corresponding to a calculated radius (r_4) of 6.26 µm. The optical fibre tip used in this case, has a MM segment with a length of ~30 µm. Figure 3(a) depicts the amplitude mask used in this particular case. The beam current and dwell time were 0.1 nA and 1 µs, respectively. The overall milling time was approximately 15 minutes. The resultant plate can be seen in Fig. 3(b,c). Once again, the radius of the fabricated plate (r_4=6.11 µm, relative error=2.35%) was slightly different from the projected one. From the SEM images, the height of the phase discontinuities is estimated to be around 534 nm. After the FIB milling, the optical fibres were cleaned with aqua regia, to remove the metallic film.

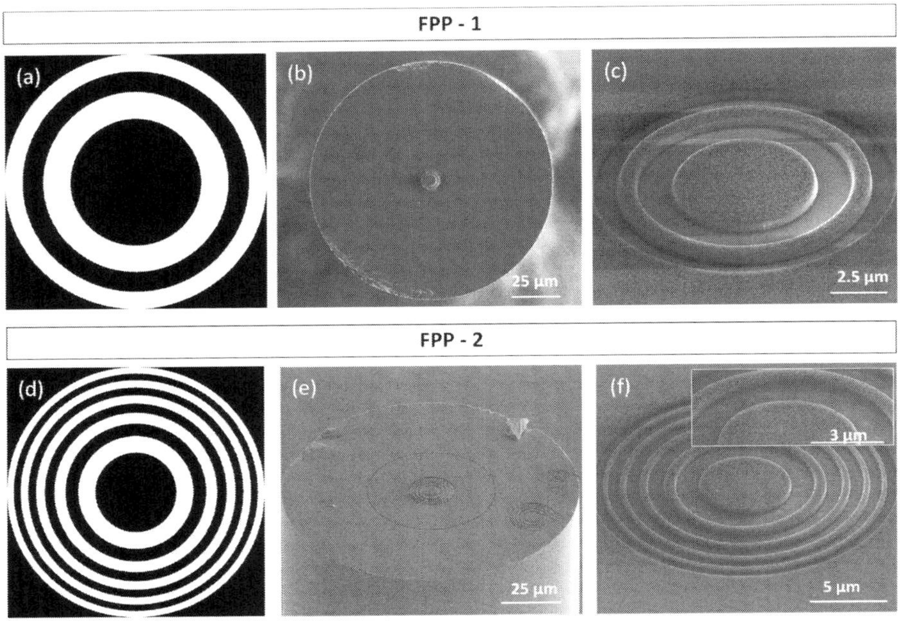

Figure 3: Fresnel phase plates: (a) amplitude mask used to fabricate FPP-1; (b) image of the optical fibre with FPP-1; (c) zoom of FPP-1; (d) amplitude mask used to fabricate FPP-2; (e) image of the optical fibre with FPP-2; (f) zoom of FPP-2, (inset) details of the first three zones of FPP-2.

In order to verify the influence of the number of rings a second zone plate is here presented. FPP-2 is composed by 10 zones, and is also dimensioned to have a focal distance of 10 µm, corresponding to a calculated radius (r_{10}) of 9.90 µm. The length of the MM segment is ~50 µm. Figure 3(d) depicts the amplitude mask used in this particular case. The beam current and dwell time were 0.1 nA and 1 µs, respectively. The overall milling time was approximately 18 minutes. The phase plate can be seen in Fig. 3(e,f). The radius was estimated to be 10.14 µm with a relative error of 2.43%. The height of the phase steps is 419 nm. The inset of Fig. 3(f) shows some imperfections on the surface of the plate, due to some constrains of the FIB gun. However, the dimensions are very reduced (Fig. 3(f), inset), and no effects were observed on the output beam profile.

Characterization of Fresnel plates

The characterization of the output beams generated by the Fresnel zone and phase plates was performed using the setup presented in Fig. 4. First, the

This is demonstrated by the plot of the intensity curves, along the propagation direction, in Fig. 5(a,b). Besides the main focal point, the plate exhibits two secondary focusing positions, at 10 μm and 20 μm (experimental values). The number of extra focal points is supported by the simulations, although the locations are not exactly the same.

To analyse the second zone plate, FZP-2, an analogous process was followed. Hereof, Fig. 5(c,d) depicts the experimental and computational results, respectively. In this case, FZP-2 was dimensioned to have a focal distance of 6 μm, which is verified by the simulations. Regarding the experiments, the focal distance is situated at 5 μm. Similarly to FZP-1, this structure also causes secondary focal points, which are also corroborated by the simulations.

In Fig. 5(e,f), the experimental transversal profiles at the main focal points for each zone plate are depicted. In this case, this allows to calculate the full width at half maximum (FWHM) of the central peak of the resultant patterns. For FZP-1, the inner peak has a dimeter of 0.9 μm, whereas for FZP-2 is 0.6 μm. This shows the influence of the number of zones, i.e., although the higher focal distance of FZP-2, the focusing of the light is stronger, since the plate is composed by twice the number of diffractive rings.

The analysis of the output beams of the phase plates (FPP-1 and FPP-2) follows a similar approach. Hereupon, Fig. 6(a,b) depict the results for FPP-1, which is projected to have a focal distance of 10 μm. Overall, the experimental and computational results match with the projected value. Looking at the measured intensity profile plotted in Fig. 6(a), the maximum of intensity should be located at ~10 μm. In this regard, this is in line with the computational results, Fig. 6(b), where the intensity maximum is located at 10 μm, and slightly decreases after this point.

Figure 6(c,d) depict the results for FPP-2, which was also projected to have a focal point at 10 μm, however, containing more rings than the previous structure. Once again, from the experimental results (Fig. 6(c)) one may conclude that the maximum of intensity is located at ~10 μm. The simulations also present a maximum at ~10 μm, that slowly loses intensity along a few micrometres. This results are coherent with the previous ones, presented for FPP-1.

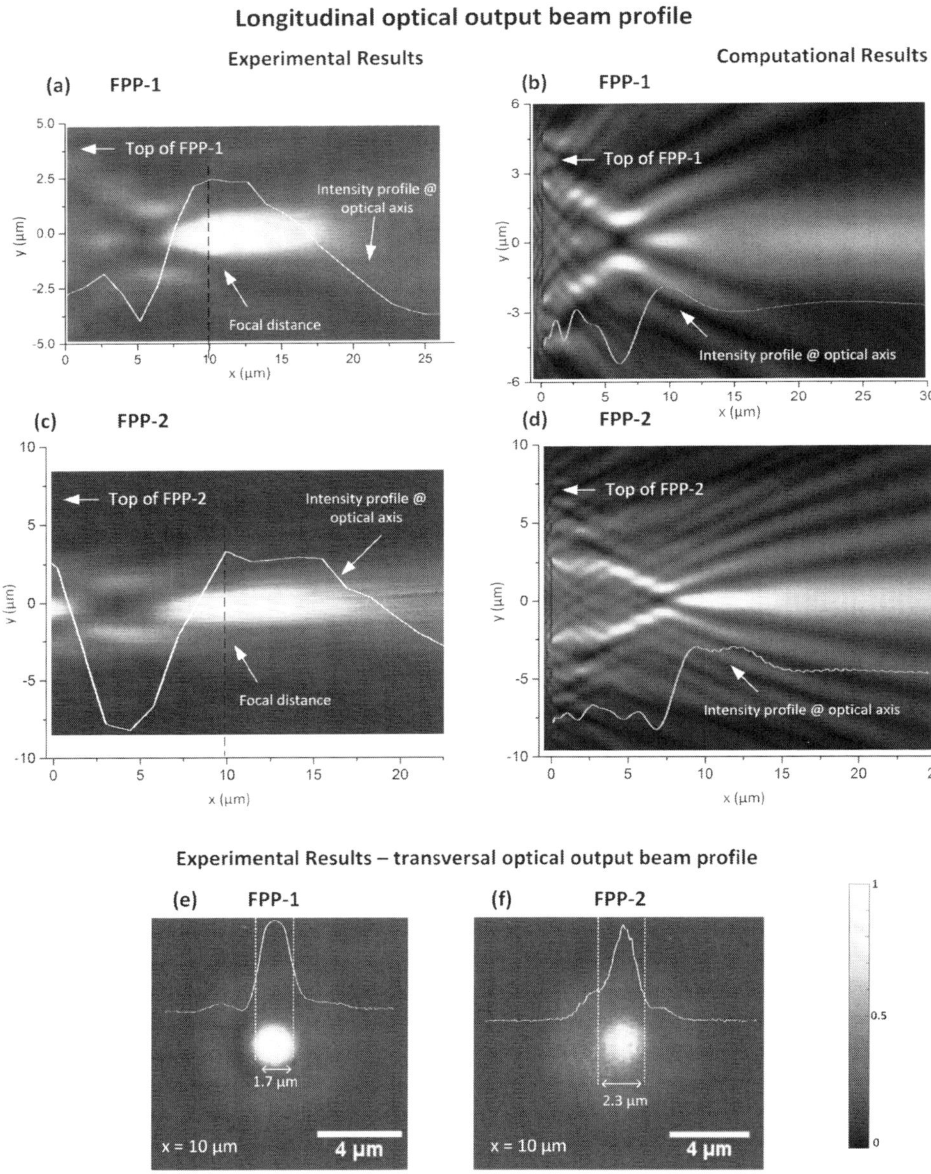

Figure 6: Longitudinal optical output beam profile for FPP-1: (a) experimental; (b) computational. Longitudinal optical output beam profile for FPP-2: (c) experimental; (d) computational. Transversal optical output beam profile at the main focal point: (e) FPP-1; (f) FPP-2.

Figures 6(e,f) show the transversal profiles experimentally acquired at the maximum intensity positions for each phase plate. The FWHM of the central peak for FPP-1 is ~1.7 μm and for FPP-2 is ~2.3 μm. In this case, increasing the number of diffractive rings caused a slightly broader peak.

DISCUSSION

The studies of the optical properties of the zone and phase plates presented in the previous sections demonstrated some important features that differentiates both Fresnel diffractive structures. Table 1summarizes some of the measured parameters. The zone plates cause a narrower central peak at the focal point, however, present two extra focal points in addition to the main one. In contrast to the zone plates, the phase plates only originate one focal point, nevertheless, the central peak is wider in the transversal direction. The efficiency of the devices was calculated based on the ratio between the power detected at the output of a cleaved optical fibre and the output detected at the output of each device. The measured values were: 33% and 38% for the FZP-1 and 2 and 60% and 67% for FPP-1 and 2, respectively. This proves that the losses are higher for the case of the zone plates, since the light is blocked by alternated zones, while the phase plates are more efficient in the conversion of the optical profiles.

Table 1: Fresnel Plates parameters (distances in μm).

Fresnel Plate	N	f (theo.)	r_N (theo.)	r_N (exp.)	$E(r_N)$	f (simul.)	f (exp.)
FZP-1	10	3	5.42	5.57	2.73%	3	5
FZP-2	22	6	11.37	12.18	7.05%	6	5
FPP-1	4	10	6.26	6.11	2.35%	10	10
FPP-2	10	10	9.90	10.14	2.43%	10	~10

Besides this, a general comment concerning the differences between the computational and experimental optical field profiles should be done. Looking at Figs 5(a,c) and 6(a,c), the fabricated tips generate an output optical field that maintains some degree of collimation, while in the simulations (Figs 5(b,d) and 6(b,d)) it is more evident that the resulting output beams diverge. Although the simulated structures have been dimensioned to reproduce the fabricated ones, there are always some parameters that may deviate from the intended ones. For instance, the fabricated structures may lack absolute circular symmetry, causing an impact on the structure of the output beam. Furthermore, during the scan of the ion beam, the milling is more pronounced in the edges that in the centre, inducing a lensing effect41. On the other hand, during the fabrication there is also a possibility to have

Gallium ions implanted on the fibre surface, as well as carbon residues, due to the contamination of the FIB chamber. These can be responsible for blockage or absorption of the light, as addressed by Janeiro et al.42. In the particular case of the zone plates, the simulations have taken into account the dominant effect of the structure produced by the pattern on the fiber tip, but have left out second order corrections. These are associated with the dispersion of the metal and the possibility of producing localized surface plasmons that might result in a small contribution of the spatial distribution of light of the output beam. Although most of the electric field is parallel to the surface of the metal rings, and thus is not effective in producing surface plasmons, at the edges of the rings it is indeed possible to have plasmonic excitation, which changes the effective dimensions of the rings and introduces a minor correction in the spatial distribution of the beam structure. As a result, some spatial components of the beam dispersed by the metallic structure may gain slightly different phase delays than predicted by the simulations, thus justifying the contrast and definition between simulations and experimental results. However, including more realistic parameters at such level would add to much complexity to the model, which at present was simply intended to guide the design. Nevertheless, these properties will be explored in futures works as the plasmonic features may provide interesting mechanisms for further focusing and sensing abilities.

Overall, from the comparison of the two types of devices it can be said that FPP are definitely more easily fabricated to fit a simple modeling approach providing interesting features as micromanipulation devices. Instead, the FZP show a more complex behavior, that requires a more complex modeling approach to explore its full potential.

In the following section, FPP-1 will be tested for optical trapping of dielectric particles and yeast cells. On the one hand, the match between the computational and experimental results, regarding Fresnel phase plates, proves that the methodology presented in this work is adequate and trustworthy. Besides this, the phase plates have better optical conversion efficiency than the zone plates. On the other hand, the zone plates experimental results differ from the projections, due to the facts above mentioned: implantation of Ga ions, defects on the ring size and possible excitation of plasmons. At the same time, the zone plates are also covered with a metallic film, which might act as a source of heating, that can compromise the trapping effect, and the sample integrity. In this regard, choosing a phase plate not only fits the need for optical trapping but also represents less constrains from the fabrication to the application. Regarding the choice of FPP-1 rather than

FPP-2, this relies on the dimension of the central peak, which is narrower at the focal distance, contributing to an increase of the optical force responsible for the trapping.

Optical Trapping using Fresnel Plates

Manipulation Setup

To test the Fresnel plates for optical trapping, the setup depicted in Fig. 7 was used. This is composed by two main parts: an image acquisition system and a 4-axis motorized micromanipulator to precisely handle the optical probes. The imaging system is essentially an inverted microscope, composed by a CMOS camera (EO-2018C, Edmund Optics) and an objective. To assemble the optical probes at the stage, first they were spliced to a 980 nm pigtailed laser diode (500 mW, Lumics) and then carefully inserted into a metallic capillary. After this, the capillary was attached to the stage and adjusted at a suitable angle. Then, the fibre was inserted into the sample that was placed over the glass slide.

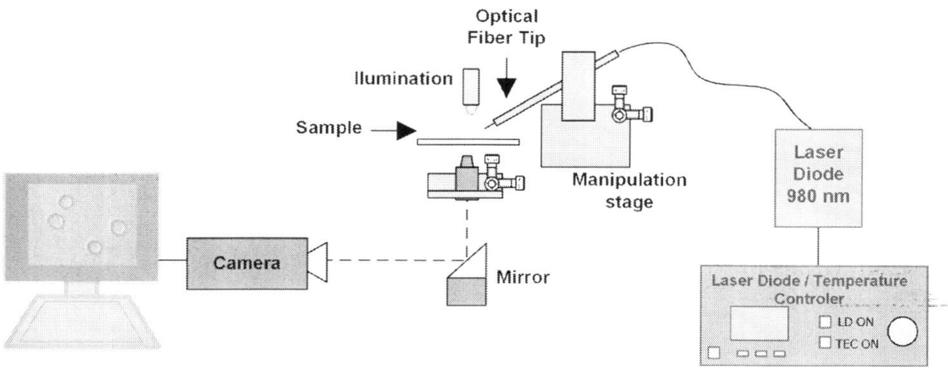

Figure 7: Optical manipulation setup.

Optical Trapping of PMMA Particles

To understand how the optical beam is affected by the media, and how do the optical forces act, some simulations based on the FDTD method, earlier mentioned, were performed. The fibre was modelled using the experimental features of FPP-1, nevertheless, the refractive index of the media was set to 1.32 (instead of 1,00), since the experiments are done in an aqueous medium. The fibre was tilted at 45°, and a glass slide was included in the computational configuration, to better reproduce the experimental conditions. Figure 8(a) represents the electric field intensity, simulated in the mentioned conditions. Generally speaking, the focal distance increased, being located near the glass slab surface, and the beam spot became wider. This is caused by the increase in the solution refractive index to 1.32, resulting in a reduced contrast and lower focusing power. The optical forces acting on the PMMA particles are calculated based on the Lorentz model, considering the beads composed by dipoles15, 47. Thus, the force is given by:

$$\vec{f} = \frac{1}{2}\varepsilon_0(\varepsilon_p - \varepsilon_m)\nabla I \quad (4)$$

where ε_0 is the vacuum permittivity, ε_p is the particle relative permittivity, ε_m is the relative permittivity of the surrounding media and **I** is the electric field intensity. The calculations of the optical forces are performed for different positions of the particle, allowing to obtain a map with the distribution of the net forces, as depicted in Fig. 8(b). The available results show that when the bead is located aside the stable trapping position, it is first driven to the optical axis, and then towards the optical trap, close to the glass slab. Consequently, the expected trapping position is no longer at 10 μm from the optical fibre, but at the new focal position, as shown by the simulations, and corroborated by the following experiments.

The optical fibre probe, containing FPP-1, was assembled accordingly to the procedure previously described, with an inclination angle of 45°. After this, a drop of an aqueous sample containing 8 μm diameter PMMA beads (1.4843 refractive index48) was placed on the glass slide. Using the motorized micromanipulator, the fibre was then immersed in the sample. Through the imaging system, the fibre was adjusted to be close to the glass slide, and when in the vicinity of some PMMA beads, the trapping was tested. Please see Fig. 9, and Supplementary Video 1. At t=0.05 s the laser was off, and the fibre was positioned so that a bead was very close to the trapping point found in the computational simulations.

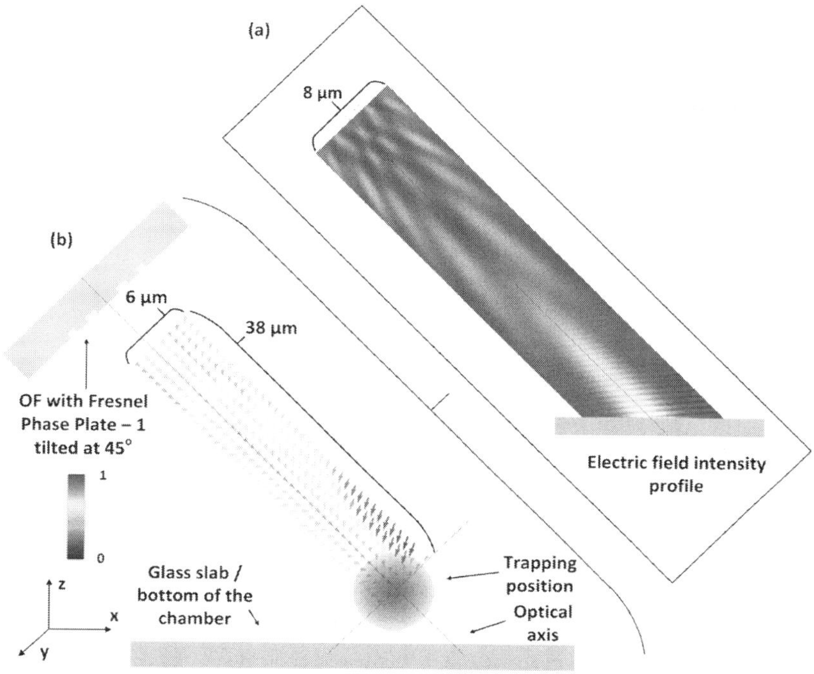

Figure 8: (a) Field intensity profile for FPP-1 tilted at 45° immersed in water; (b) corresponding force distribution map, for the particular case of 8 µm PMMA beads.

After this, the laser was turned on (2.50 s) and the fibre was moved in the −x direction (12.35 s). The frame acquired at t = 17.30 s shows that the bead (within the white circle) was moved towards the trapping position, while the particle delimited by the white square remained in the same position since the beginning of the experiment. This was followed by the displacement of the fibre in the −y direction (23.30 s), and simultaneous movement of the particle. The probe was lastly moved in the +x (31.09 s) and +y (33.09 s) directions driving the particle with it. This demonstrates that the phase plate is able to trap and move PMMA particles in 2-dimensions, i.e., the xy plane. To make the actual probe configuration more clear to the reader, close attention should be paid to the scheme in Fig. 9, where it becomes evident that the fibre does not touch the beads during the trapping process. In the acquired frames the bead and the upper border of the fibre are superposed in the image but are located at different planes, explaining the defocusing of the fibre, while the particle is focused.

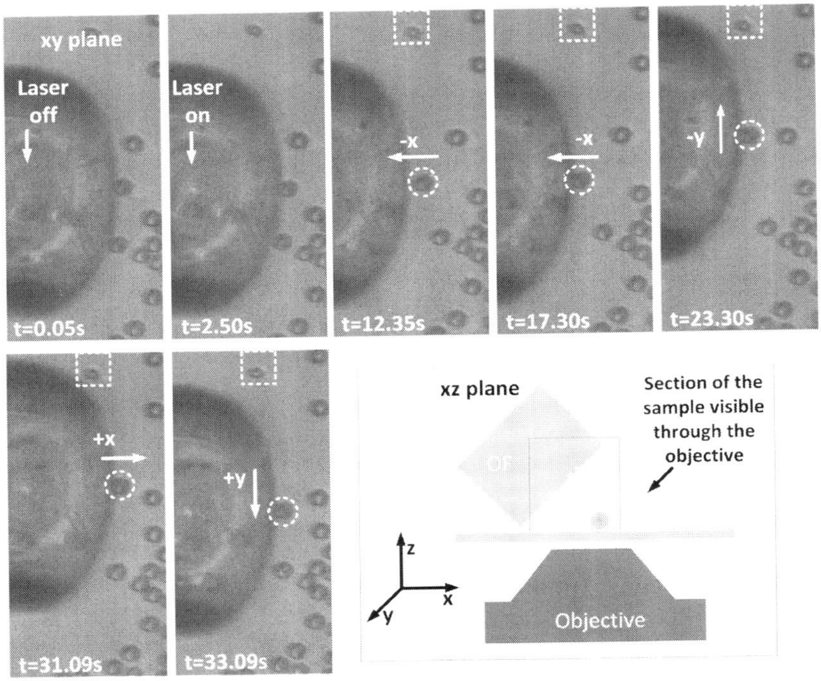

Figure 9: Demonstration of optical trapping of a PMMA particle in the xy plane. Scheme depicting the observer view, to demonstrate that the optical fibre does not touch the trapped particle (see Supplementary Video 1).

The optical forces acting on the particle can be described by the scheme depicted in Fig. 10. Since the fibre is tilted, the optical axis does not match the x direction in the xyz referential. Instead, it is rotated by θ. Consider Fig. 10(a) where the particle is located nearby the trapping point. In this case, the particle feels two forces, the axial and the transversal force. When they balance with each other, the particle is stably trapped. Otherwise, if the particle is beyond the trapping position, Fig. 10(b), the transversal force acts as a restoring force, and the particle is again driven to the trap. At last, when the bead is before the trap, the transversal and axial forces will positively contribute to guide the particle towards the steady position. These components of the force contribute to a stable trap in the xy plane, corresponding to a 2D trapping. In the meantime, in the z direction, the particle is confined in the vicinity of the glass slab. To take this into account, the revised Stokes equation, that is, Faxen's law, is used to calculate the optical forces in the xy plane14. This form of the drag equation considers the particle close to a boundary, in this case, the glass slab. The total optical force acting on the particle is given by the sum of the inertial force (1^{st}term) and the drag force (2nd term):

$$F_{opt} = F + F_{drag} = m\frac{\partial^2 s}{\partial t^2} + 6\pi\eta r\varsigma\frac{\partial s}{\partial t} \quad (5)$$

where η is the viscosity of the medium, r is the radius of the particle, ς is a correction factor that considers the proximity of the particle to the cover glass surface taking the value of 3.084 for a gap nearby zero, and s(t) denotes the trajectory of the particle. This is decomposed into the x and y component for the 2D case. The inertial force is not considered since the Reynolds number is very small for micrometric particles. Consequently, when the particle is stably trapped, the optical force has to be strong enough to balance with the drag force. The dynamics of the particle in the trap was studied following the next procedure. First a bead was trapped, then the laser was turned off, and the fibre was moved. After this, the laser was again turned on, and since the fibre is only a few micrometres away from the bead, it was attracted towards the stable position. This was performed several times for each direction, −x, +x, −y and +y and the process was repeated for a range of powers with the fibre tilted at 45° and 30°. Using a particle tracking software from ImageJ49, the trajectories of the particles were attained. Whit this information the position versus time graphs of the trajectories were plotted. Doing a fit to this graphs and employing the first derivative, the optical forces were calculated according to Eq. 5.

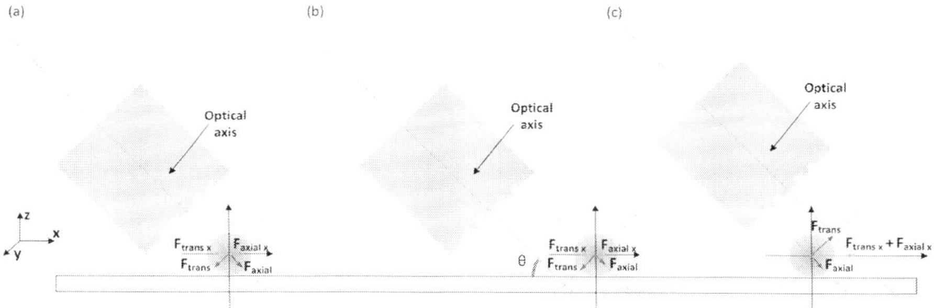

Figure 10: Optical forces acting on the particle, when it is at different locations: (a) at the trapping position $F_{trans\ x}$ and $F_{axial\ x}$ balance in the xy plan; (b) after the trapping point $F_{trans\ x}$ exceeds $F_{axial\ x}$ moving the particle towards the stable point; (c) before the trapping point, both $F_{trans\ x}$ and $F_{axial\ x}$ contribute positively to drive the particle to the stable position.

The plots presented in Fig. 11 show the calculated values for the optical force along the y and x directions considering the particle tilted at 30° and

45°. As expected, the net optical force increases with the power, and as commonly measured, the values are within the pN range. Also, the insertion angle of the fibre affects the force. Concerning the force along the y axis, the maximum values measured were 2.99×10^{-12} N and 1.81×10^{-12} N, for 45° and 30°, respectively. From Fig. 11(a) one can also say, that the trap is quite symmetric. At 45° the PMMA particle reached a maximum velocity of 10.5 ± 0.7 µm/s, while moving towards the trapping point. This means that if the stage is moved with a constant velocity along y inferior to 10.5 µm/s, the particle will remain trapped. Regarding the force along the x direction, it is stronger along the +x direction than −x. This is due to the positive contribution of the axial force, pushing the particles towards the trap. When the particle is located beyond the trapping point, the optical force is weaker, reaching a maximum value of 8.02×10^{-13} N. In this case the particle reached a maximum velocity of 3.76 ± 0.54 µm/s. The graphs of Fig. 11 also show that both in transversal and axial directions, the optical trapping net force generally increases for higher tilting angles. The exception occurs along the direction of positive x, where the forces are very similar for both angles. This is likely related with the nature of the optical forces in this direction. In the direction of negative x the optical forces are the least intense, since the gradient force has to surpass the scattering, while in the direction of positive x, both scattering and gradient add positively to drive the particle to the trapping position. In this case, the contribution of the scattering and gradient components are possibly accountable for masking the effect of the tilt angle.

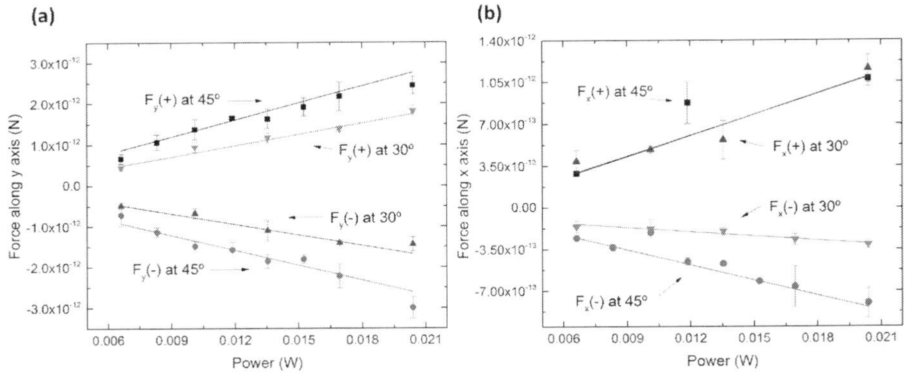

Figure 11: (a) Force along the y axis, measured with the optical fibre positioned at 45° and 30°; (b) Force along the x axis, measured with the optical fibre positioned at 45° and 30°.

Optical Trapping of Yeast Cells

In a similar fashion, the trapping of yeast cells (refractive index of 1.49–1.53[50, 51]) was also tested using FPP-1. Keeping the fibre tilted at 45°, the yeasts could be stably trapped in 2D. The range of measured forces is in line with the previous calculations, reaching the maximum trapping force of 7.96×10^{-13} N experienced along y direction, and 3.36×10^{-13} N in x direction. The ability to trap in 2D was already verified with PMMA beads. Nevertheless, it is also important to demonstrate the capacity to move a specific target to an exact point. This is here demonstrated by Fig. 12. In this particular case, yeasts A and B, are separately driven to a defined point, indicated by the arrows. These end up aligned and between two other yeast cells. The full rearrangement of the yeasts can be seen in Supplementary Video 2.

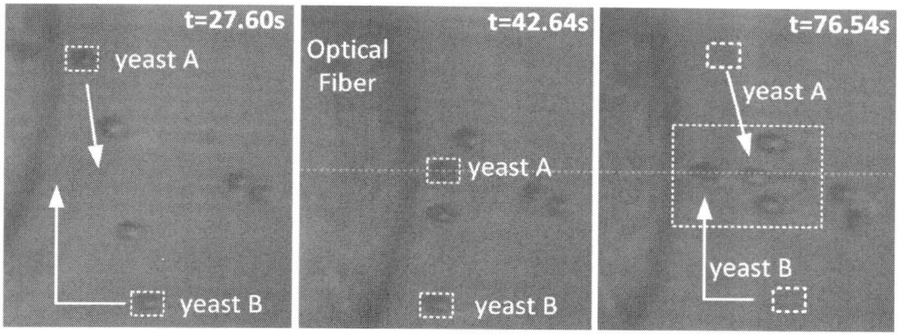

Figure 12: Rearrangement/sorting of two yeast cells, A and B, to a specific location (see Supplementary Video 2).

Detection of Trapped Particles

With a slight modification of the setup it is straightforward to automatically detect and coarsely identify that a target particle is in the trapping area. Such particle detection system is composed by a photodetector (PDA 36A-EC, Thorlabs), connected to the optical fibre probe, using an optical fibre coupler. Thus, the laser can be injected in the probe and the backscattered light can be read by the PD. The 980 nm laser was modulated with a sinusoidal signal (frequency 1 MHz, amplitude 4 V) and the scattered light signal was then filtered with a second order Butterworth filter. An example of the acquired signal can be seen in Fig. 13(a). The presence or absence of an 8 μm bead near the trapping zone is made clear through the different response signals, as indicated in the graph. Acquiring multiple data for different targets (yeast cells, PMMA beads, and clusters of beads with a maximum size of 15 μm) allowed to identify significant differences in the scattered light, which could be the base of an identification system. To study this possibility, the data was statistically analysed, and each interval of information was fitted to a normal distribution. For every test, the standard deviation (σ) of the signal in the presence (σ_{trap}) or absence ($\sigma_{no\ trap}$) of targets, considered as the reference signal, was computed. Figure 13(b) shows the fit done to the first set of data (Fig. 13(a), trapping of a PMMA particle) while Fig. 13(c) show the fit in the absence of particle. Looking at these plots, one can see that $\sigma_{no\ trap}$ should be smaller than σ_{trap}. From the difference of these quantities, $\Delta\sigma = |\sigma_{no\ trap} - \sigma_{trap}|$, one can estimate ranges where particles of different sizes will belong to. This is depicted in Fig. 13(d). In this case, it is visible that yeast cells, with an average diameter of 4 to 5 μm are characterized by an average Δσ of 0.00387, while for PMMA particles the value of Δσ is 0.04686, and for clusters/aggregates of particles the value of Δσ is 0.06473. These results show that larger particles do scatter more light and by this method it is possible to differentiate them.

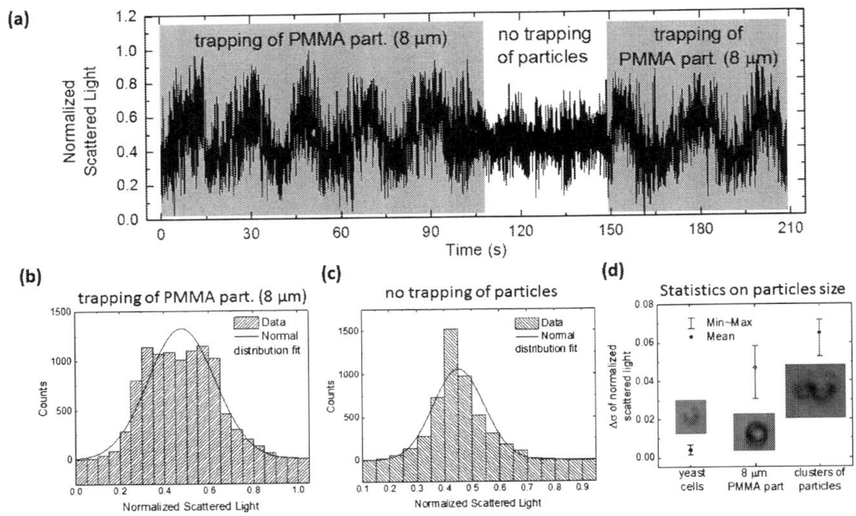

Figure 13: (a) Measurement of the light scattered by an 8 μm PMMA bead, and during the absence of trapping. (b) Normal distribution fit correspondent to the first trapping event shown in (a). (c) Normal distribution fit of the data representative of the absence of trapped particles. (d) Study of Δσ for yeast cells, 8 μm PMMA beads and clusters of particles. The images inside the graph are examples of the targets (scale photos).

DISCUSSION

The optical fibre probe allows to have full 2D trapping, allowing to push and pull dielectric particles as well as yeast cells. This unveils possible applications in biology related fields. Beyond this, the data gathered also proves that it is possible to select and arrange targets in a precise manner, which can be particularly useful in studies dedicated to single targets, where their isolation from the remaining sample population is required.

The lack of evidences of z trapping (3D), computational and experimentally verified, may be a constraint. However, for cell sorting and arrangement, the available capabilities hold great potential. To increase the trapping capabilities, as well as efficiency, further developments on the fabrication of the phase plates, that may lead to higher phase steps, is needed. In this case, the focal spot needs to be more confined, so that the electric field intensity gradient becomes steeper. According to Wright et al., beam spot sizes larger than 0.7 μm are not able to form 3D stable traps52. Such reduced spot sizes have recently been accomplished using binary phase plates fabricated by electron-beam lithography and dry etching. This validates the possibility to improve Fresnel plates using FIB milling, since both fabrication methods have resolutions of a few nanometers53.

The particle detection system, by analysis of the scattered light, enhances the applicability of the trapping setup, since sorting based on the specimen average size can be attained in an automatic fashion, if adequate feedback systems are implemented. With such approach, fibre tweezers enable immobilization, manipulation and detection all in the same platform, reducing the system cost and footprint, and greatly improving its potential applicability. Similar features, such as trapping and detection, have been recently reported by Yu-Chao Li et al., using nanojets to trap sub-wavelength targets54, 55. In this regard, this indicates the current trends on developing probes suitable for trapping but also having detection and sensing capabilities, that cover a wide range of scales from micro to nano.

CONCLUSIONS

This paper reports the fabrication of Fresnel zone and phase plates by FIB milling on optical fibre tips. The design and fabrication of the structures was supported by computational simulations. In the case of the zone plates, the experimental data differs from the simulations. This reveals that the metallic rings might produce second order effects, that lead to the excitation of plasmons. Contrary to this, the phase plates experimental and computational data agree, demonstrating that this fabrication method allows to produce diffractive structures, that despite presenting some minimal error, still comply with the projections. With this in mind, the capacity of phase plates for trapping of cells and particles was tested and characterized. This is the first time, to our knowledge, that Fresnel phase plates fabricated by FIB milling on optical fibre tips are used to trap micrometric particles and cells. The use of such diffractive structures have been tested in the past for manipulation of sub-wavelength particles, but not for targets resembling mammalian cells. Beyond this, the optical fibre probe was simultaneously used to size-detect the particles.

To summarize, this paper explores new applications for Fresnel plates fabricated on optical fibre tips, beyond their common use as coupling devices. In the future, such devices may enable advanced monitoring and manipulation devices equipped with a feedback system, for automatic single particle/cell sorting according to their size.

REFERENCES

1. Ashkin, A. Acceleration and Trapping of Particles by Radiation Pressure. Phys. Rev. Lett. 24, 156–159 (1970).
2. Tassieri, M. et al. Microrheology with Optical Tweezers: Measuring the relative viscosity of solutions 'at a glance'. Sci. Rep.5, 8831 (2015).
3. Raab, E. L., Prentiss, M., Cable, A., Chu, S. & Pritchard, D. E. Trapping of Neutral Sodium Atoms with Radiation Pressure. Phys. Rev. Lett. 59, 2631–2634 (1987).
4. Estève, J. Cold atoms: Trapped by nanostructures. Nat. Nanotechnol. 8, 317–318 (2013).
5. Nshii, C. C. et al. A surface-patterned chip as a strong source of ultracold atoms for quantum technologies. Nat. Nanotechnol. 8, 321–324 (2013).
6. Lee, K. et al. Optical tweezers study of red blood cell aggregation and disaggregation in plasma and protein solutions. J. Biomed. Opt. 21, 35001 (2016).
7. Lai, C.-W., Hsiung, S.-K., Yeh, C.-L., Chiou, A. & Lee, G.-B. A cell delivery and pre-positioning system utilizing microfluidic devices for dual-beam optical trap-and-stretch. Sensors Actuators B Chem. 135, 388–397 (2008).
8. Yang, T., Bragheri, F. & Minzioni, P. A Comprehensive Review of Optical Stretcher for Cell Mechanical Characterization at Single-Cell Level. Micromachines 7, 90 (2016).
9. Bowman, R. W. & Padgett, M. J. Optical trapping and binding. Rep. Prog. Phys. 76, 26401 (2013).
10. Bustamante, C., Macosko, J. C. & Wuite, G. J. L. Grabbing the cat by the tail: manipulating molecules one by one. Nat. Rev. Mol. Cell Biol. 1, 130–136 (2000).
11. Svoboda, K., Schmidt, C. F., Schnapp, B. J. & Block, S. M. Direct observation of kinesin stepping by optical trapping interferometry. Nature 365, 721–727 (1993).
12. Wang, M. D. et al. Force and velocity measured for single molecules of RNA polymerase. Science 282, 902–7 (1998).
13. Lee, G.-H., Kim, S.-H., Ahn, K., Lee, S.-H. & Park, J. Y. Separation and sorting of cells in microsystems using physical principles. J. Micromechanics Microengineering 26, 13003 (2016).

14. Neuman, K. C. & Block, S. M. Optical trapping. Rev. Sci. Instrum.75, 2787–809 (2004).
15. Ribeiro, R. S. R., Soppera, O., Oliva, A. G., Guerreiro, A. & Jorge, P. A. S. New Trends on Optical Fiber Tweezers. J. Light. Technol 33, 3394–3405 (2015).
16. Hu, Z., Wang, J. & Liang, J. Manipulation and arrangement of biological and dielectric particles by a lensed fiber probe. Opt. Express 12, 4123 (2004).
17. Mohanty, K. S., Liberale, C., Mohanty, S. K. & Degiorgio, V. In depth fiber optic trapping of low-index microscopic objects. Appl. Phys. Lett. 92, 151113 (2008).
18. Liu, Z., Guo, C., Yang, J. & Yuan, L. Tapered fiber optical tweezers for microscopic particle trapping: fabrication and application. Opt. Express 14, 12510 (2006).
19. Taguchi, K., Ueno, H., Hiramatsu, T. & Ikeda, M. Optical trapping of dielectric particle and biological cell using optical fibre. Electron. Lett. 33, 413 (1997).
20. Xin, H., Xu, R. & Li, B. Optical trapping, driving, and arrangement of particles using a tapered fibre probe. Sci. Rep. 2, 818 (2012).
21. Xin, H., Xu, R. & Li, B. Optical formation and manipulation of particle and cell patterns using a tapered optical fiber. Laser Photon. Rev. 7, 801–809 (2013).
22. Xin, H., Li, Y., Liu, X. & Li, B. Escherichia coli -Based Biophotonic Waveguides. Nano Lett. 13, 3408–3413 (2013).
23. Xin, H., Li, Y. & Li, B. Controllable Patterning of Different Cells Via Optical Assembly of 1D Periodic Cell Structures. Adv. Funct. Mater.25, 2816–2823 (2015).
24. Xin, H. & Li, B. Optical orientation and shifting of a single multiwalled carbon nanotube. Light Sci. Appl. 3, e205 (2014).
25. Taylor, R. & Hnatovsky, C. Particle trapping in 3-D using a single fiber probe with an annular light distribution. Opt. Express 11, 2775 (2003).
26. Liberale, C. et al. Miniaturized all-fibre probe for three-dimensional optical trapping and manipulation. Nat. Photonics 1, 723–727 (2007).
27. Schonbrun, E., Rinzler, C. & Crozier, K. B. Microfabricated water immersion zone plate optical tweezer. Appl. Phys. Lett. 92, 71112 (2008).

28. Kuo, J.-N. & Hu, H.-Z. Optical Trapping of Beads and Jurkat Cells Using Micromachined Fresnel Zone Plate Integrated with Microfluidic Chip. Jpn. J. Appl. Phys. 50, 100211 (2011).
29. Koshelev, A. et al. High refractive index Fresnel lens on a fiber fabricated by nanoimprint lithography for immersion applications. Opt. Lett. 41, 3423 (2016).
30. Kim, J. K. et al. Fabrication of Micro Fresnel Zone Plate Lens on a Mode-Expanded Hybrid Optical Fiber Using a Femtosecond Laser Ablation System. IEEE Photonics Technol. Lett. 21, 21–23 (2009).
31. Kim, J. et al. Micro Fresnel Zone Plate Lens Inscribed on a Hard Polymer Clad Fiber Using Femtosecond Pulsed Laser. IEEE Photonics Technol. Lett. 25, 761–763 (2013).
32. Liu, Y., Xu, H., Stief, F., Zhitenev, N. & Yu, M. Far-field superfocusing with an optical fiber based surface plasmonic lens made of nanoscale concentric annular slits. Opt. Express 19, 20233–43 (2011).
33. Liu, Y., Stief, F. & Yu, M. Subwavelength optical trapping with a fiber-based surface plasmonic lens. Opt. Lett. 38, 721 (2013).
34. O'Shea, D. C. & Society of Photo-optical Instrumentation Engineers. Diffractive optics: design, fabrication, and test. (SPIE, 2004).
35. Hecht, E. Optics. (Addison Wesley, 2002).
36. Handbook of Lasers and Optics. (Springer International Publishing, 2013). doi:10.1017/CBO9781107415324.004.
37. Callegari, V., Iwaniuk, D., Bronnimann, R., Schmid, E. & Sennhauser, U. Optimized fabrication of curved surfaces by a FIB for direct focusing with glass fibres. J. Micromechanics Microengineering 19, 107003 (2009).
38. Verslegers, L. et al. Planar lenses based on nanoscale slit arrays in a metallic film. Nano Lett. 9, 235–8 (2009).
39. Rodrigues Ribeiro, R. S., Dahal, P., Guerreiro, A., Jorge, P. A. S. & Viegas, J. Compact solutions for optical fiber tweezers using Fresnel zone and phase lenses fabricated using FIB milling. in SPIE OPTO (eds Glückstad, J., Andrews, D. L. & Galvez, E. J.) 97640C (International Society for Optics and Photonics, 2016). doi:10.1117/12.2213667.
40. Li, X. et al. Stretchable Binary Fresnel Lens for Focus Tuning. Sci. Rep. 6, 25348 (2016).
41. Rodrigues Ribeiro, R. S., Dahal, P., Guerreiro, A., Jorge, P. & Viegas,

J. Optical fibers as beam shapers: from Gaussian beams to optical vortices. Opt. Lett. 41, 2137 (2016).

42. Janeiro, R., Flores, R., Dahal, P. & Viegas, J. Fabrication of a phase photon sieve on an optical fiber tip by focused ion beam nanomachining for improved fiber to silicon photonics waveguide light coupling. Opt. Express 24, 11611 (2016).
43. Reyntjens, S. & Puers, R. A review of focused ion beam applications in microsystem technology. J. Micromechanics Microengineering 11, 287–300 (2001).
44. Schneider, C. A., Rasband, W. S. & Eliceiri, K. W. NIH Image to ImageJ: 25 years of image analysis. Nat. Methods 9, 671–675 (2012).
45. Oskooi, A. F. et al. Meep: A flexible free-software package for electromagnetic simulations by the FDTD method. Comput. Phys. Commun. 181, 687–702 (2010).
46. Rodrigues Ribeiro, R. S., Jorge, P. A. S. & Guerreiro, A. Computational models for new fiber optic tweezers. Photonic Sensors 3, 57–60 (2012).
47. Barnett, S. M. & Loudon, R. On the electromagnetic force on a dielectric medium. J. Phys. B At. Mol. Opt. Phys 39, 671–684 (2006).
48. Sultanova, N., Kasarova, S. & Nikolov, I. Dispersion Properties of Optical Polymers. Acta Phys. Pol. A 116, (2009).
49. Sbalzarini, I. F. & Koumoutsakos, P. Feature point tracking and trajectory analysis for video imaging in cell biology. J. Struct. Biol.151, 182–95 (2005).
50. Rines, D. R., Thomann, D., Dorn, J. F., Goodwin, P. & Sorger, P. K. Live cell imaging of yeast. Cold Spring Harb. Protoc. 2011, (2011).
51. Sigrist-Photometer AG. Refractive Index. at http://www.photometer.com/en/abc/show.html?q=Refractive index.
52. Wright, W. H., Sonek, G. J., Tadir, Y. & Berns, M. W. Laser trapping in cell biology. IEEE J. Quantum Electron. 26, 2148–2157 (1990).
53. Yu, A. et al. Creation of Sub-diffraction Longitudinally Polarized Spot by Focusing Radially Polarized Light with Binary Phase Lens. Sci. Rep. 6, 38859 (2016).
54. Li, Y. et al. Trapping and Detection of Nanoparticles and Cells Using a Parallel Photonic Nanojet Array. ACS Nano, doi:10.1021/acsnano.5b08081 (2016).
55. Li, Y.-C. et al. Manipulation and detection of single nanoparticles and biomolecules by a photonic nanojet. Light Sci. Appl. 5, e16176 (2016).

Chapter 4

Review on Fabrication Technologies for Optical Mold Inserts

Marcel Roeder [1,2], **Thomas Guenther** [2], **and André Zimmermann** [1,2]

[1]Hahn-Schickard, Allmandring 9b, 70569 Stuttgart, Germany

[2]Institute for Micro Integration (IFM), University of Stuttgart, Allmandring 9 b, 70569 Stuttgart, Germany

ABSTRACT

Polymer optics have gained increasing importance in recent years. With advancing requirements for the optical components, the fabrication process remains a challenge. In particular, the fabrication of the mold inserts for the replication process is crucial for obtaining high-quality optical components. This review focuses on fabrication technologies for optical mold inserts. Thereby, two main types of technologies can be distinguished: fabrication methods to create mold inserts with optical surface quality and methods

Citation: Roeder, Marcel; Guenther, Thomas; Zimmermann, André. 2019. "Review on Fabrication Technologies for Optical Mold Inserts." Micromachines 10, no. 4: 233. https://doi.org/10.3390/mi10040233

to create optical microstructures. Since optical mold inserts usually require outstanding form accuracies and surface qualities, a focus is placed on these factors. This review aims to give an overview of available methods as well as support the selection process when a fabrication technology is needed for a defined application. Furthermore, references are given to detailed descriptions of each technology if a deeper understanding of the processes is required.

Keywords: optical mold inserts, micro machining, micro structuring, ultra-precision machining, mold fabrication

INTRODUCTION

Polymer optics have gained increasing importance in recent years. They compete with traditional glass lenses in various fields of applications. One of the most critical points in the fabrication of polymer optical components is the mold insert required for injection molding or injection compression molding, respectively. There is a broad range of technologies which can be used to produce optical mold inserts and/or micro-structuring techniques. Which methods should be employed depends mainly on the application. This review aims to support decision making when selecting the most suitable fabrication technology by providing an overview of available technologies. The scope of this review is to describe the technologies, their advantages and limitations, and possible applications. Since the review focuses on optical mold inserts, special attention is given to achievable surface quality, accuracy and, in the case of micro-structuring techniques, the minimal structure size.

The market of polymer optics is growing rapidly, finding its way into more and more sophisticated applications [1]. Technological advantages in the fabrication process of polymer optics enable fast replication of optical elements with a wide range of geometries as well as micro-structures. This is a major advantage compared to glass optics, allowing for more freedom in optical design. Free-form optics are just one example of optical element that can be produced at a significantly lower price than traditional glass lenses. Therefore a wide range of applications emerges with increasing opportunities for the optical design. Applications range from illumination [2] and imaging [3] to automotives [4]. Another upcoming trend in polymer optics are micro-structured components. The combination of lenses with micro-structured features can be used to increase their performance, reduce the weight of optical systems, correct aberrations, and shape beams. Examples of micro-structures used in polymer optics are micro-lens arrays [5], diffractive optical elements [6], Fresnel lenses [7], prism arrays [8] and blazed structures [9]. Examples of applications of micro-structured optical components are concentration structures for solar panels using micro-lens arrays [10] or Fresnel lenses [11], beam shaping and homogenization [12,13], measurement systems [14] and sensors [15,16].

One of the main advantages of polymer optics is their fast and low-cost replication by means of hot embossing [17] or injection (compression) molding [18]. Furthermore, mounting and alignment features can be integrated into the optical components, which eliminates the need for additional holding components and assembly steps [19]. Roll-to-roll processes enable the fast replication of large areas with an accuracy even appropriate for micro-structured features [20,21]. While this opens further technological possibilities, this paper does not focus on replication technologies, but on the fabrication of mold inserts for the replication. A comprehensive overview on injection molding of polymer optics is given by Bäumer [22]. Furthermore, methods for the replication of micro- and nano-structured surface geometries are summarized by Hansen et al. [23].

The most important material properties for polymer optics are the refractive index and the Abbe number [1]. Comparing polymer optics to glass optics, the refractive index is a limiting factor since no materials with high refractive indices are available. The most commonly used materials for injection molded optics are acrylic (PMMA), polycarbonate (PC), cyclic olefin copolymer (COC) and cyclic olefin polymer (COP) [24], which provide good technical properties regarding internal stresses, reduced water absorption, optimized resistance against environmental influences, and many more. In combination with microstructured features, their optical properties can be enhanced to overcome the limitations concerning refractive index.

In the following section, technologies for the fabrication of optical mold inserts are described. The technologies are divided in form-giving methods and micro-structuring techniques. First, form-giving machining technologies for optical mold inserts are described, where ultra-precision machining presents a special case as it presents a combination of a form-giving and micro-structuring technology. The methods are investigated regarding their achievable surface quality and accuracy. Subsequently, micro-structuring methods are described, focusing on the achievable structure size. A summary is provided in the last section of the paper including a guide for aiding decision making when choosing the right technology for the required application.

FABRICATION METHODS

Throughout this section, various fabrication methods are described. Figure 1 provides an overview of the methods and their achievable surface quality and structure dimensions. More details of each technology are described below.

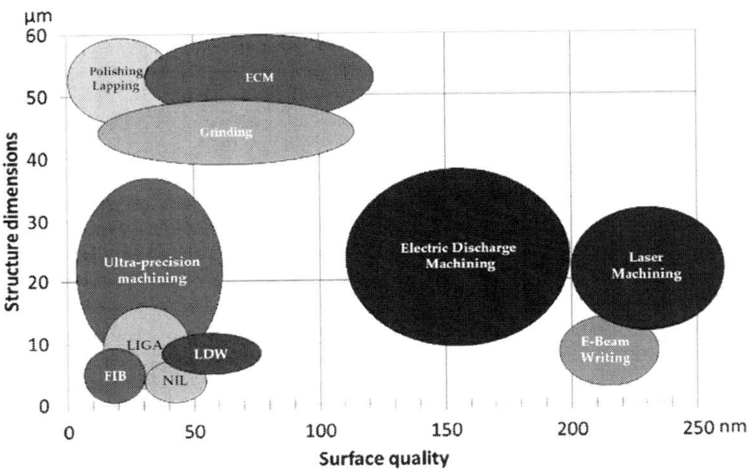

Figure 1: Structural dimensions and achievable surface quality of fabrication technologies for optical mold inserts.

FORM-GIVING TECHNOLOGIES

Ultra-Precision Machining (UPM)

Ultra-precision machining (UPM) was first introduced in the 1960's by Bryan from the Lawrence Livermore National Laboratory [25]. It is the most common method for the fabrication of optical mold inserts. Ultra-precision machines achieve a positioning accuracy in the nanometer range [26], which leads to outstanding surface quality and form accuracy. The surface roughness of diamond machined parts is usually smaller than Ra < 10 nm. Hence, post-processing of components to achieve mirror finished surfaces is not required. To obtain high-quality parts, the machine components have to be pushed to their limits. Diamond machining systems use a granite block as a foundation. High-precision positioning systems, high-speed spindles, and accurate fixture and handling equipment are needed in these systems [27]. The current state of the art in spindle technology was reviewed by Abele et al. [28]. Air bearing spindles and oil hydrostatic bearings are used for accurate movement of the tools and parts. Position control is assured by glass scales with a resolution of less than 1 nm. Furthermore, vibration suppression and temperature control is very important. Temperature should be kept constant in a range of ±0.1 K or less. When machining at the microscale, mechanics change significantly. Effects which have little to no influence at macroscale become dominant when the chip size decreases. The achievable surface roughness of diamond-machined parts is influenced by a multitude of factors like cutting conditions [29], tool vibration [30,31,32], material properties [33,34,35,36,37,38,39] and spindle vibration [40]. These factors can be separated into process factors and material factors [29]. The understanding of the effects and their impact on surface roughness is most important to improve part quality and support further development of the technology. Cheung and Lee [29,32,41,42,43] investigated the cutting dynamics and surface generation in ultra-precision machining, mainly using diamond turning as the cutting technology.

The achievable part quality and accuracy very much depends on the quality of the diamond tool. Monocrystalline diamonds are used to form the cutting tip of the tool because of their outstanding hardness and the ability to create very sharp edges with less than 50 nm edge roundness [44]. Hence, the surface finish does not depend on the cutting speed [45]. Ultra-precision machining can also be used as a micro-structuring technique. The achievable structure size is thereby limited by the nose radius of the available diamond tools to about 5 μm [46].

Diamond machining is limited to non-ferrous materials. Due to this fact, nickel-phosphorus (NiP) coatings became the industry standard as the material to machine with diamond tools for optical mold inserts. NiP can be diamond machined with negligible tool wear. The coatings are deposited onto steel molds by electroless or galvanic plating processes. The preparation of the mold inserts before the diamond machining requires three steps. First, a steel insert is fabricated using traditional milling or turning processes to fabricate the geometry roughly. Afterwards, the nickel phosphorous coating is deposited and another rough machining process is needed to remove surplus NiP since the diamond machining process only removes a couple of micrometers of the material. The necessity of the coating process makes the fabrication of optical mold inserts by means of diamond machining costly and time consuming. Therefore, efforts are made to machine tool steel inserts directly with optical surface quality. Different methods of UPM as well as methods to machine steel-based materials will be discussed in subsequent sections. Machining configurations of UPM are shown in Figure 2 and will be explained in the following sections.

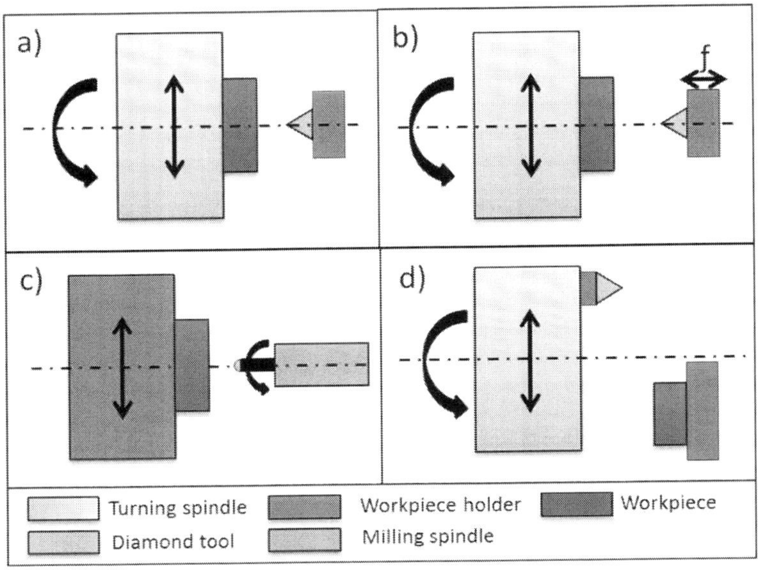

Figure 2: Ultra-precision machining (UPM) methods. (a) Diamond turning; (b) Slow-tool-servo/fast-tool-servo (c) Diamond milling; (d) Fly cutting.

Diamond Turning

Early developments of ultra-precision machining where driven by the demand for large size lenses with high-quality surfaces, mainly produced by diamond turning. Diamond turning is used to fabricate rotationally symmetrical components with high accuracy and a surface roughness Ra < 10 nm [47]. Ikawa et al. [48] reported the minimum chip thickness achievable in diamond turning is 1 nm in an experimental setup. Due to the machine configuration, possible part geometries are limited. Diamond turning is a standard process to fabricate optical mold inserts for spherical and aspherical lenses. Riedl [49] and Blough et al. [50] reported the fabrication of a diffractive optical element by means of diamond turning. When diamond turning is used to create micro-structures, the structure size is limited by the available diamond tools to about 5 μm [46]. Beside cutting conditions and machine properties the achievable surface quality of diamond-turned parts very much depends on process and material factors. The influence of process factors can be reduced or even eliminated by optimizing the operation settings. The main factors influencing the part quality are the spindle speed, tool tip radius and feed rate. Cheung and Lee [29] reported that high spindle speed, large tool tip radius and slow feed rate generally improve surface roughness. Most machines are able to rotate at a maximum speed of about 5000–6000 RPM (rounds per minute). To achieve high form accuracy the vertical and horizontal positon of the tool tip is very sensitive. Deviations lead to a residual cone in the center or an error in the surface radius. An example of a diamond-turned optical mold insert and the resulting form accuracy are shown in Figure 3.

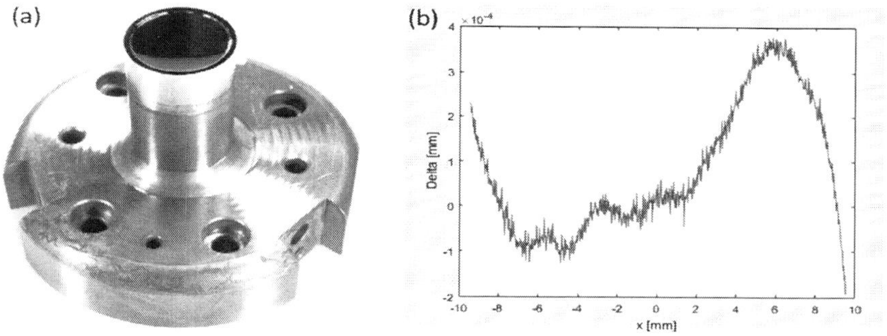

Figure 3: (a) Diamond turned mold insert, (b) form deviation of the optical aspheric surface with P-V < 1 μm (Peak to Valley).

Slow Tool Servo

Due to the high demand for non-symmetrical optics, slow tool servo (STS) was developed. Additional to the classical diamond turning setup, the z-axis oscillates during the process. Thus, the contact of the tool tip is intermittent. The slow tool servo is able to oscillate in the area of about 25 mm. Furthermore, the c-axis has to be controlled to coordinate the tool and workpiece position [51]. A slow tool servo is able to produce very accurate asymmetrical parts without any additional machine equipment. The spindle speed in these processes is typically lower compared to regular diamond turning with a rotation speed of maximum 2000 RPM. For a good result, the position accuracy and the coordination of the axis are very important [52].

Similar to the diamond turning process, an accurate tool tip position is very sensitive to the fabrication result. The machining time is long compared to fast tool servo (FTS) or diamond milling due to the fact that the z-axis is massive and can only achieve limited speed [53]. Using the STS method, surface roughness better than 10 nm is achievable [54]. The technology can also be used to fabricate micro-structured components. The structure sizes are limited by the available diamond tools and their nose radius to about 5 μm. STS is used to fabricate different optical components or optical mold inserts like micro lens arrays [55], prism arrays [56], diffractive optical elements [54], off-axis aspheric surfaces [57], freeform optical surfaces [58,59] and molds for compound eye lenses [60].

Fast Tool Servo

The machine setup for a fast tool servo is very similar to the STS configuration with a rotating workpiece and an oscillating tool. In contrast to STS, for the FTS machining an additional actuator for the tool is necessary, which oscillates the tool tip. The fast tool servo allows an accurate positioning of the tool but is limited by the stroke which is significantly smaller compared to STS technology [57]. Strokes usually are in the range of a few micrometers to a few hundred micrometers. Some FTS systems are optimized for either very short strokes <1 μm or long strokes up to 1 mm [61,62]. Different FTS technologies are reviewed by Trumper and Lu [63]. In a FTS setup, the spindle has an encoder which feeds the positon to the FTS, but does not put the spindle in position control [57]. FTS is commonly used for the fabrication of diamond-turned surfaces with structures like micro prisms, lens arrays, torics and off-axis sags with small sags [64]. FTS is also a suitable technology for the fabrication of diffractive optical elements as shown in Figure 4. Brinksmeier et al. reported the fabrication of a diffractive optical element with submicron structure size [62]. Resulting surfaces with roughness Ra <10 nm are possible [65].

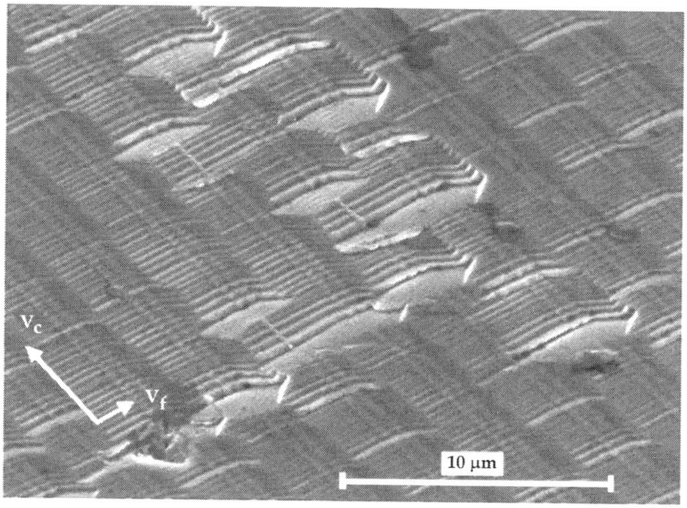

Figure 4: Scanning electron microscope (SEM) image of a diffractive surface generated by fast tool servo (FTS) in nickel silver. Reproduced with permission from [62].

Diamond Milling

In the diamond-milling process a diamond ball end mill tool with one cutting edge is used. The tool rotates at very high speed and thereby removes chips in the micrometer range. In contrast to the diamond turning process, the cutting speed is not accomplished by the rotation of the workpiece but by a fast rotation of the milling tool. Furthermore at least three controlled axes are needed. For the tool rotation, usually an air bearing spindle with very low error motion in the nanometer range is used. The rotational speed can be up to 100,000 RPM and more. For the positioning of the tool, three or more axes are used. Compared to diamond turning, the milling process is significantly slower but offers great freedom in the design. For the diamond-milling process the quality of the milling tool is very important, Therefore, the diamond has to be centered perfectly on a cylindrical shank [53,66].

Compared to the diamond-turning process, milling is much slower due to a small material removal rate. Hence, diamond milling is mainly used for non-smooth surfaces where turning processes are not applicable. Diamond milling is often used for the fabrication of micro lens arrays [66], as shown in Figure 5 and free-form surfaces [64]. Milling experiments conducted in nickel-phosphor for the fabrication of a micro lens array resulted in surface quality Ra <10 nm.

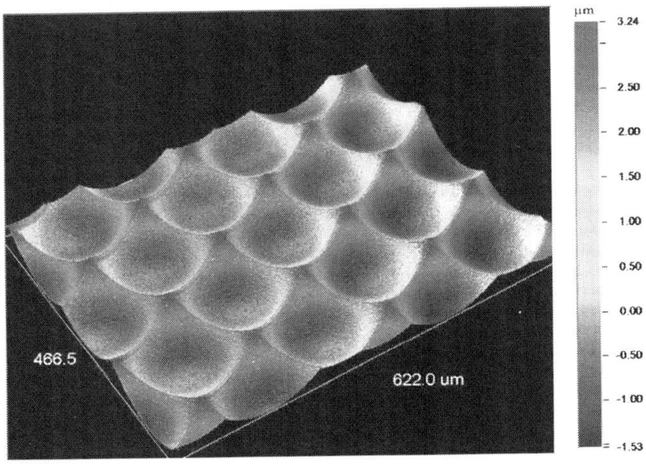

Figure 5: Micro lens array fabricated by means of diamond milling measured by white light interferometry (WLI).

Fly Cutting

Fly cutting processes use a rotating tool whereby the diamond is placed off-axis on the tool. Therefore, the diamond tool is not permanently in contact with the material.

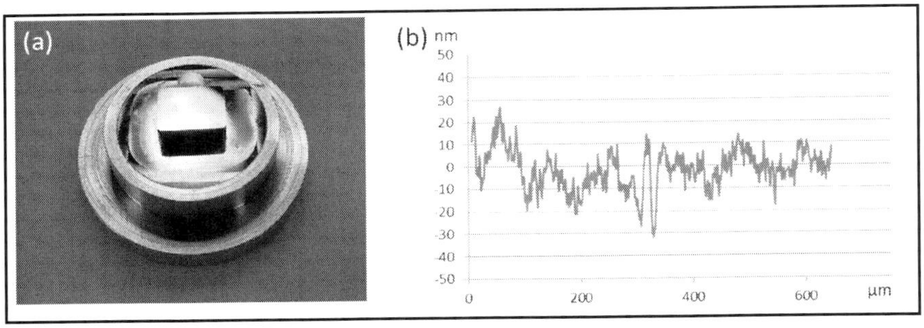

Figure 6: (a) Optical flat surface on a mold insert fabricated by fly cutting; (b) Resulting surface roughness of Ra = 8 nm measured by WLI.

Fly cutting can be used to create planar surfaces efficiently with optical surface quality, also on large areas. Furthermore, fly cutting is a suitable method to create microstructures and free form optics. Example for micro-structures are Fresnel lenses [67], micro pyramid arrays [68] and diffractive gratings [69]. Brecher et al. used the fly-cutting process for the fabrication of a complex free form mold [70]. The achievable form accuracy is in the sub-micrometer range and the resulting surface quality in the nanometer range [71]. The fly-cutting technology can also be used to create large plane surfaces quickly with a surface roughness Ra <10 nm. Figure 6 shows a flat optical surface fabricated by fly cutting with a surface roughness of Ra = 8 nm.

UPM of Steel

As already mentioned, diamond machining is usually limited to polymers and non-ferrous metals. Since hardened steels are the most popular engineering material, a lot of research has been conducted to achieve the machinability of ferrous materials with diamond tools in order to use the benefits of diamonds as a cutting material. A good review on diamond cutting of ferrous metals is given by Li et al. [72]. The review not only focuses on different methods for diamond cutting of ferrous materials but also on the wear mechanisms that are at work during the cutting process. The following wear mechanisms can be distinguished [73]:

- Adhesion and formation of a built-up edge;
- Abrasion, microchipping, fracture and fatigue;
- Tribothermal wear;
- Tribochemical wear.

The main effects for tool wear are chemical mechanisms. Paul et al. further divided the chemical mechanisms into diffusion, oxidation, graphitization and carbide-formation and linked the wear to the presence of unpaired d-electrons in the metal [74]. Metals with unpaired d-electrons lead to strong wear on the diamond tool while tool wear is negligible when machining metals without unpaired d-electrons. Diffusion of carbon atoms into the ferrous material takes place after the diamond tool graphitizes on the surface [75]. Molecular dynamics simulation supported these investigations and helped to gain a deeper understanding of the mechanisms on the atomic and nanometer scale [76,77].

To avoid severe tool wear a lot of research has been conducted to create methods which reduce the tool wear of diamond when cutting ferrous materials. Proposed methods are:

- Using ultrasonic vibration cutting;
- Optimizing cutting conditions;
- Modifying the cutting tool;
- Using binderless cubic boron nitride (cBN) tools.

Ultrasonic vibration cutting is the most promising method for the machining of ferrous materials using a diamond tool. Moriwaki and Shamoto were the first to propose a vibrating cutting tool to reduce tool wear [78]. The cutting tool is elliptically vibrating and, therefore, significantly reducing friction force and contact time of the diamond with the substrate [79]. Because of the elliptical movement of the diamond tool the technology is often referred to as elliptical vibration cutting. Different variations of the technology and applications are reviewed by Zhang et al. in [80]. Ultrasonic vibration cutting can be used in a turning process or with a moving x/y/z stage. The technology is not just very useful for the machining of ferrous materials with a diamond tool but also enables the micro-structuring of surfaces with a broad range of structures like V-grooves and pyramids as well as free forms [81,82]. At the same time, optical surface quality with Ra <10 nm is achievable [83].

Optimized Cutting Conditions

Another approach to reduce diamond wear is to modify the cutting conditions. Research teams tried different cutting conditions like reduced temperature and machining under a gaseous environment. Evans and Bryan performed diamond turning under cryogenic conditions and reported significantly reduced tool wear and achieved a surface roughness better than 25 nm [73]. Due to the low temperature during the cutting process, chemical reactions are slowed down significantly which leads to reduced tool wear. Casstevens examined diamond tool wear when machining steel under a carbon-rich atmosphere and found that tool wear under a methane environment reduces tool wear significantly and allows to achieve optical surface finish with Ra <12.5 nm [84]. However, other research groups report contrary results. Machining under inert gas environment like argon and nitrogen does not reduce the tool wear compared to air, as reported by Brinksmeier and Gläbe [85].

Tool Modification

Another method to increase wear resistance of diamond tools are protective coatings. Thus, a direct contact between the diamond and the ferrous material can be prevented. Klocke et al. discuss different coated tools for metal cutting, but did not focus on diamond tools [86]. Brinksmeier and Gläbe used TiN and TiC coatings on diamond tools to reduce tool wear and found, that flank wear is reduced as long as the protective coating is still in place but the layers are removed abrasively during the cutting process [85]. This leads to the conclusion, that coatings can prevent chemical wear but suffer from abrasive wear.

Binderless Cubic Boron Nitride (cBN)

One of the most promising methods to obtain optical surfaces in ferrous materials is the use of binderless cubic boron nitride tools (binderless cBN). cBN shows a very good heat and chemical resistance and has the second-highest hardness after diamond [87]. Thereby, very small grain sizes are needed to achieve low surface roughness. Binderless cBN tools show a strong wear resistance against adhesion, abrasion and attrition [88]. Uhlmann et al. reported a surface roughness of Ra <10 nm using a binderless cBN tool in a turning process of stainless steel with a hardness of H = 52 HRC [89]. While cBN tools are already widely used as a cutting material the availability of binderless cBN tools is still very limited. Regular cBN tools are usually sintered with binder metals, which are mainly responsible for reduced wear resistance [87]. A scanning electron microscope (SEM) image of a binderless cBN tool is shown in Figure 7.

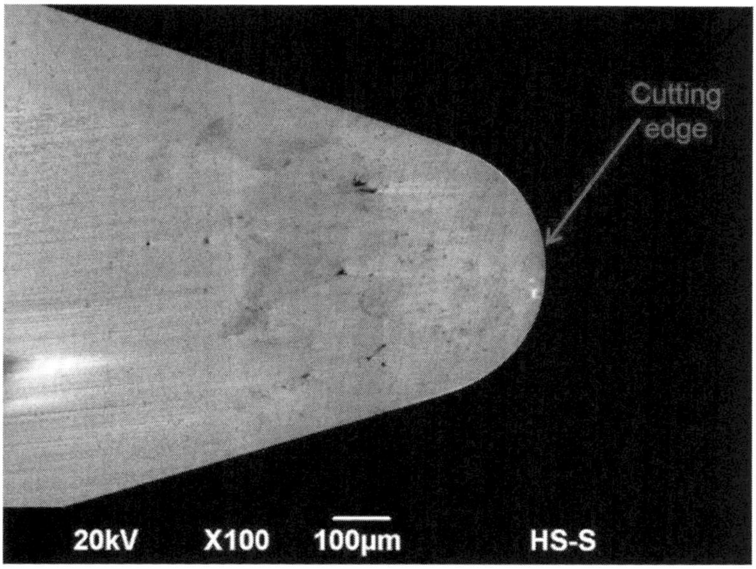

Figure 7: Scanning electron microscope (SEM) image of binderless cubic boron nitride (cBN) cutting tool.

Electric Discharge Machining (EDM)

Electric discharge machining (EDM) is a thermo-electric machining process in which a series of electrical discharges between the tool electrode and the workpiece removes material [90]. It is important to mention that the discharge sparks produced in the gap between tool electrode and workpiece remove material on both parts by melting and evaporation. The whole process is performed in a dielectric medium. The EDM process is limited to conductive or semi-conductive materials.

There are a number of different process variants for the EDM technology. A good overview is given by Uhlman et al. [91]. Using EDM for the fabrication of optical mold inserts is a relatively new approach. Takino and Hosaka proposed the use of EDM for the fabrication of a micro lens mold insert [92]. Thereby a ball-type electrode is used for shaping the mold insert made of stainless steel. In [93] they propose to use a nine-ball electrode, so that multiple lenses can be fabricated simultaneously. The form accuracy was reported to be 10 μm with a surface roughness of 0.85 μm. For optical applications this is not sufficient and post-processes like grinding, cutting or polishing are necessary to obtain smooth and accurate optical surfaces. But with further optimization of the EDM process the surface quality could be improved. Other publications report an achievable surface roughness of >0.3 μm using the EDM technology but in those cases the aim was not to produce optical mold inserts [91,94].

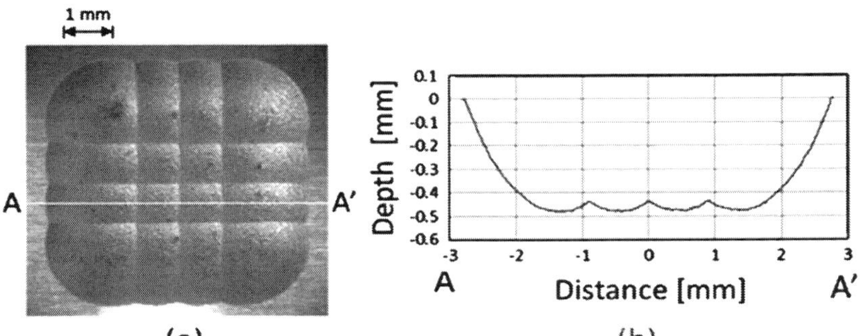

Figure 8: Micro lens array mold insert fabricated by electric discharge machining (EDM). (a) Mold surface observed with optical microscope, (b) cross section of the surface along line AA'. Reproduced with permission from [92].

Micro-EDM can also be used to fabricate micro-structures. The technology is especially powerful, when micro-structures with high aspect ratios are needed [95]. Structure sizes down to 3 μm are possible and aspect ratio can be as high as 100 [96]. An extensive review about issues with the micro-EDM process is given by Pham et al. [97]. An example of an optical mold insert fabricated by EDM technology is shown in Figure 8.

Electrochemical Machining (ECM)

Electrochemical machining (ECM) uses the anodic dissolution of metals during an electrolysis process for the material removal [98]. Similar to the EDM process, the machining shape is defined by the shape of the electrode and the workpiece has to be conductive. The resulting surface is very smooth and can be used for the smoothing of micro-metallic products [99]. Compared to traditional machining technologies, ECM has several advantages like high material removal rate, applicability regardless of the material hardness, no tool wear and smooth surfaces [98]. The technology can be used for post-processing a regularly machined workpiece. The technique is then referred as electrochemical polishing. The quality of the resulting surface is depending severely on the type of the electrolyte and the workpiece material and can reach optical quality [100]. Using a modified ECM process, Kurita and Hattori achieved a surface roughness of 0.06 μm [101]. But at this point, optical applications remain very rare. The process is more common for the elimination of burr in molds [102].

To decrease the removal rate to a minimum, short pules and a low current are required [99]. When a pulsed current is used instead of DC current, the technology is referred as PECM. Thus, the machining accuracy and process stability can be improved [90]. De Silva et al. used this process on chrome steel and achieved a surface roughness of 0.03 μm with a form accuracy better than 5 μm [103]. The main factors affecting the process are the electrolyte, the electrolyte flow conditions in the inter-electrode gap and the gap size [90].

When ECM is combined with the STM (scanning tunneling microscope) technology, microgrooves with submicron width can be achieved [99].

Grinding

Grinding is commonly used for the fabrication of optical molds. Therefore a fine grinding process is used to achieve high form accuracy. Since the achievable roughness during the grinding process is not sufficient for optical applications a post treatment process like polishing is mandatory. Recent developments aim to improve surface quality to overcome the necessity of a subsequent polishing step in the future. Polishing will be discussed in this review in a following section. Using a grinding process, surface roughness in the nanometer range can be achieved [104,105], but the results depend strongly on the tool wear, workpiece material and process parameters. In this paper, focus is placed on ultra-precision grinding since for optical applications high accuracy and low surface roughness are required. An extensive review of ultra-precision grinding has been provided by Brinksmeier [106]. In contrast to polishing and lapping, grinding uses fixed abrasives which are in interrupted contact with the workpiece. A main driving factor for grinding is the possibility to machine brittle and hard materials like ceramics, glass, carbides, glasses, hardened steels and semiconductor materials which can with difficulty be machined with ultra-precision machining [106]. Grinding processes can produce very accurate surfaces but need long machining times [57].

For optical applications, grained diamond wheels or cBN wheels are often used to achieve good form accuracy and surface roughness with Ra < 10 nm [107]. An important factor to produce high quality surfaces by grinding is to ensure a stable condition of the grinding wheel. A suitable method is the electrolytic in-process dressing (ELID) process, first proposed by Nakagawa and Ohmori [108].This in-process dressing prevents high wheel wear resulting in stable grinding conditions.

The main applications in the optical field are the manufacturing of spherical glass lenses or molds for glass injection molding [104].

MICRO-STRUCTURING TECHNOLOGIES

Lithographie, Galvanik and Abformung (LIGA)

The LIGA technology was developed at Karlsruhe Nuclear Research Center by Becker and Ehrfeld in the 1980s [109]. LIGA stands for the three German words Lithographie, Galvanik and Abformung. The LIGA technology is broadly used for the fabrication of injection molding tools [110,111]. For parts with high aspect ratio structures, the technology has particular advantages compared to other fabrication technologies [112]. Micro-structures <1 μm can be produced using this technology [113]. The LIGA technology describes a process chain of three consecutive processes. The first step is a lithography process for the structuring of a substrate. Afterwards, a nickel electroplating process is performed to create a mold using the structured substrate as a master. A detailed description of this process can be found in [114]. In the final step, parts can be produced using injection molding or hot embossing. The process chain is described in detail in [109].

A main application of the LIGA process in the optical field is the fabrication of DOE (diffractive optical elements) [115,116,117,118,119]. However, lithography for diffractive optical elements is expensive and time-consuming when the DOE level increases to over 4 [30]. A modified LIGA process can also be used for the fabrication of micro lens arrays with a resulting surface roughness of 1 nm [120], which is shown in Figure 9. Further optical components that can be fabricated with the LIGA technology are micro-prisms [8], micro-mirrors [121] and waveguides [122].

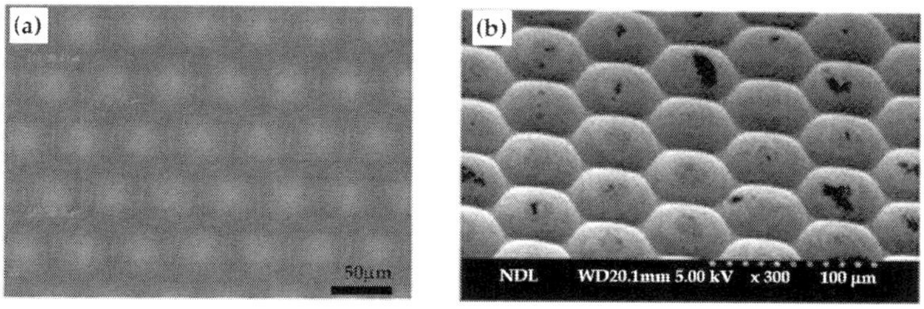

Figure 9: (a) Mold insert with a micro lens array fabricated using the Lithographie, Galvanik and Abformung (LIGA) process and (b) replicated micro lenses in a polymer film. Reproduced with permission from [123].

Nanoimprint Lithography (NIL)

Nanoimprint lithography (NIL) is a lithographic technique that allows high-throughput patterning of polymer nanostructures. The structures can be produced with a high accuracy at low costs. The process was first described by Chou et al. in 1995 [124]. NIL describes a process consisting of three main steps. First, a master mold is fabricated using a micro structuring technology like electron-beam lithography or interference lithography. In a second step the master structure is duplicated into a mold. The mold can be hard, soft or a hybrid of those [125]. The final step is the imprint process. Thereby the mold structure is imprinted into a resist on a substrate which leads to a transfer of the microstructures. Afterwards, the resist has to be cured. There are two variations of the NIL process. Thermal NIL uses heating of the resist above the glass transition temperature during the imprint step with subsequent cooling to room temperature. UV-NIL uses ultraviolet light to cure the resist. Therefore, the mold hast to be transparent. An extensive review on NIL focusing on different process variations, required material properties and nanostructure replication is given by Guo [126]. A schematic of the nanoimprint process is shown in Figure 10.

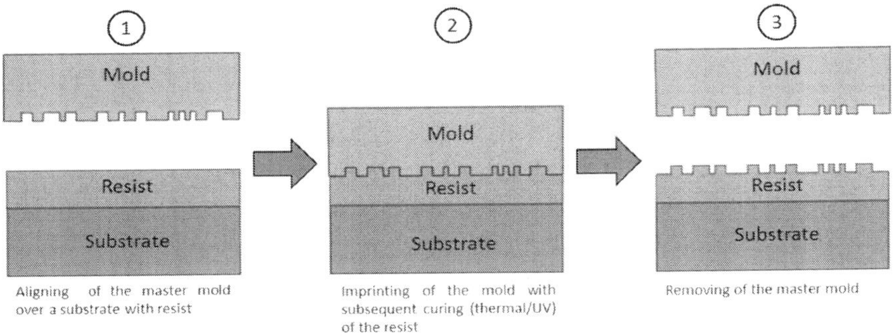

Figure 10: Nanoimprint lithography (NIL) process chain.

Using the NIL technology nanostructures with features sizes <10 nm can be produced and replicated [127]. Therefore, NIL is often used for photonics applications since the optical surface properties of substrates can be controlled precisely. Applications are holograms, diffractive structures, anti-reflective structures, micro-lens arrays and roll-to-roll applications. Since nanoimprint lithography is a method for patterning nanostructures, surface roughness is not a focus of the research.

Laser Direct Writing

In contrast to laser machining, laser direct writing (LDW) uses a laser beam for the structuring of a photoresist, comparable to a lithography process used in semiconductor manufacturing. A thin film of photoresist is deposited on a substrate. Afterwards, the LDW process is used for the structuring of the photoresist. Therefore, the substrate is scanned by a focused laser beam whereby the laser intensity is synchronously modulated [114]. In the final step the photoresist has to be developed. The LDW process is described in detail in [128].

LDW allows the fabrication of binary and continuous structures [114]. LDW is very often used for the fabrication of Fresnel or diffractive structures, mostly on planar substrates [129] where continuous structuring is beneficial for the optical performance. Compared to a lithographic approach, LDW avoids the need for submicron alignment for consecutive exposure steps. For the replication of these kinds of structures, a mold insert has to be fabricated. Therefore, nickel electroplating can be used. The produced structures in a photoresist represent the master which is afterwards cast [130]. The resulting nickel form can then be used as a mold insert for the replication of the structure with the injection molding process. An example of an electroplated mold insert with a diffractive structure is shown in Figure 11.

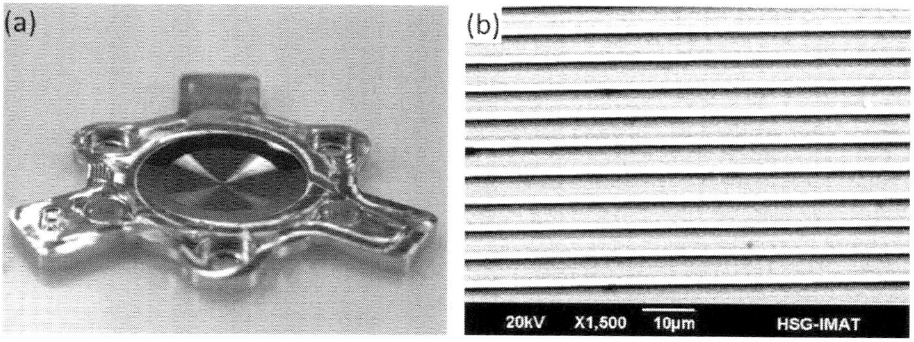

Figure 11: (a) Electroplated mold insert with diffractive structure, (b) SEM image of the diffractive structure.

Recent developments in LDW made it possible to structure curved surfaces overcoming the limitation of planar substrates [131]. This provides additional freedom for the optical design of specialized optical systems. Structure sizes usually tend to be around 5 μm to obtain sufficient diffraction efficiency of about 70% but can also go down to 1–3 μm with further reduction of the efficiency [132]. The achievable surface roughness is reported to be 25 nm, naming the raster scanning and positioning errors as the main influences [114]. An example of a curved substrate with diffractive micro-structures produced by LDW is shown in Figure 12.

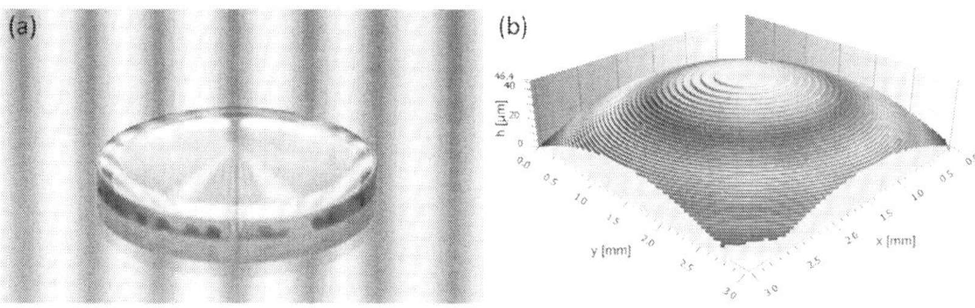

Figure 12: (a) Curved glass master with diffractive structure produced by laser direct writing (LDW), (b) confocal measurement of the diffractive structure. Reproduced with permission from [133].

E-Beam Writing

Electron beam writing is an alternative method for the structuring of a photoresist. Similar to the LDW technology, it is used for the fabrication of a master structure with a subsequent nickel electroplating process. The technology was originally developed for semiconductor mask writing but can also be used for the fabrication of micro optics. E-beam writing is especially suitable for the generation of Fresnel and diffractive structures [134,135]. Figure 13 shows a Fresnel zone plate fabricated by e-beam writing. By scanning the electron beam over a substrate, continuous relief microstructures can also be produced. Thereby the writing area of the electron beam is limited to a few millimeters. For the fabrication of larger substrates, the parts can be moved on a controlled xy-stage. To obtain high-resolution structures, the positioning has to be very accurate. The scanning approach allows a high flexibility for the writing process but results in long writing times.

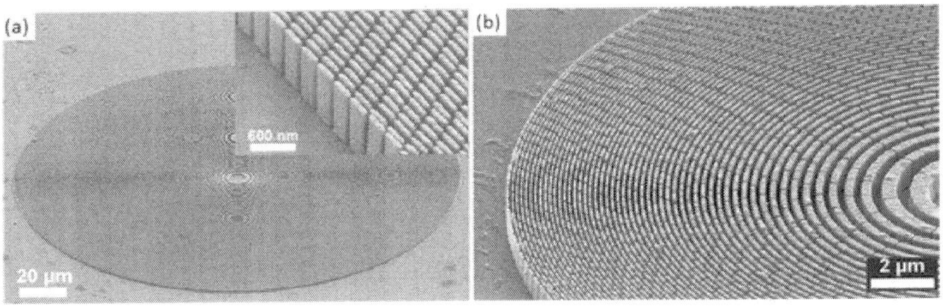

Figure 13: (a) Fresnel zone plate fabricated by e-beam writing, (b) magnified fiew of the fresnel zone plate. Reproduced with permission from [138].

It is worth mentioning that for the e-beam process additional arrangements are necessary compared to LDW. To avoid charging effects, an additional conducting film under the photoresist is necessary. Furthermore the process has to be performed under vacuum conditions. A detailed description of the electron beam process can be found in [136]. Since the technology is used in semiconductor processes, a lot of effort has been made to push the achievable resolution to the limits. The resolution of e-beam writing in a PMMA-based photoresist can be lower as 10 nm [137]. Similarly to the LDW process, the nickel electroplating process can be used to transfer the structures into a mold insert.

The technology can also be used as a polishing process for metal surfaces. Therefore, a defocused electron beam is used with a spot of some hundred microns and scanned over the surface [139]. Thereby, the metal surface melts which leads to a reduced surface roughness. Using this method a large area can be smoothed within minutes. Uno et al. reported a reduction of the surface roughness from Rz = 6 μm to Rz = 1 μm using this method [140]. Another polishing method using the e-beam technology uses an explosive electron emission applied on a surface without focusing the e-beam. Thereby, the surface roughness can be reduced from Ra = 1 μm to Ra = 0.2 μm [139]. Moreover the e-beam polishing treatment improves corrosion and oxidation resistance significantly [141].

Ion Beam Lithography

Ion beam lithography uses a focused ion beam (FIB) to scan a surface and, thereby, very small structures can be created. The technology is very similar to e-beam writing, but the ions are much heavier and heavier charged. The ion beam has a smaller wavelength than electrons and, therefore, the resolution is even higher than in case of e-beam lithography [142]. Like e-beam lithography, ion beam lithography is a direct writing process. Therefore, there is a lot of freedom in the structuring design but also processing time tends to be long. Structure sizes smaller than 5 nm have been reported using FIB [143]. Early applications for FIB were mainly failure analysis of integrated microelectronics, which is still an important field of activity [144]. Nowadays further applications are the fabrication of micro-channels, micro-structured optics and waveguides [145,146,147]. A subsequent electroplating process is necessary to transfer the micro-structures into a mold insert, mostly provided as sheet metal as a nickel shim.

The technology is also used as a polishing method for lithographic optics. These optics require ultra-smooth surfaces on lenses with 170 mm diameter and larger. Using the FIB method with low energy ions to remove form errors and reduce roughness achieves surface roughness Ra <1 nm [148].

Laser Machining

The use of short and ultrashort laser pulses is an upcoming technology for different micromachining applications. The method can be used for structuring molding tools [141,149]. Thereby, high-intensity short or ultrashort laser pulses are used for creating micro-features as shown in Figure 14. Pulse width can reach from nanoseconds down to femtoseconds. Examples for applied lasers are excimer, CO_2 or Nd:YAG lasers. An extensive review of laser beam micro machining is given in [150]. A main advantage of laser machining is the fact that virtually all materials can be machined [151]. The resulting quality very much depends on the combination of laser, workpiece and process parameters. When all these parameters are optimized, laser machining can even be used as a polishing treatment resulting in surface qualities Ra <1 μm [152,153]. Structures with dimensions down to 10 μm can be produced by laser machining [99]. Since the laser spot represents a well-defined energy input, complex structured workpieces can be polished and machined as well. Laser polishing is reviewed by Bordachev et al. in [154].

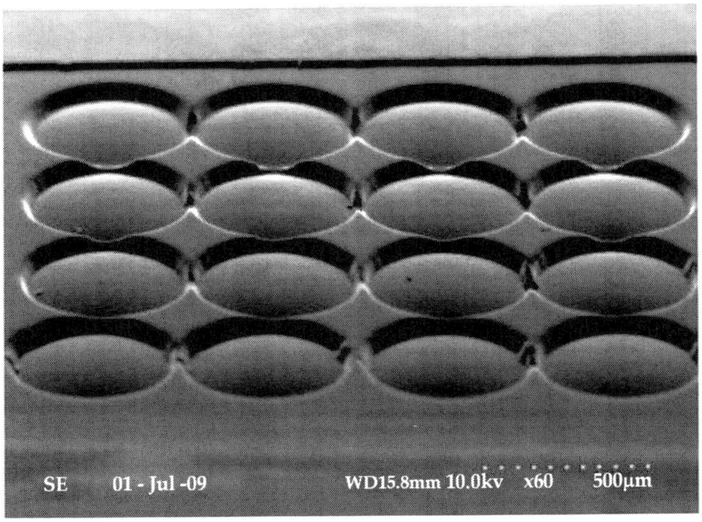

Figure 14: Laser-machined micro lens array with aspheric shape using an excimer laser Reproduced with permission from [155].

Polishing/Lapping

Polishing is a finishing treatment with undefined cutting edge to create smooth surfaces with very low roughness. Polishing is not a structuring process, therefore the geometry has to be formed by a different technology in advance. There is a broad range of process variations, but all have in common the fact that an abrasive material is used to smooth a surface. An overview on existing polishing processes is given by Yuan et al. [156]. The abrasives are suspended in a fluid. The suspended abrasive is called slurry. Polishing can create very high surface qualities in the nanometer and sub-nanometer range [157], but the removal rate is typically very low. To create this kind of high quality surface the polishing process, grain size of the abrasive, and polishing time have to be chosen carefully. Polishing can be used to machine plane, spherical, aspherical and freeform workpieces as well as structured surfaces [158,159].

Similar to polishing, lapping uses an abrasive to smooth the surface. The abrasive is rubbed between two surfaces, which can be done by hand or using machines. Lapping is mainly used when high form accuracy is needed. In contrast to polishing, the removal rate is comparatively high. Therefore, the applied grain sizes of the abrasives are usually larger. However, the transition between lapping and polishing is blurry. Both technologies are based on the same material removal mechanism. Since polishing and lapping are mainly used as finishing treatments, previous machining steps have to be performed for the form shaping. Those machining steps are usually done with cutting technologies like (UPM-) turning, milling or variations of these technologies.

An overview of all previously described technologies is shown in Table 1. All technologies are listed with their achievable surface quality and structure size as well as their advantages and limitations.

Table 1: Overview of technologies to fabricate optical mold inserts and micro-structured molds

Method	Surface Roughness	Micro-Structuring	Advantages	Limitations	Ref
Ultra-Precision Machining (UPM)	-	-	-	Available diamond tools limit size and shape of micro-structures	[46]
Diamond Turning	<5 nm	5 µm	Very high accuracy and surface quality	Limited to symmetrical parts and non-ferrous materials	[47]
Slow Tool Servo (STS)	<10 nm	5 µm	Fabrication of asymmetrical parts	Geometries are limited due to the slow stroke of the tool	[54]
Fast Tool Servo (FTS)	<10 nm	<1 µm	Fabrication of asymmetrical parts, fast and accurate positioning of the tool	Geometry has to be within the scope of the FTS stroke	[63]

Diamond Milling	<10 nm	50 μm	Fabrication of free-form structures	Long machining time especially when good surface quality is required	Exp. data
Fly Cutting	<10 nm	<1 μm	Fabrication of complex microstructures like prisms and pyramids	Limited to flat substrates	Exp. data
UPM of steel	<10 nm	5 μm	Machining of ferrous materials with high accuracy	Ultrasonic vibration cutting is limited to a turning process, other methods have problems with wear	[80]
Electric Discharge Machining	<0.1 μm	<10 μm	Large material removal rate	Only conductive workpieces, surface roughness not sufficient for optical applications	[91,96]
Electrochemical Machining	30 nm	Not suitable as a micro-structuring technique	No tool wear, high removal rate also in hardened materials	Only conductive workpieces, electrodes can be complex and expensive	[98]
Grinding	<10 nm	Not suitable as a micro-structuring technique	Machining of hardened steel	Long machining time	[107]
Lithographie, Galvanik and Abformung (LIGA)	<10 nm	<1 μm	Micro-structures with high aspect ratio are possible, broad range of micro-structures is possible	Limited to flat substrates, expensive and complex when multiple lithography steps are necessary	[113]
Nanoimprint Lithography	-	<10 nm	Fabrication and replication of very small micro- and nano-structures, high throughput	Quality is very much depending on the stamp which has to be fabricated by a micro-structuring technology, limited to 2D substrates	[126]
Laser Direct Writing	25 nm	1–3 μm	Suitable for curved substrates, fabrication of continuous structures	Limited to structuring of a photoresist	Exp. data

E-Beam Writing	0.2 µm	<100 nm	Machining of all materials, suitable for large area smoothing	Limited to small areas due to long process time	[139]
Ion Beam Lithography	<1 nm	<10 nm	Machining of all materials except for magnetic materials, fabrication of nano- and micro-structures	Limited to small areas when used as a structuring method due to long process time	[148]
Laser Machining	0.2 µm	10 µm	Processing of every material	Resulting surface quality not sufficient for optical applications	[99]
Polishing/Lapping	<1 nm	Not suitable as a micro-structuring technique	Very high surface quality	Limited form accuracy especially in free-form parts	[156]

SUMMARY AND DISCUSSION

The available technologies for the fabrication of optical mold inserts enable the production of high-quality optical parts that can compete with glass optics and replace them in a multitude of applications. To obtain high-quality optical polymer components, deep process knowledge and control during the fabrication of the mold insert as well as during the replication process are necessary. Expanding the know-how in the areas of mold fabrication and polymer replication will open new possibilities for polymer optics and replace glass lenses in further applications.

The challenge for an engineer is to determine the best suitable option for the mold fabrication, which requires a complete understanding of all technologies and fabrication methods and their limitations. This review on different available technologies provides an overview of the technologies as well as a structure to categorize the methods and their capabilities. In the following section, the advantages and disadvantages are summarized in order to support the selection process.

Form-Giving Technologies

The most common technology to produce regular optical mold inserts for spheric and aspheric optical components is ultra-precision machining. It combines high accuracy with optical surface quality in one machining process without necessity of a post-treatment. However, material removal rate is very small, usually down to 1–2 μm for the finishing cut. Therefore, the geometry has to be pre-machined using a regular machining process like turning or milling.

For complex geometries, slow- and fast-tool servo processes as well as diamond milling can be used. Being able to machine complex geometries like non-rotational symmetric parts and free forms is a major advantage of the UPM technology. The fabrication of optical mold inserts by means of ultra-precision machining is time- and cost-consuming since a coating process of the mold insert is necessary. However, improvements in the UPM technology now allow the machining of steel materials, overcoming the limitations of diamond tools. In particular, ultrasonic vibration cutting as well as alternative cutting materials like binderless cBN are powerful technologies to fabricate optical mold inserts in steel-based materials. UPM remains the most promising technology when accurate form-giving is required in an optical mold insert. When complex geometries are required, no other technology offers as much freedom in the design as UPM.

EDM offers the possibility to produce very accurate forms at comparatively high removal rates. However, achievable surface quality is not sufficient for optical applications. Further technological improvements are necessary to enable form-giving fabrication of optical mold inserts without additional post-treatment. EDM is a powerful machining process when mold inserts with high aspect ratios need to be fabricated, however improvements in surface quality are mandatory.

ECM is a relatively new approach as a machining technology for optical mold inserts. The possibility to machine even hardened materials is a main advantage of the technology. Optical surface quality is reported, but a lot of process know-how is necessary to achieve this kind of surface quality. Still, compared to ultra-precision machining, the surface roughness is significantly higher.

Grinding is a form-giving machining technology applied to create very accurate forms. However, the technology is limited concerning the achievable geometry and surface quality. The use of optimized grinding wheels like cBN-wheels as well as in-process dressing methods achieves surface qualities in the sub-micrometer range. Post treatments like polishing or lapping can be used to optimize surface quality when a regular grinding process is used or surface roughness has to be further improved.

Micro-Structuring Technologies

To fabricate micro-structured mold inserts, a broad range of technologies is available. Finding the best method is even more challenging compared to form-giving technologies. Critical factors to consider are the size of the structured area, continuous or binary structures, aspect ratio and substrate material. UPM is not just a suitable method to shape a mold insert but can be also used to create micro-structures. In particular, the fly-cutting process can fabricate large structured areas in the centimeter range in a fast and cost-efficient way, but the achievable geometry is limited. To fabricate rotationally symmetrical structured mold inserts, diamond turning is a perfectly suited method. For all micro-structures created by UPM, the available diamond tools limit the structure geometry as well as the structure size. A big advantage of the UPM technology for the fabrication of micro-structures is the fact that structuring happens directly in the mold insert and no subsequent casting process like electroplating is necessary. This reduces fabrication costs and time as well as possible inaccuracies due to a multitude of process steps.

Similar to UPM, the EDM process can also be used to create micro-structures. EDM is applied for micro structures with several micrometer or tens of micrometer in size and the aspect ratio can be high. The technology can be used to structure steel molds directly, which is a major advantage of the process compared to other micro-structuring methods.

The LIGA process is a well-established technology to produce optical mold inserts with micro features. The fabrication of diffractive gratings has been done by LIGA for a very long time. The technology allows the structuring of areas in the centimeter range but is limited to flat substrates. Part of the LIGA process chain is a galvano-copying step, creating a hard mold insert made out of nickel. Geometries of the micro-structures range from diffractive gratings, micro lenses and micro prisms to waveguides. The application of the LIGA process is advisable when high aspect ratio and accuracy is required and the micro structures are very small (<1 μm). However, the process becomes very complex and expensive, when many hierarchic levels are needed. Then, the lithography process becomes challenging.

NIL is a micro-structuring technology to produce very small structures at the highest possible accuracy. Part of the process is the fabrication of a mold insert. For the fabrication of the mold, micro-structuring technologies like lithography or e-beam writing are used. NIL can be used to produce micro-structured polymer parts, so it is not a typical technology for producing mold inserts. Nevertheless, electroplating can be used to create a mold insert by either casting the master mold or a structured photoresist. Similar to the LIGA process, high aspect ratios are possible. The technology is very well suited for applications where very small (<1 μm) and accurate micro-structures are needed.

LDW is a technology that is very well suited for the fabrication of Fresnel and diffractive structures. A big advantage of the technology is the possibility of creating continuous micro-structures. Especially for diffractive optical elements the efficiency of the elements can be increased significantly. Another advantage of the technology is the possibility to create micro-structures on curved substrates. Since the structure is written into a photoresist, the fabrication of an optical mold insert needs a subsequent electroplating process. The achievable structure size is mainly limited by the spot size of the laser. The process is advisable when feature sizes are larger than 1 μm.

E-beam writing is a micro-structuring technology that can be used when very small micro-structures in the submicron range are required. A major limitation of the process is the processing time, limiting the structured area to the micro-meter/millimeter range. In particular, for the fabrication of small Fresnel or diffractive structures with structure sizes <1 μm, e-beam writing can make use of its technological advantages. Since the process is usually applied to a photoresist, a subsequent electroplating process is necessary.

Laser machining comes with the same obstacle as the LDW process, with the limitation of the structure size to the laser sport size. However, a major advantage of the laser-machining process is the freedom to choose the substrate material. Micro-structures can be directly machined into a mold insert, eliminating a subsequent electroplating process. Since the process is a scanning method, the structuring area is limited to small areas, otherwise processing time increases significantly. Suitable applications for the laser machining process are diffractive gratings and waveguides.

To support the decision on finding a suitable fabrication method, a guide is given in Figure 15. Thereby, three categories are distinguished, namely form-giving, micro-structuring and post-treatment. For each category, the technologies are ordered according to the most important factor of the category. Grinding and UPM enable high accuracy and good surfaces as form-giving methods, however the material removal rate decreases significantly compared to ECM and EDM. For micro-structuring technologies, the achievable structure sizes are an important factor. As a rule of thumb it can be said that with decreasing structure size and increasing form accuracy, the area that can be structured decreases due to long processing time. For all machining methods where the resulting surface quality is not sufficient for optical applications, post-treatments enable a subsequent enhancement of the surface quality. Especially, polishing and lapping enable the fabrication of optical surfaces. However it needs to be considered that the overall form and form accuracy might be affected by a post-treatment process.

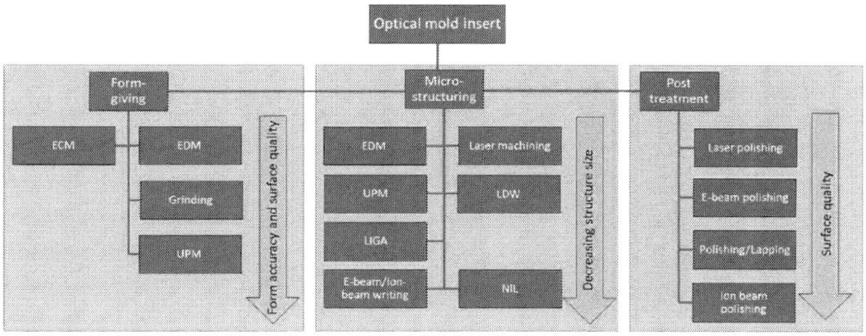

Figure 15: Available technologies for form-giving machining, micro-structuring and post-treatment.

As presented in this review, a broad range of technologies is available. Improvements in replication techniques like hot embossing, injection molding, injection compression molding and roll-to-roll replication enable the fabrication of high-performance optical components. Further developments in this area will open new possibilities and applications for polymer optics. In particular, in combination with precise micro-assembly technologies, these optical components can be integrated into optical systems to increase performance and reduce weight.

Increasing requirements of the optical components in accuracy and structure size also needs improvements in the available metrology. The combination of sub-micron resolution with large measurement areas as well as complex geometry remains a challenge for existing measurement systems [160]. Merging high-end mold-fabrication technologies, accurate replication methods and precise micro-assembly methods will enable new applications of polymer optics in optical systems as well as upcoming fields like quantum sensors and plasmon effects.

AUTHOR CONTRIBUTIONS

M.R. reviewed the literature, collected the data and wrote the manuscript. T.G. and A.Z. critically revised the manuscript, supervised the project and provided the funding.

REFERENCES

1. Doushkina, V.; Fleming, E. Optical and mechanical design advantages using polymer optics. In Proceedings of the 2009 SPIE Optical Engineering + Applications, San Diego, CA, USA, 2–6 August 2009; p. 74240Q.
2. Zhenrong, Z.; Xiang, H.; Xu, L. Freeform surface lens for LED uniform illumination. *Appl. Opt.* 2009, *48*, 6627–6634.
3. Thompson, K.P.; Rolland, J.P. Freeform optical surfaces: A revolution in imaging optical design. *Opt. Photonics News* 2012, *23*, 30–35.
4. Hicks, R.A. Controlling a ray bundle with a free-form reflector. *Opt. Lett.* 2008, *33*, 1672–1674.
5. Lee, B.-K.; Kim, D.S.; Kwon, T.H. Replication of microlens arrays by injection molding. *Microsyst. Technol.*2004, *10*, 531–535.
6. Roeder, M.; Schilling, P.; Hera, D.; Guenther, T.; Zimmermann, A. Influences on the fabrication of diffractive optical elements by injection compression molding. *J. Manuf. Mater. Process.* 2018, *2*, 5.
7. Lin, C.-M.; Hsieh, H.-K. Processing optimization of Fresnel lenses manufacturing in the injection molding considering birefringence effect. *Microsyst. Technol.* 2017, *23*, 5689–5695.
8. Brenner, K.-H.; Frank, M.; Kufner, M.; Kufner, S.; Moisel, J.; Mueller, A.; Sinzinger, S.; Testorf, M.E.; Goettert, J.; Mohr, J. Micro-optical setup with microlenses and microprisms based on refractive optics. In Proceedings of the Topical Meeting on Optical Computing, Minsk, Belarus, 29 June–1 July 1992; pp. 228–234.
9. Lee, C.; Kuriyagawa, T.; Woo, D.-K.; Lee, S.-K. Optimizing the fabrication process of a high-efficiency blazed grating through diamond scribing and molding. *J. Micromech. Microeng.* 2010, *20*, 055028.
10. Karp, J.H.; Tremblay, E.J.; Ford, J.E. Planar micro-optic solar concentrator. *Opt. Express* 2010, *18*, 1122–1133.
11. Sierra, C.; Vázquez, A.J. High solar energy concentration with a Fresnel lens. *J. Mater. Sci.* 2005, *40*, 1339–1343.
12. Schreiber, P.; Kudaev, S.; Dannberg, P.; Zeitner, U.D. Homogeneous LED-illumination using microlens arrays. In Proceedings of the Optics & Photonics 2005, San Diego, CA, USA, 31 July–4 August 2005; p. 59420K.
13. Zimmermann, M.; Lindlein, N.; Voelkel, R.; Weible, K.J. Microlens

laser beam homogenizer: from theory to application. In Proceedings of the 2007 Photonic Devices + Applications, San Diego, CA, USA, 26–30 August 2007; p. 666302.

14. Dobson, S.L.; Sun, P.; Fainman, Y. Diffractive lenses for chromatic confocal imaging. *Appl. Opt.* 1997, *36*, 4744–4748.
15. Räsänen, J.; Peiponen, K.-E. On-line measurement of the thickness and optical quality of float glass with a sensor based on a diffractive element. *Appl. Opt.* 2001, *40*, 5034–5039.
16. Kuivalainen, K.; Peiponen, K.-E.; Myller, K. Application of a diffractive element-based sensor for detection of latent fingerprints from a curved smooth surface. *Meas. Sci. Technol.* 2009, *20*, 077002.
17. Becker, H.; Heim, U. Hot embossing as a method for the fabrication of polymer high aspect ratio structures. *Sens. Actuators Phys.* 2000, *83*, 130–135.
18. Beich, W.S. Injection molded polymer optics in the 21st Century. In Proceedings of the Optics & Photonics 2005, San Diego, CA, USA, 31 July–4 August 2005; p. 58650J.
19. Beich, W.S. *Plastic Optics: Specifying Injection-Molded Polymer Optics*; Photonics Media: Pittsfield, MA, USA, 2010.
20. Ahn, S.H.; Guo, L.J. Large-area roll-to-roll and roll-to-plate nanoimprint lithography: A step toward high-throughput application of continuous nanoimprinting. *ACS Nano* 2009, *3*, 2304–2310.
21. Kooy, N.; Mohamed, K.; Pin, L.; Guan, O. A review of roll-to-roll nanoimprint lithography. *Nanoscale Res. Lett.* 2014, *9*, 320.
22. Bäumer, S. *Handbook of Plastic Optics*; John Wiley & Sons: Hoboken, NJ, USA, 2011; ISBN 978-3-527-63545-0.
23. Hansen, H.N.; Hocken, R.J.; Tosello, G. Replication of micro and nano surface geometries. *CIRP Ann.* 2011, *60*, 695–714.
24. Beich, W.S.; Fendrock, L.; Smock, C.; Turner, N. Recent trends in precision polymer optics fabrication. In Proceedings of the Optical Fabrication and Testing 2008, Rochester, NY, USA, 21–24 October 2008; p. OTuB5.
25. Ikawa, N.; Donaldson, R.R.; Komanduri, R.; König, W.; Aachen, T.H.; McKeown, P.A.; Moriwaki, T.; Stowers, I.F. Ultraprecision metal cutting—the past, the present and the future. *CIRP Ann. Manuf. Technol.* 1991, *40*, 587–594.
26. Taniguchi, N. Current status in, and future trends of, ultraprecision

machining and ultrafine materials processing. *CIRP Ann. Manuf. Technol.* 1983, *32*, 573–582.

27. Dornfeld, D.; Min, S.; Takeuchi, Y. Recent advances in mechanical micromachining. *CIRP Ann. Manuf. Technol.* 2006, *55*, 745–768.
28. Abele, E.; Altintas, Y.; Brecher, C. Machine tool spindle units. *CIRP Ann. Manuf. Technol.* 2010, *59*, 781–802.
29. Cheung, C.F.; Lee, W.B. Study of factors affecting the surface quality in ultra-precision diamond turning. *Mater. Manuf. Process.* 2000, *15*, 481–502.
30. Zhang, S.J.; To, S.; Zhang, G.Q.; Zhu, Z.W. A review of machine-tool vibration and its influence upon surface generation in ultra-precision machining. *Int. J. Mach. Tools Manuf.* 2015, *91*, 34–42.
31. Wang, H.; To, S.; Chan, C.Y.; Cheung, C.F.; Lee, W.B. A theoretical and experimental investigation of the tool-tip vibration and its influence upon surface generation in single-point diamond turning. *Int. J. Mach. Tools Manuf.* 2010, *50*, 241–252.
32. Cheung, C.-F.; Lee, W.-B. An investigation of cutting dynamics in single point diamond turning. *JSME Int. J. Ser. C* 2000, *43*, 116–126.
33. Furukawa, Y.; Moronuki, N. Effect of material properties on ultra precise cutting processes. *CIRP Ann. Manuf. Technol.* 1988, *37*, 113–116.
34. Moronuki, N.; Liang, Y.; Furukawa, Y. Experiments on the effect of material properties on microcutting processes. *Precis. Eng.* 1994, *16*, 124–131.
35. Kong, M.C.; Lee, W.B.; Cheung, C.F.; To, S. A study of materials swelling and recovery in single-point diamond turning of ductile materials. *J. Mater. Process. Technol.* 2006, *180*, 210–215.
36. To, S.; Cheung, C.F.; Lee, W.B. Influence of material swelling on surface roughness in diamond turning of single crystals. *Mater. Sci. Technol.* 2001, *17*, 102–108.
37. To, S.; Lee, W.B.; Cheung, C.F. Orientation changes of aluminium single crystals in ultra-precision diamond turning. *J. Mater. Process. Technol.* 2003, *140*, 346–351.
38. Masuda, M.; Maeda, Y.; Nishiguchi, T.; Sawa, M.; Ikawa, N. A Study on diamond turning of Al-Mg alloy—Generation mechanism of surface machined with worn tool. *CIRP Ann. Manuf. Technol.* 1989, *38*, 111–114.

39. Moriwaki, T. Machinability of copper in ultra-precision micro diamond cutting. *CIRP Ann. Manuf. Technol.*1989, *38*, 115–118.
40. Zhang, S.J.; To, S.; Cheung, C.F.; Wang, H.T. Dynamic characteristics of an aerostatic bearing spindle and its influence on surface topography in ultra-precision diamond turning. *Int. J. Mach. Tools Manuf.* 2012, *62*, 1–12.
41. Lee, W.B.; Cheung, C.F.; Chiu, W.M.; Leung, T.P. An investigation of residual form error compensation in the ultra-precision machining of aspheric surfaces. *J. Mater. Process. Technol.* 2000, *99*, 129–134.
42. Cheung, C.F.; Lee, W.B. Characterisation of nanosurface generation in single-point diamond turning. *Int. J. Mach. Tools Manuf.* 2001, *41*, 851–875.
43. Lee, W.B.; Cheung, C.F. A dynamic surface topography model for the prediction of nano-surface generation in ultra-precision machining. *Int. J. Mech. Sci.* 2001, *43*, 961–991.
44. Yuan, Z.J.; Zhou, M.; Dong, S. Effect of diamond tool sharpness on minimum cutting thickness and cutting surface integrity in ultraprecision machining. *J. Mater. Process. Technol.* 1996, *62*, 327–330.
45. Brinksmeier, E.; Preuss, W. Micro-machining. *Phil. Trans. R Soc. A* 2012, *370*, 3973–3992.
46. Gläbe, R.; Riemer, O. Diamond machining of micro-optical components and structures. In Proceedings of the SPIE Photonics Europe, Brussels, Belgium, 12–16 April 2010; p. 771602.
47. Uhlmann, E.; Mullany, B.; Biermann, D.; Rajurkar, K.P.; Hausotte, T.; Brinksmeier, E. Process chains for high-precision components with micro-scale features. *CIRP Ann. Manuf. Technol.* 2016, *65*, 549–572.
48. Ikawa, N.; Shimada, S.; Tanaka, H. Minimum thickness of cut in micromachining. *Nanotechnology* 1992, *3*, 6.
49. Riedl, M.J. Diamond-turned diffractive optical elements for the infrared: suggestion for specification standardization and manufacturing remarks. In Proceedings of the SPIE's 1995 International Symposium on Optical Science, Engineering, and Instrumentation, San Diego, CA, USA, 9–14 July 1995; pp. 257–270.
50. Blough, C.G.; Rossi, M.; Mack, S.K.; Michaels, R.L. Single-point diamond turning and replication of visible and near-infrared diffractive optical elements. *Appl. Opt.* 1997, *36*, 4648–4654.
51. Luttrell, D. Machining non-axisymmetric optics. In Proceedings of

the ASPE Fifth Annual Conference, Rochester, NY, USA, September 1990; pp. 31–34.

52. Tohme, Y.E.; Lowe, J.A. Machining of freeform optical surfaces by slow slide servo method. In Proceedings of the ASPE Annual Meeting 2004, Orlando, FL, USA, 24–29 October 2004.
53. Davis, G.E.; Roblee, J.W.; Hedges, A.R. Comparison of freeform manufacturing techniques in the production of monolithic lens arrays. In Proceedings of the 2009 SPIE Optical Engineering + Applications, San Diego, CA, USA, 2–6 August 2009; p. 742605.
54. Li, L. Fabrication of diffractive optics by use of slow tool servo diamond turning process. *Opt. Eng.* 2006, *45*, 113401.
55. Yi, A.Y.; Li, L. Design and fabrication of a microlens array by use of a slow tool servo. *Opt. Lett.* 2005, *30*, 1707–1709.
56. Li, L.; Allen, Y.Y. Design and fabrication of a freeform prism array for 3D microscopy. *JOSA A* 2010, *27*, 2613–2620.
57. Yin, Z.Q.; Dai, Y.F.; Li, S.Y.; Guan, C.L.; Tie, G.P. Fabrication of off-axis aspheric surfaces using a slow tool servo. *Int. J. Mach. Tools Manuf.* 2011, *51*, 404–410.
58. Yi, A.Y.; Raasch, T.W. Design and fabrication of a freeform phase plate for high-order ocular aberration correction. *Appl. Opt.* 2005, *44*, 6869.
59. Chen, C.-C.; Cheng, Y.-C.; Hsu, W.-Y.; Chou, H.-Y.; Wang, P.-J.; Tsai, D.P. Slow tool servo diamond turning of optical freeform surface for astigmatic contact lens. In Proceedings of the 2011 SPIE Optical Engineering + Applications, San Diego, CA, USA, 21–25 August 2011; p. 812617.
60. Zhang, X.; Fang, F.; Yu, L.; Jiang, L.; Guo, Y. Slow slide servo turning of compound eye lens. *Opt. Eng.* 2013, *52*, 023401.
61. Elfizy, A.T.; Bone, G.M.; Elbestawi, M.A. Design and control of a dual-stage feed drive. *Int. J. Mach. Tools Manuf.* 2005, *45*, 153–165.
62. Brinksmeier, E.; Riemer, O.; Gläbe, R.; Lünemann, B.; Kopylow, C.v.; Dankwart, C.; Meier, A. Submicron functional surfaces generated by diamond machining. *CIRP Ann. Manuf. Technol.* 2010, *59*, 535–538.
63. Trumper, D.L.; Lu, X. Fast tool servos: advances in precision, acceleration, and bandwidth. In *Towards Synthesis of Micro-/Nano-Systems*; Springer: London, UK, 2007; pp. 11–19.
64. Fang, F.Z.; Zhang, X.D.; Weckenmann, A.; Zhang, G.X.; Evans, C. Manufacturing and measurement of freeform optics. *CIRP Ann. Manuf.*

Technol. 2013, *62*, 823–846.

65. Zhu, L.; Li, Z.; Fang, F.; Huang, S.; Zhang, X. Review on fast tool servo machining of optical freeform surfaces. *Int. J. Adv. Manuf. Technol.* 2017, 1–22.
66. Holme, N.C.R.; Berg, T.W.; Dinesen, P.G. Diamond micro-milling for array mastering. In Proceedings of the 2008 SPIE Optical Engineering + Applications, San Diego, CA, USA, 10–14 August 2008; p. 70620J.
67. Takeuchi, Y.; Maeda, S.; Kawai, T.; Sawada, K. Manufacture of multiple-focus micro fresnel lenses by means of nonrotational diamond grooving. *CIRP Ann. Manuf. Technol.* 2002, *51*, 343–346.
68. Zhao, Q.; Guo, B.; Yang, H.; Zhang, X. A mechanistic cutting force model for diamond fly-cutting of microstructured surface. In Proceedings of the AOMATT 2008-4th International Symposium on Advanced Optical Manufacturing, Chengdu, China, 19–21 November 2008; p. 728204.
69. Fang, F.Z.; Liu, Y.C. On minimum exit-burr in micro cutting. *J. Micromech. Microeng.* 2004, *14*, 984.
70. Brecher, C.; Weck, M.; Winterschladen, M.; Lange, S.; Wetter, O.; Pfeifer, T.; Dörner, D.; Brinksmeier, E.; Autschbach, L. Manufacturing of free-form surfaces in optical quality using an integrated NURBS data interface. In Proceedings of the ASPE Winter Topical Meeting 2004, Chapel Hill, NC, USA, 4–5 February 2004.
71. Zhang, S.J.; To, S.; Zhu, Z.W.; Zhang, G.Q. A review of fly cutting applied to surface generation in ultra-precision machining. *Int. J. Mach. Tools Manuf.* 2016, *103*, 13–27.
72. Li, Z.J.; Fang, F.Z.; Gong, H.; Zhang, X.D. Review of diamond-cutting ferrous metals. *Int. J. Adv. Manuf. Technol.* 2013, *68*, 1717–1731.
73. Evans, C.; Bryan, J.B. Cryogenic diamond turning of stainless steel. *CIRP Ann. Manuf. Technol.* 1991, *40*, 571–875.
74. Paul, E.; Evans, C.J.; Mangamelli, A.; McGlauflin, M.L.; Polvani, R.S. Chemical aspects of tool wear in single point diamond turning. *Precis. Eng.* 1996, *18*, 4–19.
75. Ikawa, N. Tanaka Thermal aspects of wear of diamond grain in grinding. *CIRP Ann. Manuf. Technol.* 1971, *19/1*, 153–158.
76. Narulkar, R.; Bukkapatnam, S.; Raff, L.M.; Komanduri, R. Molecular dynamics simulations of diffusion of carbon into iron. *Philos. Mag.* 2008, *88*, 1259–1275.

77. Narulkar, R.; Bukkapatnam, S.; Raff, L.M.; Komanduri, R. Graphitization as a precursor to wear of diamond in machining pure iron: A molecular dynamics investigation. *Comput. Mater. Sci.* 2009, *45*, 358–366.
78. Moriwaki, T.; Shamoto, E. Ultraprecision diamond turning of stainless steel by applying ultrasonic vibration. *CIRP Ann. Manuf. Technol.* 1991, *40*, 559–562.
79. Moriwaki, T.; Shamoto, E. Ultrasonic Elliptical Vibration Cutting. *CIRP Ann. Manuf. Technol.* 1995, *44*, 31–34.
80. Zhang, J.; Cui, T.; Ge, C.; Sui, Y.; Yang, H. Review of micro/nano machining by utilizing elliptical vibration cutting. *Int. J. Mach. Tools Manuf.* 2016, *106*, 109–126.
81. Kim, G.D.; Loh, B.G. Direct machining of micro patterns on nickel alloy and mold steel by vibration assisted cutting. *Int. J. Precis. Eng. Manuf.* 2011, *12*, 583–588.
82. Suzuki, N. Ultraprecision sculpturing of hardened steel by applying elliptical vibration cutting. In Proceedings of the 2012 International Symposium on Ultraprecision Engineering and Nanotechnology (ISUPEN2012), Tokyo, Japan, 15 March 2012; pp. 1–6.
83. Klocke, F.; Dambon, O.; Bulla, B.; Heselhaus, M. Direct diamond turning of steel molds for optical replication. In Proceedings of the AOMATT 2008-4th International Symposium on Advanced Optical Manufacturing, Chengdu, China, 19–21 November 2008; p. 728202.
84. Casstevens, J.M. Method for machining steel with diamond tools. US Patent 17067662, 7 January 1986.
85. Brinksmeier, E.; Gläbe, R. Advances in precision machining of steel. *CIRP Ann. Manuf. Technol.* 2001, *50*, 385–388.
86. Klocke, F.; Krieg, T. Coated tools for metal cutting–features and applications. *CIRP Ann. Manuf. Technol.*1999, *48*, 515–525.
87. Fujisaki, K.; Yokota, H.; Furushiro, N.; Yamagata, Y.; Taniguchi, T.; Himeno, R.; Makinouchi, A.; Higuchi, T. Development of ultra-fine-grain binderless cBN tool for precision cutting of ferrous materials. *J. Mater. Process. Technol.* 2009, *209*, 5646–5652.
88. Polte, J.; Polte, M.; Lorenz, D.; Oberschmidt, D.; Sturm, H.; Uhlmann, E. Binderless-cBN as Cutting Material for Ultra-Precision Machining of Stainless Steel. *Adv. Mater. Res.* 2014, *1018*, 107–114.
89. Uhlmann, E.; Oberschmidt, D.; Polte, J.; Polte, M.; Huth-Herms, K.

Effect of cooling lubricant on surface roughness for turning stainless steel with binderless-cBN. In Proceedings of the 16th International Conference of the European Society for Precision Engineering and Nanotechnology, Nottingham, UK, 30 May–3 June 2016; pp. 259–260.

90. Rajurkar, K.P.; Sundaram, M.M.; Malshe, A.P. Review of Electrochemical and Electrodischarge Machining. *Procedia CIRP* 2013, *6*, 13–26.
91. Uhlmann, E.; Piltz, S.; Doll, U. Machining of micro/miniature dies and moulds by electrical discharge machining—Recent development. *J. Mater. Process. Technol.* 2005, *167*, 488–493.
92. Takino, H.; Hosaka, T. Shaping of steel mold surface of lens array by electrical discharge machining with single rod electrode. *Appl. Opt.* 2014, *53*, 8002.
93. Takino, H.; Hosaka, T. Shaping of steel mold surface of lens array by electrical discharge machining with spherical ball electrode. *Appl. Opt.* 2016, *55*, 4967.
94. Liao, Y.S.; Huang, J.T.; Chen, Y.H. A study to achieve a fine surface finish in Wire-EDM. *J. Mater. Process. Technol.* 2004, *149*, 165–171.
95. Gao, S.; Huang, H. Recent advances in micro- and nano-machining technologies. *Front. Mech. Eng.* 2016.
96. Liu, K.; Lauwers, B.; Reynaerts, D. Process capabilities of Micro-EDM and its applications. *Int. J. Adv. Manuf. Technol.* 2010, *47*, 11–19.
97. Pham, D.T.; Dimov, S.S.; Bigot, S.; Ivanov, A.; Popov, K. Micro-EDM—recent developments and research issues. *J. Mater. Process. Technol.* 2004, *149*, 50–57.
98. Rajurkar, K.P.; Zhu, D.; McGeough, J.A.; Kozak, J.; De Silva, A. New Developments in Electro-Chemical Machining. *CIRP Ann. Manuf. Technol.* 1999, *48*, 567–579.
99. Masuzawa, T. State of the Art of Micromachining. *CIRP Ann. Manuf. Technol.* 2000, *49*, 473–488.
100. Qin, Y. *Micromanufacturing Engineering and Technology*; William Andrew: Norwich, NY, USA, 2015; ISBN 978-0-323-31267-7.
101. Kurita, T.; Hattori, M. A study of EDM and ECM/ECM-lapping complex machining technology. *Int. J. Mach. Tools Manuf.* 2006, *46*, 1804–1810.
102. Schaller, T.; Heckele, M.; Ruprecht, R.; Schubert, K. Microfabrication of a mold insert made of hardened steel and first molding results. In

Proceedings of the ASPE 1999, Monterey, CA, USA, 31 October–5 November 1999; pp. 224–227.

103. De Silva, A.K.M.; Altena, H.S.J.; McGeough, J.A. Precision ECM by process characteristic modelling. *CIRP Ann. Manuf. Technol.* 2000, *49*, 151–155.

104. Tohme, Y.E. Grinding aspheric and freeform micro-optical molds. In Proceedings of the MOEMS-MEMS 2007 Micro and Nanofabrication, San Jose, CA, USA, 20–25 January 2007; p. 64620K.

105. Stephenson, D.J.; Veselovac, D.; Manley, S.; Corbett, J. Ultra-precision grinding of hard steels. *Precis. Eng.*2001, *25*, 336–345.

106. Brinksmeier, E.; Mutlugünes, Y.; Klocke, F.; Aurich, J.C.; Shore, P.; Ohmori, H. Ultra-precision grinding. *CIRP Ann. Manuf. Technol.* 2010, *59*, 652–671.

107. Zhong, Z.W.; Venkatesh, V.C. Recent developments in grinding of advanced materials. *Int. J. Adv. Manuf. Technol.* 2009, *41*, 468.

108. Ohmori, H.; Nakagawa, T. Analysis of Mirror Surface Generation of Hard and Brittle Materials by ELID (Electronic In-Process Dressing) Grinding with Superfine Grain Metallic Bond Wheels. *CIRP Ann. Manuf. Technol.* 1995, *44*, 287–290.

109. Becker, E.W.; Ehrfeld, W.; Hagmann, P.; Maner, A.; Münchmeyer, D. Fabrication of microstructures with high aspect ratios and great structural heights by synchrotron radiation lithography, galvanoforming, and plastic moulding (LIGA process). *Microelectron. Eng.* 1986, *4*, 35–56.

110. Despa, M.S.; Kelly, K.W.; Collier, J.R. Injection molding of polymeric LIGA HARMs. *Microsyst. Technol.* 1999, *6*, 60–66.

111. Ruprecht, R.; Bacher, W.; Hausselt, J.H.; Piotter, V. Injection molding of LIGA and LIGA-similar microstructures using filled and unfilled thermoplastics. In Proceedings of the Micromachining and Microfabrication, Austin, TX, USA, 23–25 October 1995; pp. 146–157.

112. Malek, C.K.; Saile, V. Applications of LIGA technology to precision manufacturing of high-aspect-ratio micro-components and -systems: a review. *Microelectron. J.* 2004, *35*, 131–143.

113. Mappes, T.; Worgull, M.; Heckele, M.; Mohr, J. Submicron polymer structures with X-ray lithography and hot embossing. *Microsyst. Technol.* 2008, *14*, 1721–1725.

114. Herzig, H.P. *Micro-Optics: Elements, Systems and Applications*; CRC

Press: Boca Raton, FL, USA, 1997; ISBN 978-0-7484-0481-0.

115. Suleski, T.J.; O'Shea, D.C. Fidelity of PostScript-generated masks for diffractive optics fabrication. *Appl. Opt.*1995, *34*, 627–635.

116. Suleski, T.J.; O'Shea, D.C. Gray-scale masks for diffractive-optics fabrication: I. Commercial slide imagers. *Appl. Opt.* 1995, *34*, 7507–7517.

117. Däschner, W.; Long, P.; Stein, R.; Wu, C.; Lee, S.H. Cost-effective mass fabrication of multilevel diffractive optical elements by use of a single optical exposure with a gray-scale mask on high-energy beam-sensitive glass. *Appl. Opt.* 1997, *36*, 4675–4680.

118. Ogura, Y.; Shirai, N.; Tanida, J.; Ichioka, Y. Wavelength-multiplexing diffractive phase elements: design, fabrication, and performance evaluation. *JOSA A* 2001, *18*, 1082–1092.

119. Lee, C.C.; Chang, Y.C.; Wang, C.M.; Chang, J.Y.; Chi, G.C. Silicon-based transmissive diffractive optical element. *Opt. Lett.* 2003, *28*, 1260–1262.

120. Kim, D.S.; Lee, H.S.; Lee, B.-K.; Yang, S.S.; Kwon, T.H.; Lee, S.S. Replications and analysis of microlens array fabricated by a modified LIGA process. *Polym. Eng. Sci.* 2006, *46*, 416–425.

121. Oh, D.-Y.; Gil, K.; Chang, S.S.; Jung, D.K.; Park, N.Y.; Lee, S.S. A tetrahedral three-facet micro mirror with the inclined deep X-ray process. *Sens. Actuators Phys.* 2001, *93*, 157–161.

122. Rogner, A.; Ehrfeld, W. Fabrication of light-guiding devices and fiber-coupling structures by the LIGA process. In Proceedings of the Micro-Optics II, The Hague, Netherlands, 11–15 March 1991; pp. 80–92.

123. Liu, K.H.; Chen, M.F.; Pan, C.T.; Chang, M.Y.; Huang, W.Y. Fabrication of various dimensions of high fill-factor micro-lens arrays for OLED package. *Sens. Actuators Phys.* 2010, *159*, 126–134.

124. Chou, S.Y.; Krauss, P.R.; Renstrom, P.J. Imprint of sub-25 nm vias and trenches in polymers. *Appl. Phys. Lett.* 1995, *67*, 3114–3116.

125. Yao, Y.; Liu, H.; Wang, Y.; Li, Y.; Song, B.; Bratkovsk, A.; Wang, S.-Y.; Wu, W. Nanoimprint lithography: An enabling technology for nanophotonics. *Appl. Phys. A* 2015, *121*, 327–333.

126. Guo, L.J. Nanoimprint Lithography: Methods and Material Requirements. *Adv. Mater.* 2007, *19*, 495–513.

127. Traub, M.C.; Longsine, W.; Truskett, V.N. Advances in Nanoimprint Lithography. *Annu. Rev. Chem. Biomol. Eng.* 2016, *7*, 583–604.

128. Bowen, J.P.; Michaels, R.L.; Blough, C.G. Generation of large-diameter diffractive elements with laser pattern generation. *Appl. Opt.* 1997, *36*, 8970–8975.

129. Gale, M.T.; Rossi, M.; Pedersen, J.; Schuetz, H. Fabrication of continuous-relief micro-optical elements by direct laser writing in photoresists. *Opt. Eng.* 1994, *33*, 3556–3566.

130. Roeder, M.; Hera, D.; Thiele, S.; Pruss, C.; Osten, W.; Zimmermann, A. 3D Laser direct-writing based master fabrication for injection compression molding of diffractive-refractive elements. In Proceedings of the 2016 European Optical Society Annual Meeting (EOSAM), Berlin, Germany, 26–30 September 2016; p. 66.

131. Häfner, M.; Reichle, R.; Pruss, C.; Osten, W. Laser direct writing of high resolution structures on curved substrates: evaluation of the writing precision. In *Fringe 2009*; Springer: Berlin, Germany, 2009; pp. 1–4.

132. Pruss, C.; Reichelt, S.; Tiziani, H.J.; Korolkov, V.P. Metrological features of diffractive high-efficiency objectives for laser interferometry. In Proceedings of the 7th International Symposium on Laser Metrology Applied to Science, Industry, and Everyday Life, Novosibirsk, Russia, 9–13 September 2002; pp. 873–885.

133. Roeder, M.; Thiele, S.; Hera, D.; Pruss, C.; Guenther, T.; Osten, W.; Zimmermann, A. Fabrication of curved diffractive optical elements by means of laser direct writing, electroplating, and injection compression molding. *Submitt. J. Manuf. Process.* 2019.

134. Aoyama, S.; Horie, N.; Yamashita, T. Micro-Fresnel lens fabricated by electron-beam lithography. In Proceedings of the Computer and Optically Formed Holographic Optics, Los Angeles, CA, USA, 14–19 January 1990; pp. 175–184.

135. Däschner, W.; Larsson, M.; Lee, S.H. Fabrication of monolithic diffractive optical elements by the use of e-beam direct write on an analog resist and a single chemically assisted ion-beam-etching step. *Appl. Opt.*1995, *34*, 2534–2539.

136. Verheijen, M.J. E-beam Lithography for Digital Holograms. *J. Mod. Opt.* 1993, *40*, 711–721.

137. Vieu, C.; Carcenac, F.; Pépin, A.; Chen, Y.; Mejias, M.; Lebib, A.; Manin-Ferlazzo, L.; Couraud, L.; Launois, H. Electron beam lithography: resolution limits and applications. *Appl. Surf. Sci.* 2000, *164*, 111–117.

138. Gorelick, S.; Vila-Comamala, J.; Guzenko, V.A.; Barrett, R.; Salomé, M.; David, C. High-efficiency Fresnel zone plates for hard X-rays by 100 keV e-beam lithography and electroplating. *J. Synchrotron Radiat.* 2011, *18*, 442–446.

139. Laperrière, L.; Reinhart, G. *CIRP Encyclopedia of Production Engineering*; Springer Berlin Heidelberg: Berlin/Heidelberg, Germany, 2014; ISBN 978-3-642-20616-0.

140. Uno, Y.; Okada, A.; Uemura, K.; Raharjo, P.; Furukawa, T.; Karato, K. High-efficiency finishing process for metal mold by large-area electron beam irradiation. *Precis. Eng.* 2005, *29*, 449–455.

141. Selada, A.; Manaia, A.; Vieira, M.T.; Pouzada, A.S. Effect of LBM and large-area EBM finishing on micro-injection moulding surfaces. *Int. J. Adv. Manuf. Technol.* 2011, *52*, 171–182.

142. Watt, F.; Bettiol, A.A.; Van Kan, J.A.; Teo, E.J.; Breese, M.B.H. Ion beam lithography and nanofabrication: A review. *Int. J. Nanosci.* 2005, *04*, 269–286.

143. Menard, L.D.; Ramsey, J.M. Fabrication of Sub-5 nm Nanochannels in Insulating Substrates Using Focused Ion Beam Milling. *Nano Lett.* 2011, *11*, 512–517.

144. Nikawa, K. Applications of focused ion beam technique to failure analysis of very large scale integrations: A review. *J. Vac. Sci. Technol. B Microelectron. Nanometer Struct. Process. Meas. Phenom.* 1991, *9*, 2566–2577.

145. Tseng, A.A. Recent Developments in Nanofabrication Using Focused Ion Beams. *Small* 2005, *1*, 924–939.

146. Fu, Y. Investigation of microlens mold fabricated by focused ion beam technology. *Microelectron. Eng.* 2001, *56*, 333–338.

147. Nellen, P.M.; Callegari, V.; Brönnimann, R. FIB-milling of photonic structures and sputtering simulation. *Microelectron. Eng.* 2006, *83*, 1805–1808.

148. Weiser, M. Ion beam figuring for lithography optics. *Nucl. Instrum. Methods Phys. Res. Sect. B Beam Interact. Mater. At.* 2009, *267*, 1390–1393.

149. Kaldos, A.; Pieper, H.J.; Wolf, E.; Krause, M. Laser machining in die making—a modern rapid tooling process. *J. Mater. Process. Technol.* 2004, *155–156*, 1815–1820.

150. Mishra, S.; Yadava, V. Laser Beam MicroMachining (LBMM)—A

review. *Opt. Lasers Eng.* 2015, *73*, 89–122.

151. Fleischer, J.; Halvadjiysky, G.; Haupt, S. Process parameter analysis in ablating micro-mold manufacturing. *Microsyst. Technol.* 2008, *14*, 1367–1372.
152. Ukar, E.; Lamikiz, A.; López de Lacalle, L.N.; del Pozo, D.; Arana, J.L. Laser polishing of tool steel with CO2 laser and high-power diode laser. *Int. J. Mach. Tools Manuf.* 2010, *50*, 115–125.
153. Trotta, G.; Volpe, A.; Ancona, A.; Fassi, I. Flexible micro manufacturing platform for the fabrication of PMMA microfluidic devices. *J. Manuf. Process.* 2018, *35*, 107–117.
154. Bordatchev, E.V.; Hafiz, A.M.K.; Tutunea-Fatan, O.R. Performance of laser polishing in finishing of metallic surfaces. *Int. J. Adv. Manuf. Technol.* 2014, *73*, 35–52.
155. Chiu, C.-C.; Lee, Y.-C. Fabricating of aspheric micro-lens array by excimer laser micromachining. *Opt. Lasers Eng.* 2011, *49*, 1232–1237.
156. Yuan, J.; Lyu, B.; Hang, W.; Deng, Q. Review on the progress of ultra-precision machining technologies. *Front. Mech. Eng.* 2017, *12*, 158–180.
157. Namba, Y.; Shimomura, T.; Fushiki, A.; Beaucamp, A.; Inasaki, I.; Kunieda, H.; Ogasaka, Y.; Yamashita, K. Ultra-precision polishing of electroless nickel molding dies for shorter wavelength applications. *CIRP Ann. Manuf. Technol.* 2008, *57*, 337–340.
158. Brinksmeier, E.; Riemer, O.; Gessenharter, A. Finishing of structured surfaces by abrasive polishing. *Precis. Eng.* 2006, *30*, 325–336.
159. Brinksmeier, E.; Riemer, O.; Gessenharter, A.; Autschbach, L. Polishing of Structured Molds. *CIRP Ann. Manuf. Technol.* 2004, *53*, 247–250.
160. Manske, E.; Jäger, G.; Hausotte, T.; Füßl, R. Recent developments and challenges of nanopositioning and nanomeasuring technology. *Meas. Sci. Technol.* 2012, *23*, 074001.

SECTION II:
SURFACE POLISHING PROCESSES

Chapter 5

Research Progress of Optical Fabrication and Surface-Microstructure Modification of SiC

Fang Jiang[1,2] , Yan Liu[1] , Yong Yang[1] , Zheng-Ren Huang[1] , Dan Li[1,2] , Gui-ling Liu[1] , and Xue-Jian Liu[1]

[1]State Key Laboratory of High Performance Ceramics and Superfine Microstructure, Shanghai Institute of Ceramics, Chinese Academy of Sciences, Shanghai 200050, China

[2]Graduate University of the Chinese Academy of Sciences, Beijing 100049, China

ABSTRACT

SiC has become the best candidate material for space mirror and optical devices due to a series of favorable physical and chemical properties. Fine surface optical quality with the surface roughness (RMS) less than 1 nm is necessary for fine optical application. However, various defects are present in SiC ceramics, and it is very difficult to polish SiC ceramic matrix with

Citation: Fang Jiang, Yan Liu, Yong Yang, et al., "Research Progress of Optical Fabrication and Surface-Microstructure Modification of SiC," Journal of Nanomaterials, vol. 2012, Article ID 984048, 9 pages, 2012. https://doi.org/10.1155/2012/984048

the 1 nm RMS. Surface modification of SiC ceramics must be done on the SiC substrate. Four kinds of surface-modification routes including the hot pressed glass, the C/SiC clapping, SiC clapping, and Si clapping on SiC surface have been reported and reviewed here. The methods of surface modification, the mechanism of preparation, and the disadvantages and advantages are focused on in this paper. In our view, PVD Si is the best choice for surface modification of SiC mirror.

INTRODUCTION

At present, mirror systems as the most important device are applied commonly in the optical system with high precision. Until now, three generations of reflector materials have been developed. The first generation is glass-ceramic; the second one is mainly made of metal, such as Beryllium metal and its alloys; the third generation of the reflector material is based on silicon carbide. SiC may be the best material available for mirror optics because of its outstanding combination of thermal and mechanical properties. It has remarkable dimensional stability even under the disturbances of temperature, humidity, and chemicals. Its specific stiffness and elastic modulus are higher than that of beryllium, which has toxicity. The density of SiC is slightly higher than aluminum and its fracture toughness is higher than glass. The remarkable properties of SiC in terms of hardness, stiffness, and thermal stability combined with a reasonable density are indeed of primary importance for all space applications. This combination of material advantages makes SiC an excellent material candidate for space optical instruments [1, 2].

Based on microstructure and processing routes, four kinds of SiC ceramics including hot-pressed SiC (HP-SiC), reaction-bonded SiC (RB-SiC), sintered SiC (S-SiC), and chemical vapor deposition SiC (CVD-SiC) were developed. The properties of different SiC materials and the brief description of various SiC component manufacturing techniques are summarized in Table 1 [2, 17]. Whatever the preparation process, it is difficult to obtain high-quality optical surface due to polishing the bare SiC very difficultly. Moreover, microstructure defects, like pores, steps at different phases, and grain boundary damages, are unavoidable under certain surfacing condition and present further difficulty in polishing this material. The AFM topography images of surfaces polished by 1 um abrasive grains of different SiC ceramics are shown in Figure 1. The steps still exist at the interfaces between two phases both in RB-SiC and S-SiC, and it cannot meet the optical requirements (<1 nm RMS). The rms surface roughness values of SiC ceramics are 6.464 nm and 3.017 nm [18]. The high hardness and complex crystal-phase structure of the SiC material indicate that the fabrication of high precision optical components is costly and slow. The key to solving this problem is to select the appropriate preparation process for surface modification of SiC matrix material, the dense SiC mirror optics plated coating, and the coating polishing. Efficiency of surface finishing for large optical components can be greatly improved as well. Moreover, microstructure defects on polished surface of SiC ceramic mentioned above can be covered up by the coating process. This paper presents the optical surface processes and the recent developments of SiC substrate by hot pressed glass, C/SiC clapping, CVD- and PVD-coated SiC and Si coating method in detail [19–21].

Table 1: Different preparation methods and properties of silicon carbide

Materials	Density (g/cm^3)	Free state of Si	Isotropic	Cost	Preparation cycle	Surface modification
HP-SiC	3.20	No	Bad	High	Short term	Requirement
RB-SiC	3.04	Yes	Good	Low	Short term	Requirement
S-SiC	3.10	No	Worse	Lower	Long term	Requirement
CVD-SiC	3.21	No	Best	High	Long term	Nonrequirement

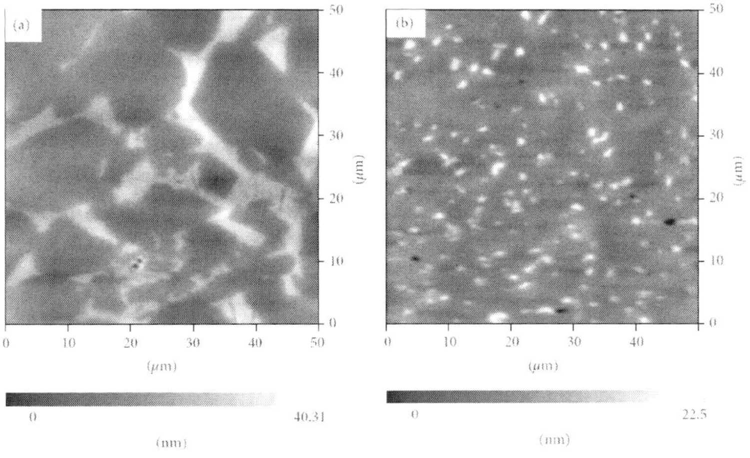

Figure 1: AFM images of SiC surfaces polished by 1 um abrasive grains within 50 um × 50 um area: (a) RB-SiC, (b) S-SiC, reprinted with permission from [18].

OPTICAL FABRICATION OF SILICON CARBIDE

However, RB-SiC is typically a difficult material to machine. SiC is harder than most other materials except diamond, cubic boron nitride (cBN), and boron carbide (B_4C), and hence available cutting tool materials for machining RB-SiC are very limited. Recent efforts have focused on the precision machining of RB-SiC by grinding, lapping, polishing, and combinations of these [24, 25]. Main features and classification of optical fabrication methods of SiC are shown in Table 2 [26–28].

Table 2: Main features of optical fabrication methods of SiC

Optical fabrication of silicon carbide	Principle	Characteristic	Example
Mechanical polishing	Mechanical friction and material removal	Low efficiency	The final surface accuracy of 1 nm RMS was reported by Paquin et al. [3]
Wet polishing	Polished mold is immersed in the slurry, RMS becomes lower by an abrasive	High surface accuracy	The final surface accuracy of 0.75 nm RMS was reported by Xu et al. [4]
Ultraprecision grinding	Ductility grinding	Equipment require high	Bifano et al. [5] has used CVD SiC to obtain the 5.5 nm RMS
Tribochemical polishing	The tribochemical reaction	Low efficiency	In [6], final surface accuracy of 1 nm RMS
Electrolytic in-process dressing (ELID)	The electrolysis process so that the grinding wheel functions	Good surface quality	In [7], final surface accuracy of 1.4 nm RMS
Chemical mechanical Polishing (CMP)	Combination of mechanical grinding and chemical etching	Good surface quality, but is corrosive	In [8], final surface accuracy of 0.5 nm RMS

Magneto rheological Finishing (MRF)	Magnetorheological polishing fluid viscosity increases the shear force generated for material removal in the magnetic field gradient	Low efficiency	Johnson et al. have used this method to fabricate CVD SiC [9]
Laser-induced photochemical polishing	Laser-induced photochemical reactions	High efficiency, but is corrosive	The final surface accuracy of 80 nm RMS was reported by Murahara [10]
Ion beam milling	High-speed ion beam hits the surface of the sample	Good surface quality, expensive equipment	The final surface accuracy of 1 nm RMS was reported by Johnson et al. [11]
Float polishing	The sample is floating on the polishing plate by the high-speed rotating fluid dynamic pressure	Good surface quality	In [12], final surface accuracy of 3 nm RMS

SURFACE MODIFICATION

Surface modification of SiC material is to add a thick film coating to the substrate of SiC for obtaining better surface quality and easier polishing. After fine polishing, the surface shows an extremely smooth surface and it can meet the optical requirement. The earliest study method is hot pressing glass, the recent developments of SiC surface modification is CVD- and PVD-coated SiC and Si, and C/SiC clapping is also applied to space optical devices.

Hot Pressed Glass

The hot pressed glass process was developed by the German IABG Company and the thin preshaped borosilicate glass plate is thoroughly pressed to a preground C/SiC substrate at elevated temperature near the glass softening temperature. After cooling down, the glass plate is bonded to the substrate without the use of any adhesive. The resulted layer of approximate 1 mm thickness is well and economically polishable. But there are some drawbacks for usage of the glass cladding on large surfaces. For example the application of the high pressure led to the CTE mismatching and residual internal stresses. Now, this process has rarely been used in the size of a large mirror [29].

Si/SiC Clapping

The Si/SiC claddings are processed by an economic deposition of SiC-containing slurry onto the preground surface. Subsequent ceramization of the deposited cladding by thermal treatment leads to the conversion of the slurry cladding into a dense and homogeneous two-phase ceramic surface layer consisting of SiC and silicon. The Si/SiC layer exhibits a strong chemical bonding on the C/SiC base material due to its related chemical nature. By optimizing the ratio of Si to SiC in the layer, its CTE can be tuned to match with that of the C/SiC bulk material over a very broad temperature range. This is important for a thermally stable mirror. Although the described two-phase Si/SiC cladding might not be suitable for superpolished surfaces below 1 nm RMS microroughness, the advantages of the Si/SiC cladding technology for conventional optical components can be summarized as that surface roughness with <2 nm RMS can be obtained with standard optical polishing techniques. A series of studies for this technology [29–31] have been reported by IABG and DSS company.

SiC Clapping

SiC clapping has the same mechanical properties and thermal physical properties as the SiC matrix, so that SiC clapping as a surface modification coating of silicon carbide mirrors attracts extensive attention of researchers. SiC coatings exhibit many outstanding properties, like good isotropic, high hardness, high thermal conductivity, and excellent optical performance characteristics. The amorphous SiC coating prepared by PVD technique with the low temperature and short cycle has also been reported.

CVD SiC Clapping

Chemical vapor deposition has been widely applied in thick-film preparation from the 1960s, and its principle is that the reaction materials resulted from the decomposition of Si-containing precursors are deposited on the substrate surface to form a thin film. The film is of good uniformity and repeatability. Such a method is applied to the preparation of SiC coating. Polished CVD-SiC (crystalline cubic α-SiC) holds also good physical properties for making mirrors such as low density, high melting point, and low expansion coefficient.

The main precursors for CVD SiC is CH_3SiCl_3, and the reaction equations are as follows: $CH_3SiCl_3(g) \rightarrow SiC(s) + 3HCl(g)$. The dense SiC coating with excellent optical properties (surface roughness <0.3 nm RMS) can be obtained. It can meet the application requirements of the mirror surface and it is one of the most effective methods for surface modification to prepare SiC-based reflection mirror. But the CVD process with high substrate temperature (>1000°C) would lead to the deformation of SiC-matrix. Another disadvantage of the CVD process is time consumption.

There are different treating methods for CVD SiC process, including atmospheric pressure chemical vapor deposition SiC (APCVD SiC), low-pressure chemical vapor deposition SiC (LPCVD SiC), and plasma enhanced chemical vapor deposition SiC (PECVD SiC) [32–34]. Different methods for preparing CVD SiC are shown in Table 3.

Table 3: Different methods for preparing CVD SiC

Methods	Advantages and disadvantages	Example
APCVD	The process is simple and has fast response, but with poor uniformity	Chung and Kim [13] use APCVD growth of single-crystalline 3C-SiC (cubic silicon carbide) thin films. The 3C-SiC film had a very good crystal quality without twins, defects, or dislocations, and a very low residual stress
LPCVD	It is possible to grow thin films with uniform and smooth morphology during the 3C-SiC film growth. LPCVD has disadvantages of low growth rate	Clavaguera-Mora et al. [14] have used Si $(CH_3)_4$ as the source materials to deposit SiC film by hot CVD device. Films grown at 900–980°C are amorphous with nanocrystals of mean grain size 10 nm and have smooth surfaces. Surface profilometry measurements give a root mean square roughness (RMS) of 6 nm
PECVD	Substrate temperature is low and deposition rate is fast. SiC films deposited by PECVD appeared to be in an amorphous state	Novi et al. [15] have used PECVD to prepare amorphous SiC, and there has been a lot of cubic structures on the surface, which indicates the existence of polycrystalline structures

CVD SiC coating has been widely applied to surface modification of SiC matrix. For example, CVD SiC with 0.2 nm RMS on the C/SiC substrates had been reported by Trex Advanced Materials [21]. The CVD SiC on S-SiC substrates in the polished roughness of less than 0.1 nm has also been reported by BOOSTEC. In China, Zhang [35] applied this method to obtain CVD-SiC with the surface roughness of 1.478 nm RMS.

PVD SiC

High temperature (typically 1300°C) in the preparation process of CVD SiC may result in stronger film stress on the SiC mirror. This is due to the unacceptable high residual stress buildup in heavy cross-sections. The cost of producing CVD SiC as a large self-supporting substrate is very high. However, PVD SiC appears to be very attractive due to its relative simplicity, low substrate temperature, and wide accessibility by industry.

Ion-Assisted Deposition SiC (IAD SiC)

The ion implantation of semiconductors rapidly became an accessible technology from the 1970s because of its ability to produce superior electronic devices. Ion beam modification of nonsemiconductor materials for enhancing surface sensitive properties has been actively pursued in the international R&D community. The advantages of this technique include high density, superior adhesion, and the ability to produce arbitrarily thick coatings. Perhaps the most important feature of the IBAD technology is the frequently demonstrated ability to control many coatings properties such as morphology, adhesion, and stress, as well as stoichiometry. The amorphous SiC coating with large area can be prepared by this method on the SiC substrate. The α-SiC coating has been prepared on the RB SiC and polished with surface roughness up to 0.2 nm RMS by U.S. HDOS [3, 36] using ion beam deposition method. Hall ion source [37] and high-energy Kaufman ion source [38] can both be used as the ion-beam resources for preparing α-SiC and its system of SiC-modified membrane and the surface roughness can be down to 0.867 nm RMS and 0.743 nm RMS.

Magnetron Sputtering SiC Clapping

Magnetron sputtering is used widely due to its low cost, high deposition rate, low deposition temperature, and good adhesion of the film feature. The SiC coatings were deposited by RF magnetron sputtering from a sintered SiC target onto commercially monopolished RB SiC substrate kept at low temperature. The deposition rate is fast and the desired temperature is low, but the film stability of its system is not sustainable [39]. Magnetron sputtering SiC clapping was prepared with the roughness down to 1.394 nm RMS [40] and 3.184 nm RMS as shown in Figure 2 by Tang et al. [22] and the surface roughness of 2 angstrom by Kortright and Windt [41].

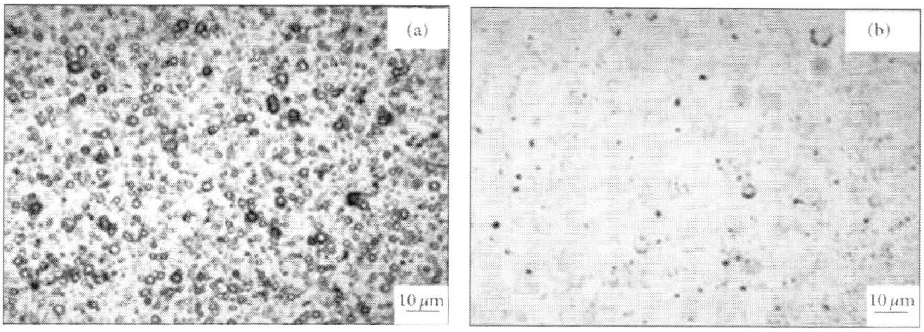

Figure 2: Photomicrograph of (a) unpolished, (b) polished PVD SiC coating on RB SiC, reprinted with permission from [22].

Pulsed Laser Deposition SiC (PLD SiC)

PLD is a relatively new technique and attracting great attention for its simplicity, reduced investment cost, and flexibility. In the PLD process [42], a high flux pulsed laser beam is focused on the target material leading to the formation of a plasma plume. The high heating rate of the target surface (≈100 K/s) due to pulsed laser irradiation leads to a congruent evaporation of the target irrespective of the evaporating point of its constituent elements or compounds, so that the target stoichiometry can be retained in the deposited films. Furthermore, because of the high heating rate of the ablated materials, PLD of crystalline films demands a much lower substrate temperature than other film growing techniques. The major drawback of this technique is that the high energy involved in the process also leads to the formation of microparticulate on the thin film surface. Vendan et al. [43] has prepared α-SiC thin films by fs-PLD and ns-PLD techniques and found that the surface roughness of SiC films by the ns-PLD technique was bigger than that by fs-PLD. Magida et al. [32] has prepared SiC thin films which exhibit high reflectivity in the ultraviolet band (40.7 nm–121.6 nm) and the surface roughness of 1 nm RMS.

Ion Beam Sputtering SiC (IBS SiC)

The ion beam engine concept was firstly developed in the United States. By the IBS technique, the SiC films being flat and smooth, a large area of dense internal stress and low defect density can be obtained. For example, the IBS SiC coatings have been prepared and polished to be less than 2A RMS by Johnson [36] for the optical system requiring ultralow scatter performance.

Si Clapping

The thermal properties of Si coating match well with that of SiC as shown in Table 4, then Si clapping can be used for the surface modification of SiC. It can be seen that their thermal performance is well matched, and Si can be used as a good reflector material. CVD and PVD are the main preparation methods.

Table 4: Thermal performance of RB-SiC and Si [16]

Materials	α/(ppm/K)	K/(W/mK)	D	α/K	α/D
RB-SiC	2.4	170	80	0.014	0.03
Si	2.6	156	89	0.017	0.029

CVD Si Clapping

The Si clapping is easier to be polished well with better surface quality and low-cost SiC-clapping. Polycrystalline Si produced by a scalable CVD process has exhibited a surface finish of 0.2 nm RMS. Polycrystalline Si was fabricated by reacting trichlorosilane ($SiHCl_3$) with excess H_2 in a hot-wall CVD reactor according to the reaction: $SiHCl_3 + 2H_2 \rightarrow Si(s) + 4HCl(g)$. However, it is not widely applied [14, 44] because columnar structure is often present in CVD Si.

Polycrystalline Si [19] was also used to clad on several advanced ceramic materials such as SiC, sapphire, pyrolytic BN, and Si by a chemical vapor deposition (CVD) process. The thickness of Si cladding ranged from 0.025 to 3.0 mm. CVD Si adhered quite well to all the above materials, where the Si cladding was highly stressed and cracked. The surface roughness can reach 0.2 nm RMS after polishing. Amorphous silicon thin films were formed by chemical vapor deposition by Choi et al. [45]. The amorphous silicon films without reflector bias voltage exhibit 0.119 nm RMS, but down to 0.171 nm RMS with reflector bias voltage of −120 V, respectively.

PVD Si Clapping

CVD Si is generally prepared with high temperature (>600°C), and the PVD Si film is widely used. The application of Si coating can reduce surface wearing resistance of SiC ceramic without changing mechanical properties of the bulk materials. The efficiency of surface finishing for large optical components can be greatly improved as well [23]. Therefore, PVD Si appears to be very attractive because of its relative simplicity, low substrate temperature, and wide accessibility by industry. Si clapping is easier to be polished well because it is single-phase material without the different heterogeneous phase in the polishing process. PVD Si coating is now becoming a preferred method of the SiC surface modification.

Vacuum Evaporation Si Clapping

Vacuum evaporation deposition of Si film is processed in a high vacuum by heating the Si gasification or sublimation condensing into Si film and thus being deposited on SiC substrate surface. The method is relatively simple and has high deposition rate. But the columnar structure is present in Si film and the physical property of Si film is not stabile. Zhao [46] has obtained amorphous Si films by thermal evaporation. In recent years, this method has been improved. Fang et al. [47] deposited Si film on steel and alumina substrates by electron beam evaporation of solid silicon.

Magnetron Sputtering Si Clapping

Magnetron sputtering (belongs to PVD) appears to be very attractive due to its relative simplicity and low substrate temperature. Many researchers have studied this method, Aoucher et al. [48] deposited amorphous silicon by DC magnetron sputtering on a quartz substrate at a rate around 1.5 nm/s. Tang et al. [22] has used RF magnetron sputtering method to prepare Si film with the surface roughness from 17.992 nm RMS to 1.183 nm RMS as shown in Figure 3.

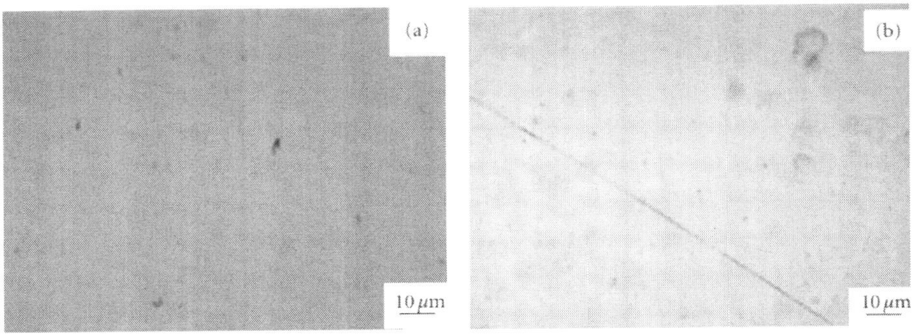

Figure 3: Photomicrograph of (a) unpolished, (b) polished PVD Si coating on RB SiC, reprinted with permission from [22].

Ion-Beam-Assisted Si Clapping

Hydrogenated amorphous silicon (a-Si:H) films are generally prepared by glow-discharge decomposition of silane or by sputtering of silicon in an argon-hydrogen mixture. The reaction temperature is low (200°C) and the preparation parameters can be controlled. So it was used to prepare amorphous silicon films.

Photoconductive hydrogenated amorphous silicon films were deposited by ion-beam-assisted evaporation using hydrogen-argon plasma. Surface modification for the RB-SiC substrate [49] is carried out using e-beam evaporation with plasma ion assisted. The surface roughness of the RB-SiC substrate is reduced to 0.0632 nm, the scattering coefficient is reduced to 2.81%, and the average reflectance from 500 nm to 1000 nm is raised to 97.05%; these data indicate that good optical quality similar to that of the fine polished glass ceramics can be obtained by the modification process.

Plasma-Ion-Assisted of Deposition Si Clapping

For the production of thin films of high quality standard, thermal evaporation techniques are applied with the assistance of ion sources which provide additional energy and momentum to influence the growth process. Larger ion current densities on increased substrate areas can be generated by employing plasma sources, and the process is plasma-ion-assisted deposition (PIAD) [50]. In PIAD, a growing thin film is bombarded by energetic ions from a plasma ion source and the columnar microstructure of the film is disrupted, resulting in the improved optical and mechanical properties of thin films [51]. Liu et al. [23] has prepared Si thin film by this method and obtained continuous, homogenous, well-bonded amorphous Si coatings on SiC ceramics. It means that the SiC substrate can be fully covered up and the effect of substrate surface defects on the surface morphology of the Si coating can be overlooked as shown in Figure 4.

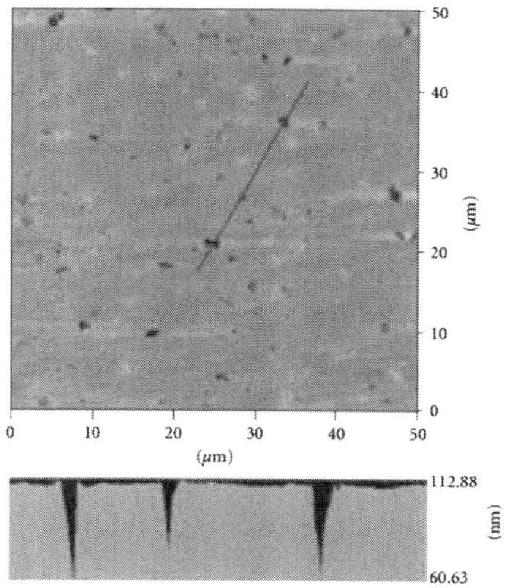

Figure 4: Surface topography of Si coating tested by AFM, reprinted with permission from [23].

Recently, the PVD Si coatings on SiC substrates have been investigated. SSG has applied PVD Si in optical systems of the GEO telescope [52], designed and constructed for an SBIR program funded by NASA's Marshall Space Flight Center (MSFC). The SiC telescope and "GOES-like" scan mirror are designed to "generic" GEO specifications, and the surface roughness decreased to 0.4 nm RMS after polishing. In China, the PVD Si coatings deposited on the surface of polished RB-SiC and S-SiC were demonstrated to improve the optical surface quality after being polished by Tang et al. [22]. Both the surfaces of PVD Si coating on RB-SiC and S-SiC are smoother than that of bare SiC. The RMS can reach to the angstrom grade, and the reflectance improves significantly.

CONCLUSION

Silicon carbide as the third generation of space mirror material has attracted more and more attention and is widely applied. Silicon carbides prepared by different preparation methods have their advantages, but still cannot meet the optical requirements (<1 nm RMS) after the current optical processing. The surface roughness and reflection rate of the SiC material can be well improved after surface modification. Especially in the CVD SiC coating and PVD-Si coating on the SiC ceramic, surface roughness of Angstrom level can be reached. All kinds of surface-modification methods have been developed and every method has its disadvantages. The hot press glass has some drawbacks for usage of the glass cladding on large surfaces. The C/SiC coating may not be suitable for low surface roughness (<1 nm RMS). The CVD process with high temperature (>1000°C) would lead to the deformation of SiC matrix. PVD method to prepare SiC-film is slower and the modified film is very difficult to be polished. In our view, PIAD Si has low reaction temperature (<300°C) and is very easy to be polished. The preparation process is relatively simple, and reproducible preparation of silicon-modified layer gives a dense structure, combined with the base firmly. Therefore, PVD Si is the best choice for surface modification of SiC mirror because of its relative simplicity, low substrate temperature, and wide accessibility by industry, especial PIAD Si.

ACKNOWLEDGMENTS

Y. Yang thanks the Century Program (One-Hundred-Talent Program) of the Chinese Academy of Sciences for special funding support. This study was also supported in part by a fund from the National Natural Science Foundation of China (no. 51071167, 51102266), the Shanghai Yangtze River Delta Science Project (no. 11495810100), and the Shanghai Pujiang Program (10PJ1410700).

REFERENCES

1. J. Robichaud, J. J. Guregian, and M. Schwalm, "SiC optics for Earth observing applications," in Earth Observing Systems VIII, vol. 5151 of Proceedings of SPIE, pp. 53–62, August 2003.
2. S. G. Johnson, "SiC coatings on RB SiC mirrors for ultrasmooth surfaces," vol. 2018 of Proceedings of SPIE, no. 1993, pp. 237–247.
3. R. A. Paquin, M. B. Magida, and C. L. Vernold, "Large optics from silicon carbide," in Large Optics II, vol. 1618 of Proceedings of SPIE, pp. 53–60, October 1991.
4. Q. L. Xu, F. Wu, and S. B. Wu, "Super smooth surfacing fabrication for lightweight silicon carbide plan mirror," Opto-Electronic Engineering, vol. 31, pp. 22–25, 2004.
5. T. Bifano, Y. Yi, and K. Kahl, "Fixed abrasive grinding of CVD SiC mirrors," Precision Engineering, vol. 16, no. 2, pp. 109–116, 1994.
6. Z. Z. Zhu, V. Muratov, and T. E. Fischer, "Tribochemical polishing of silicon carbide in oxidant solution," in Proceedings of the 12th International Conference on Wear of Materials, vol. 225–229, pp. 848–856, Atlanta, Ga, USA, 1999.
7. W. Lin, H. Ohmori, T. Suzuki, Y. Uehara, and S. Morita, "Polishing characteristics of ELID-ground surface of nano precision optical elements," Key Engineering Materials, vol. 291-292, pp. 365–370, 2005.
8. L. Zhou, V. Audurier, P. Pirouz, and J. A. Powell, "Chemomechanical polishing of silicon carbide," Journal of the Electrochemical Society, vol. 144, no. 6, pp. L161–L163, 1997.
9. J. S. Johnson, K. Grobsky, and D. J. Bray, "Rapid fabrication of lightweight silicon carbide mirrors," in Optomechanical Design and Engineering, vol. 4771 of Proceedings of SPIE, pp. 243–253, July 2002.
10. M. Murahara, "Excimer laser-induced photochemical polishing of SiC mirror," in Laser-Induced Damage in Optical Materials, vol. 4679 of Proceedings of SPIE, pp. 69–74, October 2001.
11. J. S. Johnson, K. Grobsky, and D. J. Bray, "Rapid fabrication of lightweight silicon carbide mirrors," in Optomechanical Design and Engineering, Proceedings of SPIE, pp. 243–253, July 2002.
12. D. Fan, Z. Y. Zhang, and H. Y. Niu, "Surfacing fabrication of silicon

carbide optical mirror," Journal of Inorganic Materials, vol. 31, pp. 1096–1100, 2003.

13. G. S. Chung and K. S. Kim, "Heteroepitaxial growth of single 3C-SiC thin films on Si(100) substrates using a single-source precursor of hexamethyldisilane by APCVD," Bulletin of the Korean Chemical Society, vol. 28, no. 4, pp. 533–537, 2007.
14. M. T. Clavaguera-Mora, J. Rodríguez-Viejo, Z. El Felk et al., "Growth of SiC films obtained by LPCVD," Diamond and Related Materials, vol. 6, no. 10, pp. 1306–1310, 1997.
15. A. Novi, G. Taglioni, and L. Novella, "Silicon carbide for mirrors by plasma enhanced chemical vapour deposition at low temperature," European Space Agency ESA SP, vol. 554, pp. 687–691, 2004.
16. L. D. Xu, X. J. Zhang, and X. Wang, "Surface coating technique of reaction bonded sic mirrors," Opto-Electronic Engineering, vol. 36, no. 1, pp. 120–124, 2009.
17. C. J. Shih and A. Ezis, "Application of hot-pressed silicon carbide to large high-precision optical structures," in Silicon Carbide Materials for Optics and Precision Structures, vol. 2543 of Proceedings of SPIE, pp. 24–37, July 1995.
18. G. Liu, Z. Huang, X. Liu, and D. Jiang, "Removal behaviors of different SiC ceramics during polishing," Journal of Materials Science and Technology, vol. 26, no. 2, pp. 125–130, 2010.
19. J. S. Goela and R. L. Taylor, "Chemical vapor deposition for silicon cladding on advanced ceramics," Journal of the American Ceramic Society, vol. 72, pp. 1747–1750, 1989.
20. Y. Y. Han, Y. M. Zhang, and J. C. Han, "Current status of research on silicon carbide mirror technology," Materials Review, vol. 19, pp. 5–8, 2005.
21. H. Zhou, C. R. Zhang, and Y. B. Cao, "Lightweight C/SiC mirrors for space application," Large Mirrors and Telescopes, vol. 6148, pp. L1480–L1486, 2006.
22. H. Tang, Z. Huang, and S. Tan, "PVD SiC and PVD Si coatings on RB SIC for surface modification," in Proceedings of the 2nd International Symposium on Advanced Optical Manufacturing and Testing Technologies—Advanced Optical Manufacturing and Testing Technologies, pp. 1–6, November 2005.

23. G. L. Liu, Z. R. Huang, J. H. Wu, and X. J. Liu, "Surface morphology evolution and properties of silicon coating on silicon carbide ceramics by advanced plasma source ion plating," Surface and Coatings Technology, vol. 207, pp. 204–210, 2012.
24. J. Yan, Z. Zhang, and T. Kuriyagawa, "Mechanism for material removal in diamond turning of reaction-bonded silicon carbide," International Journal of Machine Tools and Manufacture, vol. 49, no. 5, pp. 366–374, 2009.
25. H. Y. Tam, H. B. Cheng, and Y. W. Wang, "Removal rate and surface roughness in the lapping and polishing of RB-SiC optical components," Journal of Materials Processing Technology, vol. 192-193, pp. 276–280, 2007.
26. G. L. Liu, Z. R. Huang, and X. J. Liu, "Recent developments of surface coatings and optical fabrication of silicon carbide," Journal of Inorganic Materials, vol. 22, pp. 769–774, 2007.
27. A. Venu Gopal and P. V. Rao, "The optimisation of the grinding of silicon carbide with diamond wheels using genetic algorithms," International Journal of Advanced Manufacturing Technology, vol. 22, no. 7-8, pp. 475–480, 2003.
28. H. B. Cheng, Z. J. Feng, and Y. W. Wang, "Magnetorheological finishing of SiC aspheric mirrors," Materials and Manufacturing Processes, vol. 20, pp. 917–931, 2005.
29. M. Deyerler, N. Pailer, R. Wagner, and C. Mueller, "Ultra-lightweight mirrors: recent developments of C/SiC," in Optical Design, Materials, Fabrication, and Maintenance, vol. 4003 of Proceedings of SPIE, pp. 73–79, March 2000.
30. A. Novi, G. Mondello, C. Devilliers et al., "Comparison of sintered-SiC and C/SiC mirrors behaviour at cryo-temperatures," in Optical Materials and Structures Technologies, vol. 5179 of Proceedings of SPIE, pp. 246–253, August 2003.
31. B. Harnisch, B. Kunkel, M. Deyerler, and S. Bauereisen, "Ultra-lightweight C/SiC Mirrors and Structures," ESA Bulletin-European Space Agency, vol. 95, pp. 108–112, 1998.
32. M. B. Magida, R. A. Paquin, and J. J. Richmond, "Dimensional stability of bare and coated reaction bonded silicon carbide," in Dimensional Stability, vol. 1335 of Proceedings of SPIE, pp. 60–68, July 1990.
33. C. Hung Wu, C. A. Zorman, and M. Mehregany, "Growth of

polycrystalline SiC films on SiO_2 and $Si_3N_4$by APCVD," Thin Solid Films, vol. 355-356, pp. 179–183, 1999.

34. R. J. Liu, C. R. Zhang, X. G. Zhou, and Y. B. Cao, "SiC coatings for mirrors prepared by low pressure chemical vapor deposition," Journal of Materials Science Letters, vol. 22, no. 11, pp. 841–843, 2003.
35. J. H. Zhang, Y. M. Zhang, J. C. Han, X. D. He, and W. Yao, "Design, fabrication and testing of space-borne SiC mirror," Optics and Precision Engineering, vol. 14, no. 2, pp. 179–184, 2006.
36. S. Johnson, "SiC coating on RB SiC mirror for ultra-smooth surface," vol. 2018 of Proceedings of SPIE, pp. 237–247, 1993.
37. T. T. Wang, J. S. Gao, X. Y. Wang et al., "Preliminary study of reaction bonded silicon carbide surface modification," Optics and Precision Engineering, vol. 16, no. 9, pp. 1603–1607, 2008.
38. J. S. Gao, T. T. Wang, Q. Song et al., "Preparation of silicon carbide surface-modified coatings with end hall ion source assisted," Infrared and Laser Engineering, vol. 37, no. 4, pp. 710–713, 2008.
39. G. Monaco, M. Gastaldi, and P. Nicolosi, "Silicon carbide thin films for EUV and soft X-ray applications," European Physical Journal, vol. 169, pp. 159–165, 2009.
40. H. Tang, S. Tan, Z. Huang, S. Dong, and D. Jiang, "Surface morphology of α-SIC coatings deposited by RF magnetron sputtering," Surface and Coatings Technology, vol. 197, no. 2-3, pp. 161–167, 2005.
41. J. B. Kortright and D. L. Windt, "Amorphous silicon carbide coatings for extreme ultraviolet optics," Applied Optical, vol. 27, no. 14, pp. 2841–2846, 1988.
42. I. Hanyecz, J. Budai, E. Szilágyi, and Z. Tóth, "Characterization of pulsed laser deposited hydrogenated amorphous silicon films by spectroscopic ellipsometry," Thin Solid Films, vol. 519, no. 9, pp. 2855–2858, 2011.
43. M. Vendan, P. Molian, A. Bastawros, and J. Anderegg, "Ultra-short pulsed laser deposition and patterning of SiC thin films for MEMS fabrication," Materials Science in Semiconductor Processing, vol. 8, no. 6, pp. 630–645, 2005.
44. Y. F. Zhou, Y. M. Zhang, and J. C. Han, "Preparation of silicon coating," Materials Review, vol. 19, pp. 84–86, 2005.
45. S. G. Choi, S. J. Wang, H. H. Park et al., "Properties of amorphous silicon thin films synthesized by reactive particle beam assisted

chemical vapor deposition," Thin Solid Films, vol. 518, no. 24, pp. 7372–7376, 2010.

46. X. J. Zhao, "Investigation of silicon carbide mirror's reflective performance and film design," Harbin Harbin Institute of Technology, 2007.
47. P. H. Fang, P. Bai, J. H. Kinnier, Z. Huan, and C. C. Schubert, "Temperature dependent formation of microcrystal and amorphous silicon by vacuum evaporation," Journal of Non-Crystalline Solids, vol. 59-60, no. 2, pp. 819–821, 1983.
48. M. Aoucher, G. Farhi, and T. Mohammed-Brahim, "Crystallization of hydrogenated amorphous silicon deposited at high rate by dc magnetron sputtering," Journal of Non-Crystalline Solids, vol. 227–230, no. 2, pp. 958–961, 1998.
49. Z. F. Shen, J. S. Gao, H. Chen, X. Y. Wang, T. T. Wang, and X. M. Zheng, "Performance and reliability evaluation for the surface modification of space used silicon carbide mirror," Acta Photonica Sinica, vol. 38, no. 9, pp. 2353–2358, 2009.
50. J. Harhausen, I. Meyenburg, A. Ohl, and R. Foest, "Characterization of the plasma plume of a PIAD plasma source by means of optical emission spectroscopy," Surface and Coatings Technology, vol. 205, no. 2, pp. S407–S410, 2011.
51. S. H. Woo and C. K. Hwangbo, "Influence of plasma ion-beam assistance on TiO_2 and MgF_2 thin films deposited by plasma ion-assisted deposition," Surface and Coatings Technology, vol. 201, no. 19-20, pp. 8250–8257, 2007.
52. M. I. Anapol, L. R. Gardner, T. W. Tucker, and R. J. Koczor, "Lightweight 0.5-m silicon carbide telescope for a geostationary earth observatory mission," in Silicon Carbide Materials for Optics and Precision Structures, Proceedings of SPIE, pp. 164–172, July 1995.

Chapter 6

Experimental Optimization of Annular Polishing Parameters for Silicon Carbide

Yuan Liu[1], La Han[2], Haiying Liu[3], Yikai Shi[4], and Junjie Zhang[2]

[1]School of Astronautics, Harbin Institute of Technology, Harbin 150001, China
[2]Center for Precision Engineering, Harbin Institute of Technology, Harbin 150001, China
[3]Xiaguang Optical Electron Co., Ltd, Yangzhou 225127, China
[4]Science and Technology on Integrated Logistics Support Laboratory, National University of Defense Technology, Changsha 410073, China

ABSTRACT

Machined surface quality has a strong impact on the functionality of silicon carbide-based components and devices. In the present work, we first analytically investigate the complex coupling of motions in annular polishing based on the Preston equation, which derives the influential parameters for

Citation: Yuan Liu, La Han, Haiying Liu, Yikai Shi, and Junjie Zhang, "Experimental Optimization of Annular Polishing Parameters for Silicon Carbide," Advances in Materials Science and Engineering, vol. 2018, Article ID 9019848, 6 pages, 2018. https://doi.org/10.1155/2018/9019848

material removal. Subsequently, we conduct systematic annular polishing experiments of reaction-bonded silicon carbide to investigate the influence of derived parameters on polished surface quality, which yield optimized polishing parameters for achieving ultralow surface roughness of reaction-bonded silicon carbide.

INTRODUCTION

Reaction-bonded (RB) silicon carbide (SiC) is one of the preferred materials for manufacturing optical mirrors due to its unique characteristics of low density, high strength, low thermal expansion, high thermal conductivity, and high chemical inertness [1]. Machined surface quality plays an important role in determining the functionality of SiC-based components and devices. For instance, according to the theory of total integral scattering (TIS), the surface scattering capability of optical mirror is closely related to its surface roughness, as the TIS increases sharply with the increase of surface roughness. The high surface scattering coefficient will cause the system to produce stray light, which decreases the reflectivity of SiC mirrors and consequently results in the degradation of imaging quality of optical system [2]. Thus, improving machined surface quality of SiC is critical for facilitating the performance of SiC mirrors.

At present, annular polishing technology is an efficient method to obtain SiC mirrors with high surface quality [3–6]. Some typical work are as follows: Rupp et al. studied the grinding and polishing of conventional optics and found that the contact pressure between the specimen and the polishing pad changes dynamically, and is a function of time and location of points on the specimen [2]. Aspden et al. developed a CNC surface polisher that uses a small tool to achieve the desired profile by convolution to calculate the dwell time of the widget. In their topics, there is no single mathematical model that completely correlates the mechanical motion of optical surfaces with the amount of material removed during grinding or polishing process [7]. Wagner and Shannon established a model to describe the relationship between the mechanical motion and the change in surface area and to calculate the material removal of the specimen surface through the mathematical method and optical kinematics principle [8]. Gao and Cao studied the relationship between motion parameters, grinding mechanism, and the trajectory of the specimen [9]. Shen and Yuan studied the influence of applied load on pressure profile in the polishing process [10]. Liu et al. introduced the use of annular polishing technology for the manufacturing of large-diameter square optical components [11]. On the other side, the annular polishing is strongly dependent on polishing conditions. Wang et al. found that as the pressure increases, the contact area between the polishing pad and the workpiece increases, while the mechanical friction is enhanced and the material removal rate is high. As the rotation speed increases, the thickness of the polishing liquid film between the polishing pad and the workpiece gradually increases, the flow rate of the polishing liquid increases, leading to increased materials taken away from the processed surface under an increased material removal rate [12]. Sun et al. found that the oxidant, the complex agent content, and the polishing solution PH value in the polishing solution have a great influence on the material removal rate [13]. Hu et al. studied the effects of abrasive types such as SiO_2, Al_2O_3, and CeO, the concentration of abrasives and the size of abrasive grains on the polishing effect [14]. More recently, Zhang et al. performed finite element simulations and experiments of annular polishing of SiC and found that the distribution of contact pressure on the SiC specimen is significantly affected by the polishing speed, Poisson's ratio, and the elastic modulus of polishing pad [15].

Previous studies have obtained many valuable insights. However, most of previous work is mainly utilizing a sole method, either experiment or simulation. In particular, the lack of systematic experimental demonstration of theoretical analysis greatly restricts the deep understanding of the annular polishing mechanism of SiC. Therefore, based on the Preston equation and motion simulation, this paper analyzes the law of motion coupling in annular polishing and summarizes the polishing parameters that affect the annular polishing. Based on the analytical theoretical investigations, annular polishing experiments are carried out to further study the influence of different polishing parameters and ultimately obtain the optimized polishing parameters. This research has important theoretical significance and practical value to guide the annular polishing processing of SiC.

ANALYTICAL INVESTIGATION OF ANNULAR POLISHING

Kinematic Coupling of Motions

Figure 1(a) illustrates a typical annular polisher, which mainly consists of a polishing disc, a carrier disc, and a swinging bracket, respectively. The specimen is pasted on the carrier disc with paraffin, which means that the specimen has a synchronous speed with the carrier disc. The applied pressure is provided by the weight of carrier disc. Accordingly, Figure 1(b) illustrates the simplified motion diagram in the annular polishing, which indicates that the kinematic coupling of relative motions mainly includes the rotation of polishing disc and the rotation of carrier disc. As indicated in Figure 1(1), the speed of carrier disc is ω and the speed of polishing disc is δ. The distance from the center of specimen O_2 to a point A on the carrier disc is r, and the distance from the center of polishing disc O_1 to the center of the carrier disc O_2 is R. The angle between the line segments $AO_2$and O_1O_2 is θ. The speed at the point A on the polishing disc relative to O_1 is V_1, and the speed at the point A on the specimen relative to O_2 is V_2, so the relative speed of the specimen and the polishing disc at point A is V, which can be derived from the following equation [16].

$$V(r,\theta) = \left[R^2\delta^2 + r^2(\delta-\omega)^2 + 2rR\delta(\delta-\omega)\cos\theta\right]^{1/2}. \quad (1)$$

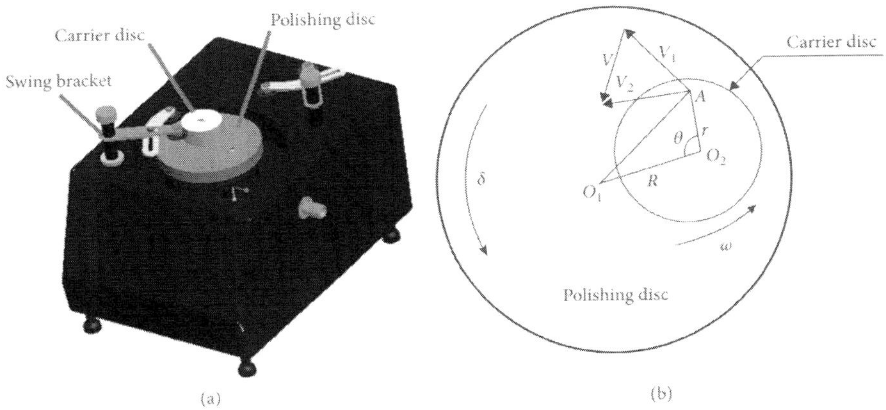

Figure 1: Illustration of annular polishing. (a) Model of annular polisher; (b) motion analysis of annular polishing. Reproduced from Zhang et al. [15] (under the Creative Commons Attribution License/public domain).

Preston Equation

Both material removal rate and surface quality of specimen in the annular polishing process are strongly affected by processing polishing parameters, which have complex interactions. Preston et al. simplified the Preston equation [17] to characterize the relationship between material removal and polishing speed V, applied pressure P, and other external factors, as shown in the following equation [17]:

$$\frac{dh}{dt} = kPV = kP\frac{ds}{dt}, \tag{2}$$

where h is the amount of material removal and k is a proportional constant that is related to various environmental factors. Therefore, the amount of material removal at one specific point can be derived according to Equation (2). However, the contact between specimen and polishing disc changes dynamically with polishing time, which induces uncertainties in the analytical investigation of annular polishing process. Therefore, three main assumptions are made to simplify the operation: (1) specimen and polishing disc are fully contacted without separation; (2) the applied pressure does not change with polishing time; and (3) the proportional constant k does not change with polishing time. According to the Preston equation, the amount of material removal within a specific polishing time can be derived by integrating over time t, as shown in the following equation:

$$h(r) = k\int_0^T P(r,t)\cdot V(r,t)\,dt. \tag{3}$$

It can be seen from Equation (3) that the material removal is only dependent on the resultant speed V, given the k and the P are constant values. By substituting the relative velocity derived from Equation (1) into Equation (3), the material removal at point A can be derived, as shown in the following equation [15]:

$$h(r) = k\int_0^T P(r,t) \cdot \left[R^2\delta^2 + r^2(\delta-\omega)^2 + 2rR\delta(\delta-\omega)\cos\theta\right]^{1/2} dt. \tag{4}$$

The amount of material removal at any point on the specimen can be calculated from Equation (4). The pressure P is affected by the type of polishing pad and polishing solution. Based on the annular polishing motion simulation and the Preston equation, four influential parameters can be determined: as polishing solution, polishing pad, polishing time, and rotation speed of polishing disc, respectively.

ANNULAR POLISHING EXPERIMENT OF RB-SIC

Effect of Polishing Solution Material

It has been demonstrated that polishing solution's particle size, particle hardness, and concentration have a strong impact on the polishing process [18]. We first investigate the influence of polishing solution material on the polishing results. Three kinds of polishing solution materials such as alumina, silica, and diamond are considered. For each polishing solution material, the average particle size is the same as 250 nm. In the polishing experiments, the polyurethane polishing pad is used with a rotation speed of 120 r/min. The polishing time is 60 min. After polishing, the surface roughness of the specimen is measured by a Taylor Hobson surface profiler.

Figure 2 presents the value of surface roughness for each polishing solution material. The surface roughness is the maximum of 7.3 nm for the alumina, followed by 6.1 nm for the silica, and the minimum of 3.2 nm for the diamond. It can be seen from Figure 2 that with the increase of abrasive hardness of the polishing solution, the surface roughness of the specimen decreases gradually, as the increase of abrasive hardness results in an improvement not only in the mechanical force between the specimen and abrasive but also in the quality of the specimen surface.

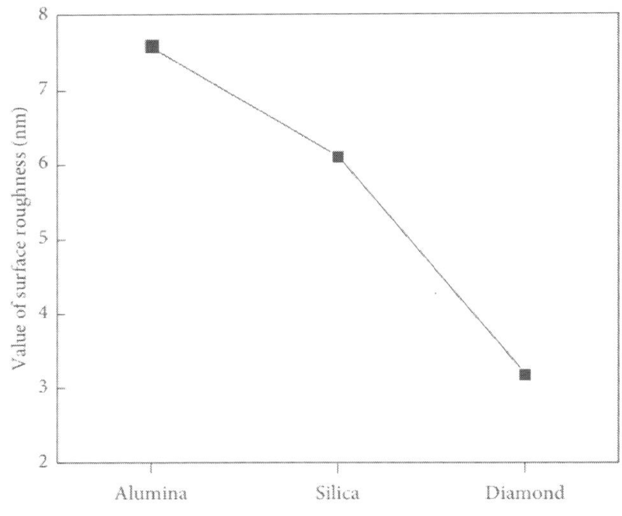

Figure 2: The relationship between the surface roughness and the type of polishing solution.

Effect of Polishing Pad Material

The polishing pad should has the following conditions: appropriate rigidity and hardness, a certain elasticity, good retention of the polishing solution, excluding the by-product of polishing process, and low impurities of the polishing pad [19]. In this paper, we use four kinds of polishing pad materials: the cot polishing pad, the matte leather polishing pad, the synthetic leather polishing pad, and the polyurethane polishing pad.

In different polishing experiments with different polishing pad materials, all the polishing parameters are kept constant. And the diamond suspension solution is used as the polishing solution for each experiment. The speed of polishing plate is 120 r/min, and the polishing time is 60 min. After the polishing experiment, the measured surface roughness of SiC is 7.6 nm, 6.9 nm, 8.1 nm, and 2.8 nm for the cot polishing pad, the matte leather polishing pad, the synthetic leather polishing pad, and the polyurethane polishing pad, respectively, as shown in Figure 3.

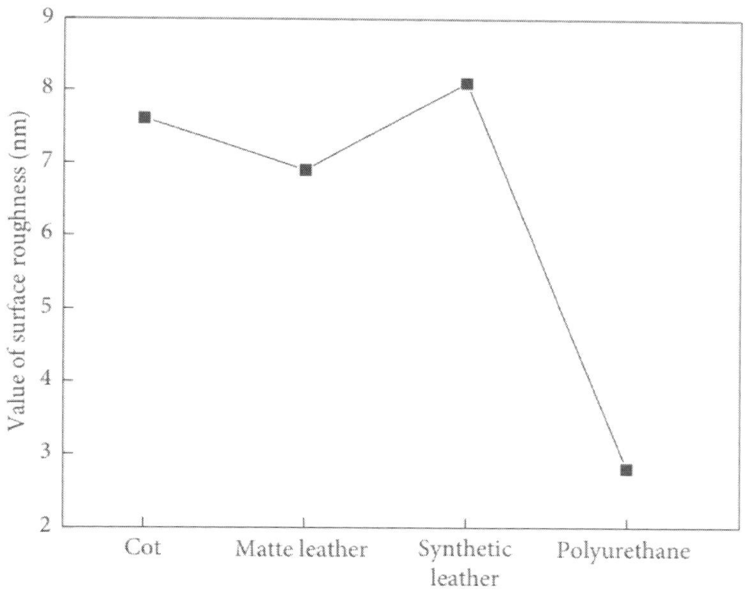

Figure 3: The relationship between the surface roughness and the type of polishing pad.

It can be seen from Figure 3 that surface roughness drops to the smallest value when using the polyurethane polishing pad. It can be concluded that due to the higher hardness of SiC specimen than the cot polishing pad, abrasive particles are pressed into the polishing pad, and thus are unable to produce effective force on the SiC surface, resulting in low processing precision and high surface roughness. The hardness of matte leather and synthetic leather polishing pad is higher, but the surface is easier to be scratched as a result of the higher interaction force, which results in higher surface roughness. The hardness of polyurethane polishing pad is moderate, which can obtain the best polishing effect with the lowest surface roughness of 2.8 nm.

Effect of Polishing Time

We further investigate the influence of polishing time on the machined surface quality of SiC. While maintaining the other polishing parameters unchanged, only the polishing time is changed. The polishing time is divided into six groups: 30 min, 60 min, 90 min, 120 min, 150 min, and 180 min, respectively. In each experiment, the diamond suspension is used as the polishing solution, and the polyurethane polishing pad is utilized. The polishing pad speed is 120 r/min. After the polishing, the measured surface roughness is 3.2 nm, 2.5 nm, 2.1 nm, 1.6 nm, 0.8 nm, and 1.2 nm for the polishing time of 30 min, 60 min, 90 min, 120 min, 150 min, and 180 min, respectively, as shown in Figure 4.

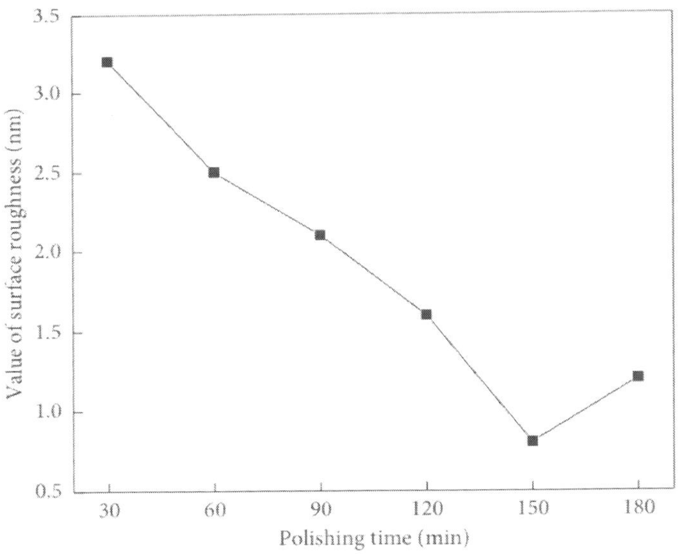

Figure 4: The relationship between the surface roughness and the polishing time.

It can be seen from Figure 4 that the surface roughness changes with the polishing time: it first decreases steadily and then starts to increase from the polishing time of 150 min, indicating a critical polishing time of 150 min. In the initial stage of the polishing experiment, large abrasive particles are embedded into the surface to accommodate large pressure, which leads to generation of surface scratches, thus affecting the surface quality. With the increase of polishing time, abrasive particles break up to small particles with sharp edges that have good cutting performance, which facilitates the formation of smooth surface of the specimen. The surface roughness reaches the lowest value of 0.8 nm at the polishing time of 150 minutes. However, with the further increase of polishing time, the increase of surface temperature produces thermal stress, which affects the machining accuracy and lowers the surface roughness.

Effect of Rotation Speed of Polishing Disc

We also investigate the influence of rotation speed of polishing disc. Six rotation speeds of polishing disc are considered: 50 r/min, 70 r/min, 90 r/min, 120 r/min, 140 r/min, and 160 r/min, respectively. For each rotation speed, all the other polishing parameters are the same. The diamond suspension is used as the polishing solution, and the polyurethane polishing pad is utilized. The polishing time is 60 min. After the polishing experiment, the measured surface roughness is 3.3 nm, 2.7 nm, 2.5 nm, 1.6 nm, 1.2 nm, and 2.6 nm for the rotation speed of 50 r/min, 70 r/min, 90 r/min, 120 r/min, 140 r/min, and 160 r/min, respectively, as shown in Figure 5.

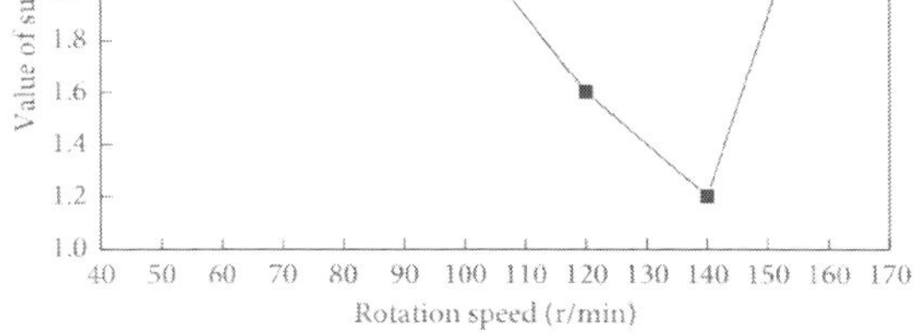

Figure 5: The relationship between the surface roughness and the rotation speed.

It can be seen from Figure 5 that with increasing rotation speed, the surface roughness first decreases and reaches the lowest value of 1.2 nm at the rotation speed of 140 r/min and then increases with a further increase of rotation speed. It is known that a high rotation speed of the polishing disk can increase the efficiency of the polishing process. However, the higher the rotation speed of polishing disc, the lower the processing stability, which result in the lower surface formation accuracy.

Based on the above analysis, the optimized parameters of annular polishing of SiC are summarized as follows: a rotating speed of polishing disc is 140 r/min, a polishing time is 150 min, using the polyurethane polishing pad and the diamond suspension solution. Under the optimized polishing conditions, a high-quality SiC specimen with a surface roughness of 1.31 nm is obtained. Figure 6 shows the characterized SiC specimen before and after polishing, indicating polished SiC with good surface quality can be obtained by using the optimized annular polishing parameters.

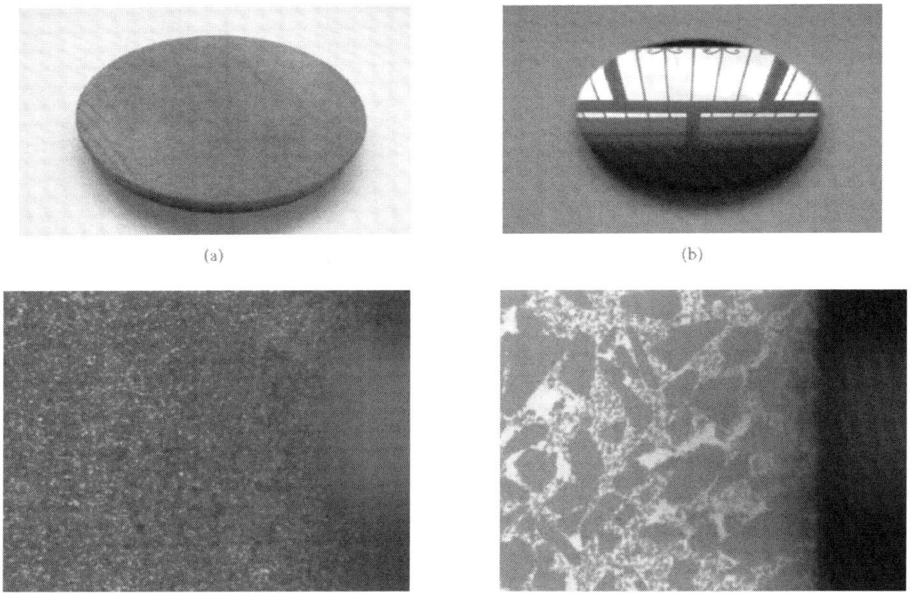

(a) (b) (c) (d)

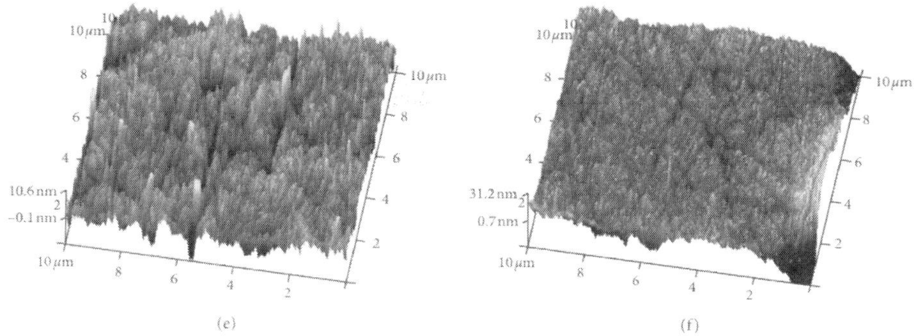

Figure 6: Morphology and microstructure of polished SiC. (a) Before polishing, (b) after polishing, (c) surface morphology before polishing, (d) surface morphology after polishing, (e) surface microstructure before polishing, and (f) surface microstructure after polishing.

SUMMARY

In this work, we experimentally investigate the optimization of annular polishing parameters of RB-SiC by first analytical investigation of influential parameters based on the Preston equation and the coupling of polishing motions, and then systematic polishing experiments and characterization. It is found that the annular polishing process can be greatly influenced by the polishing solution: the polishing pad material, the polishing time, and the rotation speed of polishing disc. The optimized annular polishing parameters are a rotating speed of polishing disc of 140 r/min, a polishing time of 150 min, using polyurethane polishing pad and diamond suspension solution, which lead to a surface roughness of 1.31 nm of RB-SiC by annular polishing.

AUTHORS' CONTRIBUTIONS

Y. L., H. L., and J. Z. conceived and designed the experiments; L. H. and Y. S. performed the analytical investigation and experiments; Y. L. and J. Z. analyzed the data and wrote the paper.

ACKNOWLEDGMENTS

The authors acknowledge financial support from the National Natural Science Foundation of China (NSFC) (51875119 and 61473096), the Fundamental Research Funds for the Central Universities, the German Research Foundation (DFG) International Joint Research Program (51761135106), and the National Key R&D Program of China (2016YFB0501203).

REFERENCES

1. R. Joseph, S. Jay, and L. David, "Recent advances in reaction bonded silicon carbide optics and optical systems," Proceeding of SPIE, vol. 5868, Article ID 586802, 2005.
2. V. Rupp, "The development of optical surfaces during the grinding process," Applied Optics, vol. 4, no. 6, pp. 743–748, 1965.
3. F. Gritti and G. Guiochon, "Gradient chromatography under constant frictional heat: Realization and application," Journal of Chromatography A, vol. 1289, pp. 1–12, 2013.
4. X. C. Wang, C. C. Wang, X. T. Shen, and F. H. Sun, "Potential material for fabricating optical mirrors: polished diamond coated silicon carbide," Applied Optics, vol. 56, no. 14, pp. 4113–4122, 2017.
5. Q. T. Fan, J. Q. Zhu, and B. A. Zhang, "Effect of the geometry of workpiece on polishing velocity in free annular polishing," Chinese Optics Letters, vol. 5, pp. 298–300, 2007.
6. Y. Hashimoto, S. Oshika, N. Suzuki, and E. Shamoto, "A new contact model of pad surface asperities utilizing measured geometrical features," in Proceedings of the International Conference on Planarization/CMP Technology, pp. 1–4, Chandler, AZ, USA, September–October 2015.
7. R. Aspden, R. Mcdonough, and F. R. Nitchie, "Computer assisted optical surfacing," Applied Optics, vol. 11, no. 12, pp. 2739–2747, 1972.
8. R. E. Wagner and R. R. Shannon, "Fabrication of aspherics using a mathematical model for material removal," Applied Optics, vol. 13, no. 7, pp. 1683–1689, 1974.
9. H. G. Gao and J. L. Cao, "Simulation of stock removal uniformity during eccentric plane polishing by a tin lap," Chinese Journal of Scientific Instrument, vol. 21, pp. 83–85, 2000.
10. Z. W. Shen and J. L. Yuan, "Ultra-precision planarization for crystals," Journal of Zhejiang University-Technology, vol. 29, pp. 20–25, 2001.
11. M. C. Liu, X. Y. Hu, and Z. S. Li, "Research of fabricating large quadrate high-precision plane optical elments," Optics Technology, vol. 27, pp. 518-519, 2004.
12. B. L. Wang, H. Gao, X. J. Teng, and R. K. Kang, "Effect of polishing parameter on material removal and surface quality of KDP crystal," Journal of Synthetic Crystals, vol. 39, pp. 29–33, 2010.

13. J. Z. Sun, G. S. Pan, Y. H. Zhu, Y. J. Dai, J. B. Luo, and W. M. Li, "Influence of slurry ingredients as particle on hard disk substrate polishing," in Advanced Tribology, J. Luo, Y. Meng, T. Shao, and Q. Zhao, Eds., pp. 993-994, Springer, Berlin, Heidelberg, 2007.
14. W. Hu, X. Wei, X. Z. Xie, and B. P. Xiang, "Study on the performance of polishing slurry in chemical mechanical polishing," Diamond and Abrasives Engineering, vol. 156, pp. 78–80, 2006.
15. J. J. Zhang, L. Han, H. Y. Liu, Y. K. Shi, Y. D. Yan, and T. Sun, "Theoretical and experimental studies of over-polishing of silicon carbide in annular polishing," Machines, vol. 6, no. 2, p. 15, 2018.
16. R. L. Goedecke and R. M. Jackson, "A fractal expansion of a three dimensional elastic-plastic multi-scale rough surface contact model," Tribology International, vol. 59, pp. 230–239, 2013.
17. F. Preston, "The theory and design of plate glass polishing machines," Journal of the Society of Glass Technology, vol. 11, pp. 214–256, 1927.
18. Y. D. Filatov, A. G. Vetrov, and V. I. Sidorko, "Polishing of optoelectronic components made of monocrystalline silicon carbide," Journal of Superhard Materials, vol. 37, no. 1, pp. 48–56, 2015.
19. D. Liu, G. Chen, and Q. Hu, "Material removal model of chemical mechanical polishing for fused silica using soft nanoparticles," International Journal of Advanced Manufacturing Technology, vol. 88, no. 9-12, pp. 3215–3525, 2016.

Chapter 7

Effects of Process Parameters on Material Removal in Vibration-Assisted Polishing of Micro-Optic Mold

Jiang Guo [1] and Hirofumi Suzuki [2]

[1]Key Laboratory for Precision and Non-Traditional Machining Technology of Ministry of Education, Dalian University of Technology, Dalian 116024, China

[2]Department of Mechanical Engineering, Chubu University, 1200 Matsumoto-cho, Kasugai, Aichi 487-8501, Japan

ABSTRACT

Process parameter conditions such as vibrating motion, abrasives, pressure and tool wear play an important role in vibration-assisted polishing of micro-optic molds as they strongly affect material removal efficiency and stability. This paper presents an analytical and experimental investigation on the effects of process parameters, aimed at clarifying interrelations between material removal and process parameters which affect polishing quantitatively. The material removal rate (MRR) and surface roughness

Citation: Guo, Jiang; Suzuki, Hirofumi. 2018. "Effects of Process Parameters on Material Removal in Vibration-Assisted Polishing of Micro-Optic Mold." Micromachines 9, no. 7: 349. https://doi.org/10.3390/mi9070349

which represent the polishing characteristics were examined under different vibrating motions, grain sizes of abrasives and polishing pressure. The effects of pressure and tool wear conditions on tool influence function were analyzed. The results showed that 2D vibrating motion generated better surface roughness with higher material removal efficiency while a smaller grain size of abrasives created better surface roughness but lower material removal efficiency. MRR gradually decreases with the increase of polishing pressure when it exceeds 345 kPa, and it was greatly affected by the wear of polisher when wear diameter on the polisher's head exceeds 300 μm.

Keywords: polishing pressure, vibration, material removal, roughness, tungsten carbide, tool wear

INTRODUCTION

There is an increasing demand for micro-optic lenses with low-cost and high optical qualities in consumer digital optical devices such as digital cameras, DVD players and microscopes. Besides that, micro-optic lenses are widely used in medical and industrial applications where their small size and renowned high-quality optical imaging are important [1,2]. Due to the large number of requirements and the complexity in structure, the molding process becomes a highly productive method for providing high accuracy and shape control of the manufactured parts. To withstand high temperature and pressure conditions in the hot-press molding process for replicating glass lenses, tungsten carbide (WC) material is usually adopted [3,4]. Due to its hard-brittle nature, a grinding process is required for generating the form. Some research has been done on grinding methods and tools to achieve high form accuracy [5,6,7,8]. However, a super-smooth surface in high form accuracy is still difficult to obtain in order to meet the requirements well. Therefore, a subsequent polishing process with loose super abrasives becomes necessary.

Several super-fine polishing technologies such as magnetorheological finishing (MRF), fluid jet polishing, and bonnet polishing, have been developed for finishing complex optics in recent years, but they are still limited to relatively large-size optics [9,10,11,12,13]. As an alternative technology, vibration-assisted machining shows strong capabilities to machine small-size parts [14,15,16]. To polish small-size optics, recently, a vibration-assisted polishing method has been proposed and successfully applied to finish the micro-optic molds with high form accuracy and low surface roughness [17,18,19,20]. Compared with the polishing methods using rotation, it has the advantages that (1) owing to the two-dimensional (2D) vibration with shorter polishing traces, better surface roughness can be obtained; (2) the polishing tool can be focused on polishing a very small area (under 0.2 mm^2) with high removal efficiency; (3) since polishing efficiency is dominated by vibration frequency and amplitude, not by the polishing tool's diameter, it can be miniaturized and arbitrarily shaped to meet specific requirements. So, vibration-assisted polishing is quite suitable to be applied to polish micro-optic molds of diameter less than 5 mm with a concave shape and large curvature. To control the polishing process well, the effects of process parameters in the material removal process needs to be well understood because it is crucial to ensuring high process repeatability. However, polishing is a very complex phenomenon which is affected by many variables. Due to the interactions between polisher, workpiece and abrasives and the physical scale of the material removal processes in polishing being difficult (practically impossible) to observe directly [21,22], up to now, the fundamental material removal mechanism is still poorly understood and a holistic knowledge of it does not exist although during the last decades efforts have been made to investigate interrelations between process parameters and removal mechanism [23].

This paper presents an analytical and experimental investigation on the effects of process parameters, aiming at clarifying interrelations between material removal mechanism and parameters which affect the polishing process qualitatively. Firstly, the effects of process parameters including different vibrating motions of the polisher and grain sizes of abrasives on material removal rate (MRR) and surface roughness are examined. Then the effects of pressure and tool wear on tool influence function are analyzed. To reveal some insights of the mechanism involved in vibration-assisted polishing, a general model of material removal is proposed to illustrate the phenomenon of MRR change by varying polishing pressure. Finally, as another parameter which greatly affects MRR, the wear of the polisher is discussed with some experimental verification.

EXPERIMENTAL SETUP AND CONDITIONS

The experiments were conducted using similar setups as previous research. The main polishing conditions are summarized as shown in Table 1. Tungsten carbide was adopted as the workpiece material since it is a hard-brittle material which can withstand high temperature and pressure conditions in the hot-press molding process to replicate the glass lenses. It was lapped prior to polishing on a lapping machine. A polisher made of polyurethane was selected for the experiments. It is a kind of soft but hard-to-wear material with an international rubber hardness degree (IRHD) of 90. The shape of the polisher was generated by precision cutting process with high form accuracy. It can be miniaturized with a radius of curvature from 0.5 mm to 2 mm. In this paper, a radius of 1 mm is used. Concerning surface roughness and material removal efficiency, diamond slurry is the best abrasive to be used to polish tungsten carbide, so a few kinds of diamond slurry with density of 1 wt % and grain size of 0.1 μm, 0.125 μm, 0.25 μm, 0.5 μm and 1 μm were employed for testing. These diamond slurries are water-based with excellent cutting performance and surface finish and they are particularly effective for materials which are hard to machine. Since loading conditions appear to be key to understanding the removal mechanism [24], this paper will focus on them. As mentioned above, the polishing force can be precisely controlled by the real-time polishing force control system, so it is set from 5–50 mN with an increment of 5 mN in the experiments. The magnetostrictive vibrating polisher vibrated at frequency of 9.2 kHz. Lateral, circular and elliptical vibration traces correspond to a phase difference of 0 deg, 45 deg and 90 deg, respectively, in the opposing pair of coils of magnetostrictive vibrating polisher [19]. The vibration amplitude in lateral vibrating motion was 43 μm. The radius of circular vibrating motion was 30 μm. The long axis and minor axis of elliptical vibration was 40 μm and 16 μm, respectively. For the lateral vibration motion, the vibration direction is along the scanning path. For the experiments to investigate interrelations between material removal and process parameters, the raster scanning path was adopted. As shown in Figure 1, the polishing tool vibrated on a workpiece surface during scanning and the material was removed in a rectangular shape. The polishing scope was set to 400 μm × 400 μm, while the polishing tool had a scanning speed of 3.5 mm/min. with a pitch size of 20 μm.

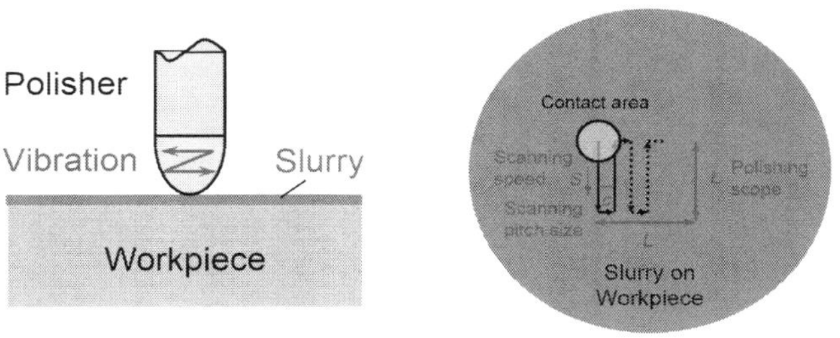

Figure 1: The scanning path of polisher on workpiece surface.

Table 1: Main polishing conditions

Workpiece Material	Tungsten Carbide
Polisher head Radius Hardness	Polyurethane 1 mm IRHD 90
Abrasive Grain size Density	Diamond slurry 0.1 μm/0.125 μm/0.25μm/0.5 μm/1 μm 1 wt %
Polishing force	W = 5–50 mN (Increment: 5 mN)
Vibrating motion	Lateral (λ = 43 μm) Circular (r = 30 μm) Elliptical (a = 40 μm, b = 16 μm)
Frequency	9.2 kHz
Polishing scope Pitch size Scanning speed	400 μm × 400 μm 20 μm 3.5 mm/min

ANALYSIS OF PRESSURE AND TOOL WEAR EFFECTS

The effects of pressure and tool wear conditions on tool influence function were analyzed to know the stress distributions. A finite element model (FEM) was developed based on contact mechanics with the commercial FEM software ABAQUS (ABAQUS6.14, Dassasult Simulia Company, Providence, RI, USA). The static contact model was built considering the boundary conditions of geometry and material properties such as hardness and Young's modulus. The material properties of the polisher and workpiece were set following the descriptions in Section 2. The slurry was not involved in the model. Figure 2a,b shows the simulated cross-sectional view of stress distribution without tool wear at 20 mN and stress distribution on the workpiece under different loads, respectively. It is noted that the simulated values in this figure may not be quantitatively accurate due to the difference between the actual and setting values of material properties, but the results are adequate for qualitative comparison. It is found that with the increase of force, the contact area was enlarged, and the center part became flatter which indicated that it will be more difficult for the abrasives to be involved in the polished area. Figure 3a,b shows the cross-sectional view of stress distribution with wear diameter of 100 μm and stress distribution on the workpiece with different wear diameters, respectively. According to the results, it can predict that the stress at the central part of contact area was less than the edge, so less material will be removed from central part and ideal tool influence function cannot be obtained.

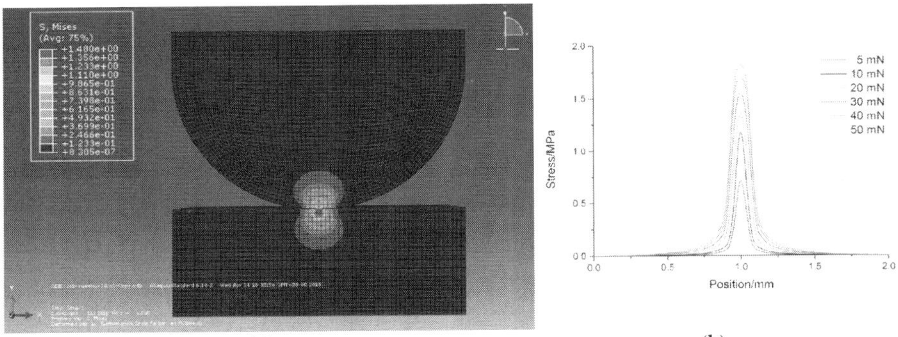

(a) (b)

Figure 2. Stress distribution w/o tool wear (**a**) cross-sectional view at 20 mN, and (**b**) under different loads on workpiece surface.

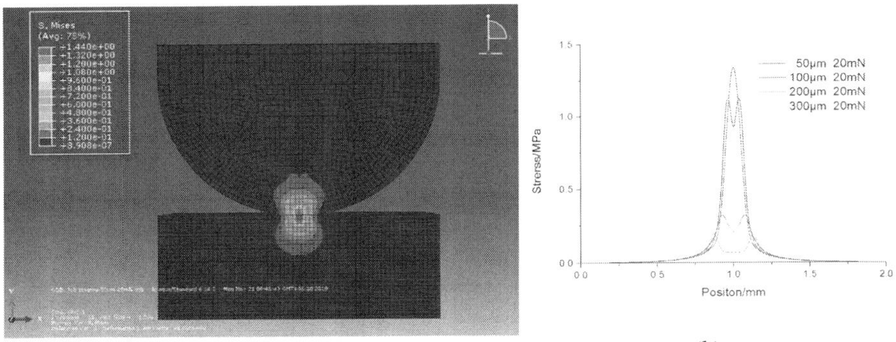

(a) (b)

Figure 3. Stress distribution with tool wear at 20 mN (**a**) cross-sectional view with wear diameter of 100 μm, and (**b**) with wear different diameters on workpiece surface.

RESULTS AND DISCUSSIONS

Effect of Vibrating Motion

Firstly, the effect of different vibration traces on the workpiece surface was observed using a microscope. The magnetostrictive vibrating polisher vibrated at a fixed position on the workpiece surface under lateral, circular and elliptical vibrating motion respectively according to the conditions in Table 1 for a few seconds at 50 mN using a grain size of 1 μm diamond slurry. The results are shown in Figure 4. The granules in diamond slurries were driven by the vibration performing two-body abrasion and three-body abrasion functions to remove material, so the movement of them follows the vibration traces of the polisher. The scratches showed how the material removal process happened on the workpiece surface under different vibrating motions, and the obvious ones were mainly generated by the two-body abrasion when abrasives became embedded and slide over the surface of the workpiece under a relatively large polishing force. Compared with the scratches on the lapped surface, scratches generated by vibration were much shorter (under 100 μm), so a better surface finish could be obtained. Also, during the polishing process, as the polisher scans on the workpiece surface, the scratch effect will be minimized.

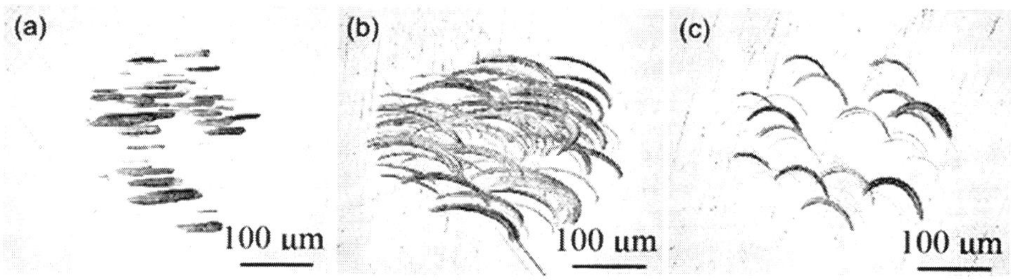

Figure 4: Observation of vibration traces under (a) lateral, (b) circular and (c) elliptical vibrating motions.

A comparison between rotational polishing and vibration-assisted polishing is shown in Figure 5. For rotational polishing, the polishing speed which is represented by the relative velocity between polisher and workpiece V is given by:

$$V = \sqrt{2}\pi r w \qquad (1)$$

where, r is the radius of the polisher and w is the rotation speed. For vibration-assisted polishing, in the case of lateral vibrating motion, the average value is adopted for ease of comparison and V is expressed as:

$$V = 4\lambda v \qquad (2)$$

where λ and v are the amplitude and frequency of vibration respectively, while in the case of elliptical vibrating motion, V is approximately changed to:

$$V = 2\pi bv + 4(a - b)v \qquad (3)$$

where a and b are the long axis and minor axis of elliptical vibration, respectively. In the case of circular vibrating motion, a equals b, so V is changed to:

$$V = 2\pi rv \qquad (4)$$

where r is the radius of circular vibration. So, the circular vibrating motion will generate a higher removal rate than a lateral vibrating motion with a gain of $\pi/2$. It is found that in the case of rotational polishing, the relative velocity has a linear relationship with the radius of the polisher, while in vibration-assisted polishing the relative velocity is independent of the radius of the polisher, demonstrating that the material removal efficiency is not affected by the size of the polisher.

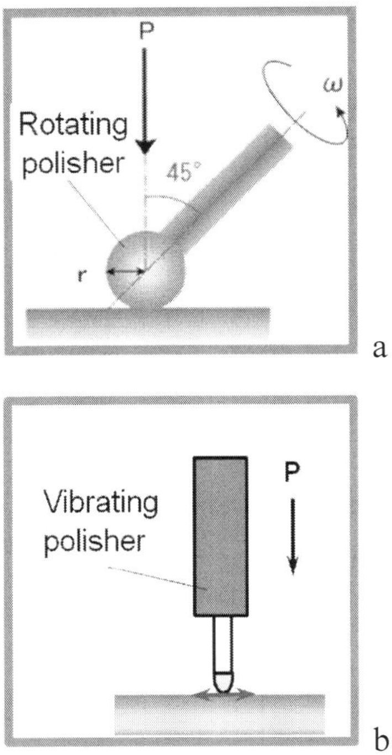

Figure 5: Mathematical models of polishing using (a) rotating and (b) vibrating tools.

Further investigations were conducted to compare material removal depth and surface roughness under different vibrating motions according to the conditions in Table 1. The magnetostrictive vibrating polisher scanned a raster path on the workpiece for one cycle. According to Preston's equation [25], the removal depth δ should be proportional to the average value of V when other parameters are fixed. The polishing force of 5 mN and the grain size of diamond slurry of 0.1 μm is adopted. The results are summarized in Figure 6. This shows the average value of experiments in 10 times with measurement size of surface roughness of 70 μm × 50 μm. It is found that circular vibrating motion has the highest removal depth up to 123 nm and the gain of circular vibrating motion by lateral vibrating motion shows an agreement with the theoretical value. The surface roughness is reduced over 50% by the 2D vibrating motion (elliptical and circular) than that of the 1D (lateral) vibrating motion. Figure 7 shows the measurement data by circular vibrating motion.

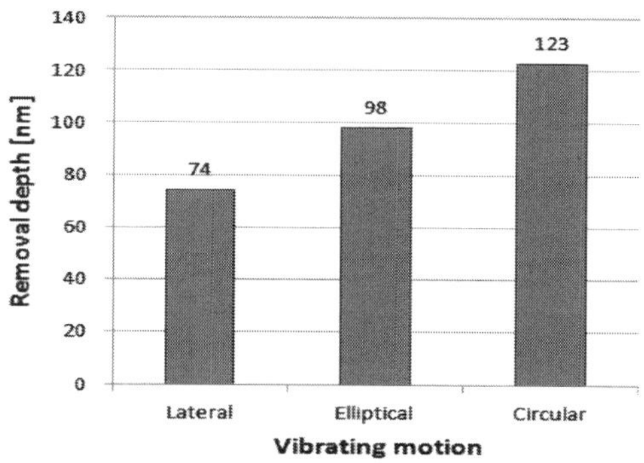

(a)

a

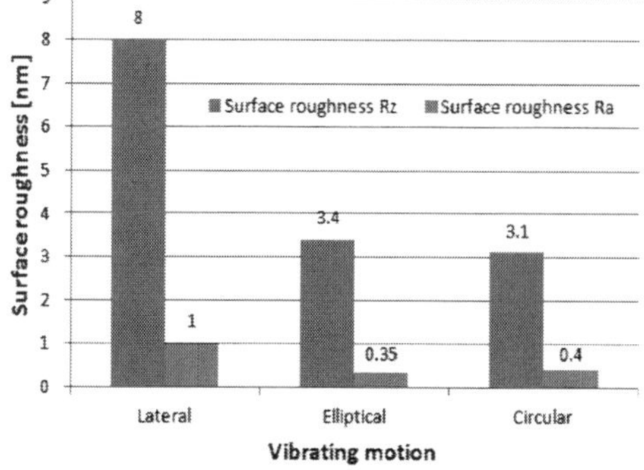

(b)

b

Figure 6: Effects of vibrating motions on (a) material removal and (b) surface roughness.

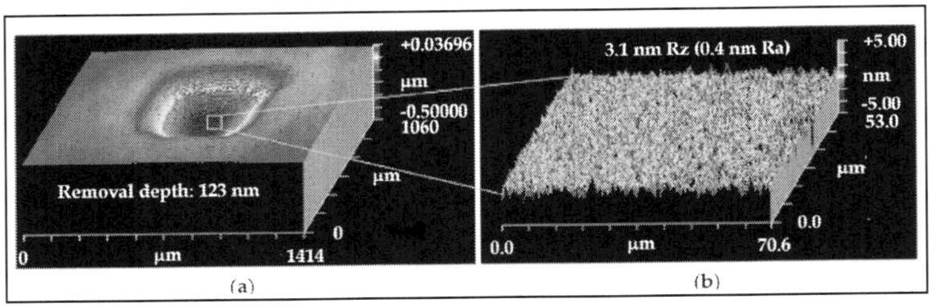

Figure 7: (a) Material removal depth and (b) surface roughness by circular vibrating motion.

Effect of Abrasives

The polishing characteristics of diamond slurries with different grain size were compared under circular vibrating motion with a polishing force of 5 mN. The grains of diamond slurries which have sharp edges demonstrated that the material removal process was mainly performed in nano- and subnano-cutting. The experiment results (see Figure 8) show that both removal depth and surface roughness decrease when grain size of diamond slurry becomes smaller. For practical application in micro-optic mold finishing, the grain size of 0.5 µm and 0.25 µm will be more suitable concerning both removal depth and surface roughness.

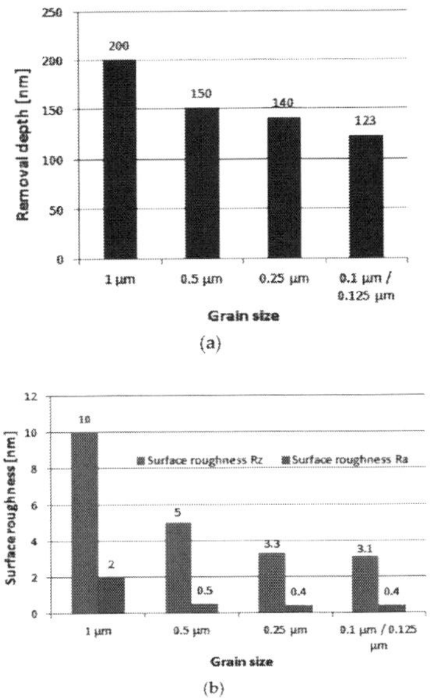

Figure 8: Effects of abrasive grain size on (a) material removal and (b) surface roughness.

Effect of Pressure Conditions

Experiments were conducted to examine the relationship between polishing pressure and MRR. Circular vibrating motion with conditions shown in Table and diamond slurry with grain size of 0.1 μm of were adopted. The average polishing pressure was obtained by a polishing force dividing the contact area. The results are shown in Figure 9. It is found that the MRR showed agreement with Preston's equation when the polishing pressure was under 345 kPa. But when the polishing pressure exceeded 345 kPa, the removal rate decreased gradually with the increase of polishing pressure.

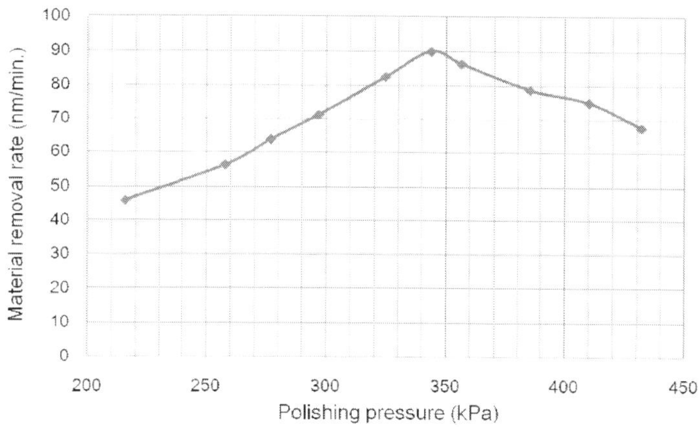

Figure 9: Material removal rate (MRR) change as a function of polishing pressure.

To explain this phenomenon, a model of material removal in general for micro-optic mold polishing is presented in Figure 10. Although the physical scale of material removal processes in polishing is difficult (practically impossible) to observe directly, there is a general view that there are two kinds of abrasive motion, which are two-body abrasion and the three-body abrasion effect, on the workpiece during loose abrasive polishing process. Two-body abrasion happens when abrasives become embedded and slide over the surface, while three-body abrasion is generated when abrasives become freely rolling abrasives [20]. In this experiment, since the urethane polisher is very soft and the polishing spot is very small (under 0.2 mm^2), it is easy to generate a high polishing pressure. When the polishing pressure is under 345 kPa, the concentration of granules in diamond slurry between polisher and workpiece keeps constant, and the MMR increases with the increase of polishing pressure although some two-body abrasives transform to the three-body abrasives. However, when the polishing pressure is over 345 kPa, due to the high pressure, some of two-body abrasives will be dropped out from the contact area of the polisher and workpiece while some of them will transform to the three-body abrasives, so the total number of acting granules reduces and the MRR decreases.

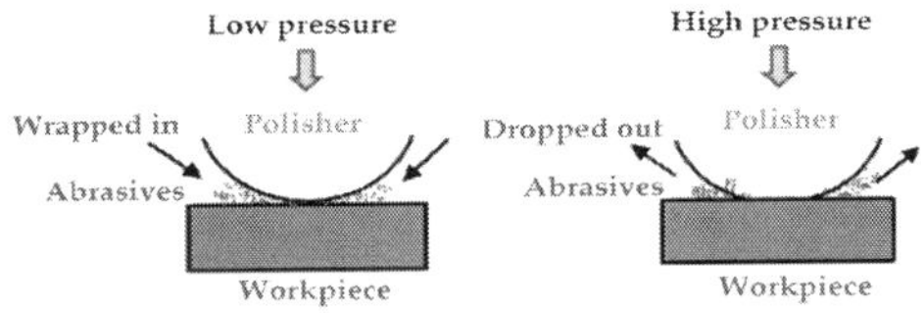

Figure 10: A general model of the material removal mechanism in vibration-assisted polishing.

Effect of Tool Wear

The aforementioned experiments were conducted under the conditions of no significant tool wear. However, during the polishing process, wear of the polisher occurs while the material is removed, and the MRR will be changed as wear persists. In vibration-assisted polishing by loose abrasive, the wear rate of the polisher is correlated with vibration amplitude and frequency, material and concentration of slurry and polishing pressure. The major fabrication process of the urethane polisher was followed by Phenol resin gluing, cutting or grinding and urethane-coating. During the polishing process, the polisher wears on the top of the head where it contacts with the workpiece, and the head becomes flat gradually.

Since the wear volume of the polisher is quite difficult to measure precisely, as an alternative method the wear diameter on the head of the polisher was tested. Figure 11 shows shape change on the polisher's head before and after polishing, and Figure 12 depicts the surface changes on top of the polisher's head before and after polishing by a microscope. A measurement result is shown in Figure 13. The wear diameter of the polisher is defined as d for the following investigations.

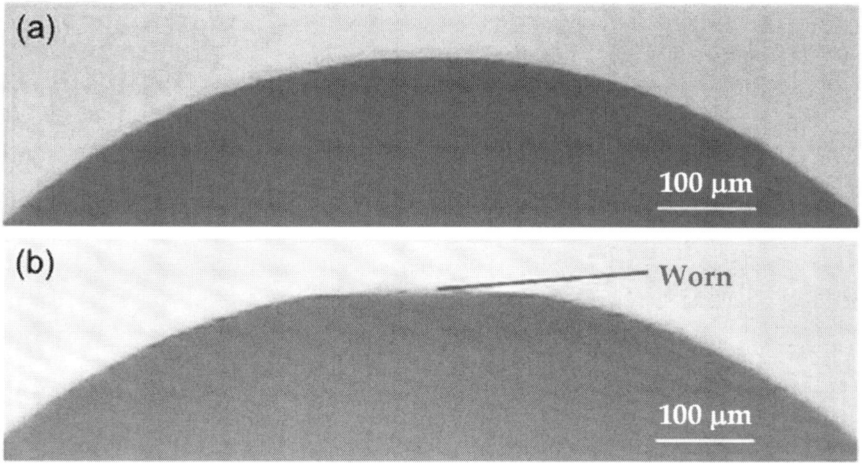

Figure 11: Shape changes on the polisher's head (a) before polishing and (b) after polishing.

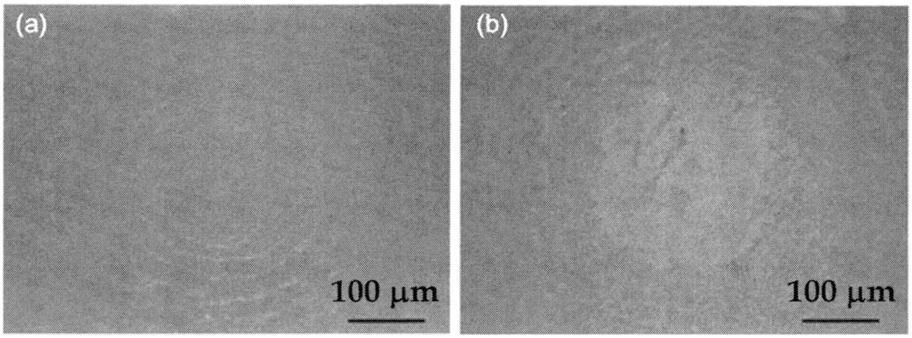

Figure 12: Surface changes on the polisher's head (a) before polishing and (b) after polishing.

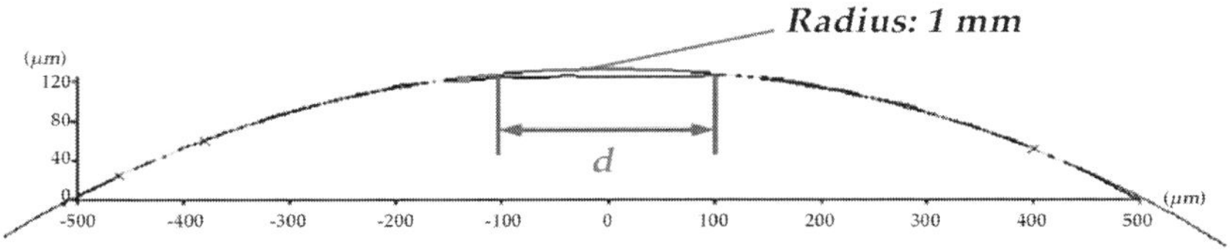

Figure 13: Wear diameter d.

To evaluate the effects of polisher's wear on material removal depth, some polishing experiments were conducted. The polisher scanned a raster scanning path and vibrated in a circular vibrating motion under a constant pressure of 10 mN on the surface of the workpiece. The experiment with the same conditions was repeated 11 times with the change of polishing areas on the workpiece surface. According to the experimental results in Figure 14, it is found that the wear process was almost divided into three steps. During the 1st step (d < 50 μm), both wear diameter and material removal depth increased, and the material removal depth goes to a maximum. At the 2nd step (50 μm ≤ d ≤ 300 μm), the polisher wears proportional to the number of experiments, but the material removal depth keeps almost constant just with a little decrease. From the 3rd step (d > 300 μm), the polisher's wear speed reduced while the material removal depth decreased dramatically and after that showed a tendency to a constant value.

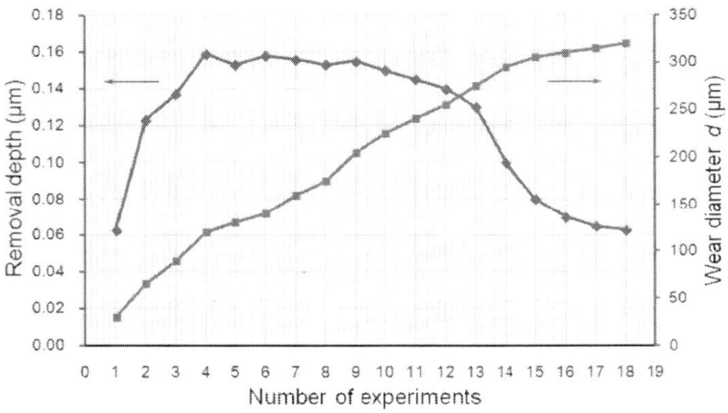

Figure 14: Relations of the polisher's wear and material removal depth.

The reason why material removal depth reduces when d > 300 μm could be explained through observing the shapes of the tool influence function to some extent. Figure 15 shows the comparison of tool influence functions under the conditions of d ≤ 300 μm and d > 300 μm at a polishing force of 20–40 mN. When d ≤ 300 μm, the shape of the removal function was center-concave and symmetrical which is near to the ideal model of removal function. But when d > 300 μm, the acting polishing abrasives became difficult to reach the contact-center of the polisher and workpiece, so the material in the center of the removal function was left and the removal depth reduced. The result also indicated that with the increase of polishing force, the center depth of the tool influence function decreased which led to a low MRR.

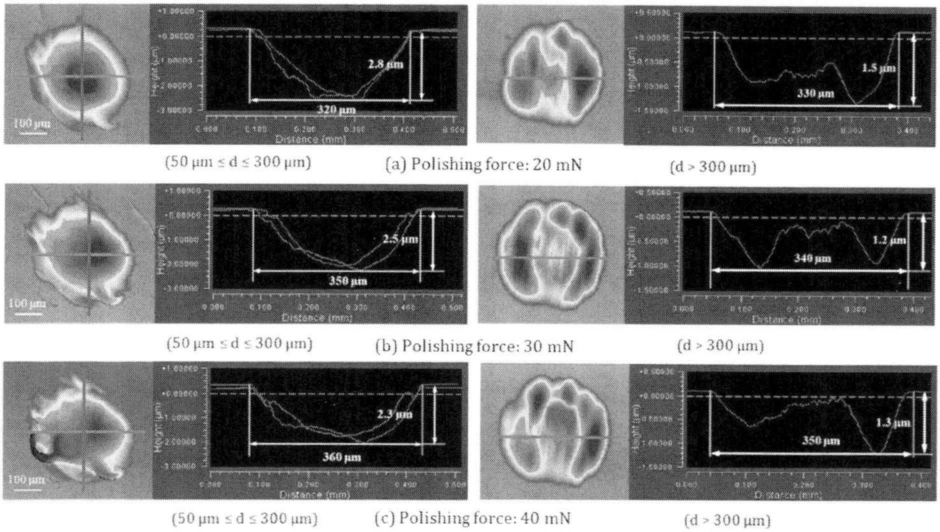

Figure 15: Comparison of tool influence functions with 50 µm ≤ d ≤ 300 µm and d > 300 µm.

CONCLUSIONS

In this paper, some insights into the material removal mechanism in vibration-assisted polishing were revealed through investigating the effects of process parameters on material removal quantitatively. According to the experimental results, some conclusions can be obtained as follows:

- 2D vibrating motion created a better surface roughness with higher removal efficiency than 1D vibration motion. The smaller the grain size of the diamond slurry, the better the surface roughness, but the MRR was reduced.
- MRR increased when the polishing pressure was under 345 kPa. However, when the polishing pressure exceeded 345 kPa, MRR decreased gradually with the increase of polishing pressure.
- MRR was greatly affected by the wear of the polisher when wear diameter on polisher's head exceeded 300 µm.

In summary, the polishing pressure condition which affects the concentration of granules in diamond slurry and transitions between two-body abrasives and three-body abrasives plays a key role in revealing insights into the material removal mechanism. Future work will be continued on the effect of polishing pressure conditions on the surface roughness to further investigate the material removal mechanism.

AUTHOR CONTRIBUTIONS

J.G. and H.S. conceived the idea and planned the experiments, J.G. carried out the experiments and took the lead in writing and revising the manuscript. H.S. supervised the project.

ACKNOWLEDGMENTS

The author J.G. appreciates T. Higuchi for his kind supervision.

REFERENCES

1. Brass, M.; Enoch, J.M.; van Stryland, E.W.; Wolfe, W.L. (Eds.) Handbook of Optics (Volume III): Vision and Vision Optics; McGraw Hill: New York, NY, USA, 2010; ISBN 0071498915/97800714.
2. Fang, F.Z.; Zhang, X.D.; Weckenmann, A.; Zhang, G.X.; Evans, C. Manufacturing and measurement of freeform optics. CIRP Ann. 2013, 62, 823–846.
3. Yi, A.; Jain, A. Compression molding of aspherical glass lenses—A combined experimental and numerical analysis. J. Am. Ceram. Soc. 2005, 88, 579–586.
4. Yan, J.; Zhou, T.; Yoshihara, N.; Kuriyagawa, T. Shape transferability and microscopic deformation of molding dies in aspherical glass lens molding press. JMTR 2009, 1, 85–102.
5. Suzuki, H.; Moriwaki, T.; Yamamoto, Y.; Goto, Y. Precision Cutting of Aspherical Ceramic Molds with Micro PCD Milling Tool. CIRP Ann.-Manuf. Technol. 2007, 56, 131–134.
6. Huang, H.; Chen, W.K.; Kuriyagawa, T. Profile error compensation approaches for parallel nanogrinding of aspherical mould inserts. Int. J. Mach. Tools Manuf. 2007, 47, 2237–2245.
7. Suzuki, H.; Okada, M.; Fujii, K.; Matsui, S.; Yamagata, Y. Development of micro milling tool made of single crystalline diamond for ceramic cutting. CIRP Ann.-Manuf. Technol. 2013, 62, 59–62.
8. Chen, F.J.; Yin, S.H.; Huang, H.; Ohmori, H.; Wang, Y.; Fan, Y.F.; Zhu, Y.J. Profile error compensation in ultra-precision grinding of aspheric surfaces with on-machine measurement. Int. J. Mach. Tools Manuf. 2010, 50, 480–486.
9. Kordonski, W.; Jacobs, S. Magnetorheological finishing. Int. J. Mod. Phys. B 1996, 10, 2837–2848.
10. Fähnle, W.; van Brug, H.; Frankena, H.J. Fluid jet polishing of optical surfaces. Appl. Opt. 1998, 37, 6771–6773.
11. Walker, D.D.; Brooks, D.; King, A.; Freeman, R.; Morton, R.; McCavana, G.; Kim, S.W. The 'Precessions' tooling for polishing and figuring flat, spherical and aspheric surfaces. Opt. Express 2003, 11, 958–964.
12. Walker, D.D.; Freeman, R.; Morton, R.; McCavana, G.; Beaucamp, A. Use of the 'Precessions'™ process for prepolishing and correcting 2D

& 2½D form. Opt. Express 2006, 14, 11787–11795.

13. Beaucamp, A.; Namba, Y.; Inasaki, I.; Freeman, R. Finishing of optical molds to λ/20 by automated corrective polishing. CIRP Ann.-Manuf. Technol. 2011, 60, 375–378.
14. Brehl, D.E.; Dow, T.A. Review of vibration-assisted machining. Precis. Eng. 2008, 32, 153–172.
15. Kumar, M.N.; Subbu, S.K.; Krishna, P.V.; Venugopal, A. Vibration assisted conventional and advanced machining: A review. Procedia Eng. 2014, 97, 1577–1586.
16. Li, Z.; Jin, G.; Fang, F.; Gong, H.; Jia, H. Ultrasonically Assisted Single Point Diamond Turning of Optical Mold of Tungsten Carbide. Micromachines 2018, 9, 77.
17. Suzuki, H.; Moriwaki, T.; Okino, T.; Ando, Y. Development of Ultrasonic Vibration Assisted Polishing Machine for Micro Aspheric Die and Mold. CIRP Ann.-Manuf. Technol. 2006, 55, 385–388.
18. Suzuki, H.; Hamada, S.; Okino, T.; Kondo, M.; Yamagata, Y.; Higuchi, T. Ultraprecision finishing of micro-aspheric surface by ultrasonic two-axis vibration assisted polishing. CIRP Ann. 2010, 59, 347–350.
19. Guo, J.; Morita, S.; Hara, M.; Yamagata, Y.; Higuchi, T. Ultra-precision finishing of micro-aspheric mold using a magnetostrictive vibrating polisher. CIRP Ann.-Manuf. Technol. 2012, 61, 371–374.
20. Guo, J.; Suzuki, H.; Higuchi, T. Development of micro polishing system using magnetostrictive vibrating polisher. Precis. Eng. 2013, 37, 81–87.
21. Evans, C.J.; Paul, E.; Dornfeld, D.; Lucca, D.A.; Byrne, G.; Tricard, M.; Klocke, F.; Dambon, O.; Mullany, B.A. Material Removal Mechanisms in Lapping and Polishing. CIRP Ann.-Manuf. Technol. 2003, 52, 611–633.
22. Komanduri, R.; Lucca, D.A.; Tani, Y. Technological Advances in Fine Abrasive Process. CIRP Ann. 1997, 46, 545–596.
23. Brinksmeier, E.; Riemer, O.; Gessenharter, A. Finishing of structured surfaces by abrasive polishing. Precis. Eng. 2006, 30, 325–336.
24. Guo, J.; Suzuki, H.; Morita, S.; Yamagata, Y.; Higuchi, T. A real-time polishing force control system for ultra-precision finishing of micro-optics. Precis. Eng. 2013, 37, 787–792.
25. Preston, F.W. The theory and design of plate glass polishing machines. J. Glass Technol. 1927, 11, 214–256.

SECTION III: SUBSURFACE DAMAGES

Chapter 8

Methods for Detection of Subsurface Damage: A Review

Jingfei Yin[1] , Qian Bai[1] and Bi Zhang[1,2]

[1] Key Laboratory for Precision and Nontraditional Machining of Ministry of Education, Dalian University of Technology, Dalian 116024, China

[2] Depart ment of Mechanical and Energy Engineering, South University of Science and Technology, Shenzhen 518055, China

ABSTRACT

Subsurface damage is easily induced in machining of hard and brittle materials because of their particular mechanical and physical properties. It is detrimental to the strength, performance and lifetime of a machined part. To manufacture a high quality part, it is necessary to detect and remove the machining induced subsurface damage by the subsequent processes. However, subsurface damage is often covered with a smearing layer

Citation: Jing-fei Yin, Qian Bai and Bi Zhang "Methods for Detection of Subsurface Damage: A Review" Chinese Journal of Mechanical Engineering 201831:41. https://doi.org/10.1186/s10033-018-0229-2

generated in a machining process, it is rather difficult to directly observe and detect by optical microscopy. An efficient detection of subsurface damage directly leads to quality improvement and time saving for machining of hard and brittle materials. This paper presents a review of the methods for detection of subsurface damage, both destructive and non-destructive. Although more reliable, destructive methods are typically time-consuming and confined to local damage information. Non-destructive methods usually suffer from uncertainty factors, but may provide global information on subsurface damage distribution. These methods are promising because they can provide a capacity of rapid scan and detection of subsurface damage in spatial distribution.

Keywords: Subsurface damage, Hard and brittle material, Taper polishing, Measurement, Laser scattering

INTRODUCTION

Hard and brittle materials are widely used because of their excellent physical and mechanical properties, such as high hardness and strength at elevated temperatures, wear and corrosion resistances, etc. Among hard and brittle materials, single crystalline silicon is the predominant substrate material for integrated circuits (IC) [1, 2, 3, 4, 5]; optical glasses are for optical windows and lens [6, 7, 8, 9, 10, 11] as well as high-power laser components [12]; ceramic materials are commonly used for bearings, cutting tools and machine parts [13, 14, 15, 16]. However, these materials are also difficult to machine due to their hard and brittle nature. Surface and subsurface damage (SSD) is often induced in a machined part made of such a material.

Due to the ever-increasing requirement for better surface integrity of mechanical and optical parts, the tolerance for surface and subsurface damage is getting tighter and tighter. It becomes necessary to detect and remove surface and subsurface damage in a machining process. Damage detection is thus important to manufacturing of hard and brittle materials. It does not only provide information on size, type, location and spatial distribution of surface and subsurface damage, but also gives a guideline for the subsequent machining processes to remove damage so as to improve efficiency and quality, and reduce cost for the manufacturing processes.

While surface damage is relatively easy to detect because many existing techniques, such as optical scanning, are available to use, subsurface damage is rather hard to detect since such a damage is often hidden in the subsurface at a certain depth. There are many types of subsurface damage, such as cracks, residual stress, porosities, phase transformation, amorphization, among which cracks are most detrimental and directly determine performance and lifetime of a part. Both destructive and non-destructive methods have also been developed for subsurface damage detections. Destructive methods include taper-polishing, cross-sectional microscopy, and etching, whereas non-destructive methods are X-ray scanning, laser scattering, and ultrasonic probing, etc. In comparison with the non-destructive methods, the destructive methods have better reliability and accuracy. However, the destructive methods need to destroy the part, do not provide global damage information of the part, and are time-consuming in sample preparation. On the other hand, non-destructive methods can realize in-process detection without the above drawbacks, and may greatly reduce cost, and improve efficiency and quality for machining. The major problem associated with the non-destructive methods is the detection accuracy and efficiency which need to be improved.

This paper provides a review of the methods for detection of subsurface damage. It discusses mechanisms of subsurface damage formation, and methods for subsurface damage detection. The paper also outlines the recent developments in subsurface damage detection. The paper compares the methods and techniques for the improvement of detection accuracy and efficiency, and further points out limitations and drawbacks of the methods and techniques.

MECHANISMS OF SUBSURFACE DAMAGE FORMATION

Zhang et al. [14, 17, 18] assessed subsurface damage induced by single diamond scratching of different ceramics and observed that subsurface damage included median (radial) and lateral cracks, as well as pulverization zone, as shown in Figure 1. During the scratching process, the pulverized material is squeezed to the two sides of the groove to form material pile-up. Under the influence of the loading stress, the region underneath the abrasive grain was turned into a pulverization zone as a consequence of pulverization and collapse of the internal pores in the material. The pulverization zone was surrounded by a zone of elastic-plastic deformation. During scratching, a median crack could be initiated at the point where the stress level exceeded the tensile strength of the material. When the abrasive grain moves away, the median crack is closed due to the elastic springback. Meanwhile, the elastic-plastic deformation zone begins to restore, lateral cracks are formed and tends to propagate to the surface. Friction between the abrasive grain and surface of the workpiece elevates the temperature in the machining area. Because the material grains in the pulverization zone are fine enough, they can be re-sintered so as to form a smearing layer that covers the scratching pass. Some lateral cracks may propagate and intersect each other and result in material removal. The residual cracks form subsurface damage.

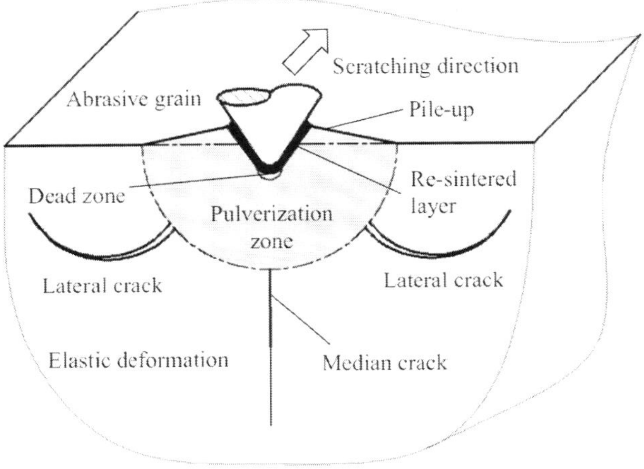

Figure 1: Schematics of subsurface damage induced in a scratching process [18].

Subsurface damage is detrimental to the strength, lifetime and performance of a part, and is necessary to be removed in subsequent processes. To detect SSD is critical to the correct adoption of the subsequent processes. Methods for detecting SSD in a part made of a hard and brittle material are classified into two categories: destructive and non-destructive. Destructive methods include tapper polishing, SEM microscopy, TEM microscopy, cross-sectional microscopy and chemical etching. On the other hand, non-destructive methods are mainly optical coherent tomography and laser scattering, etc. These methods are compared and discussed in detail in the following three sections.

DESTRUCTIVE METHODS

In the application of destructive methods for SSD detection, a machined part has to be destroyed in order to expose and detect SSD. Destructive methods are relatively mature with acceptable detection accuracies. They are widely used in academia and industry.

Polishing Techniques

Traditional Taper Polishing

The traditional taper polishing is one of the widely used methods for subsurface damage detection and observations [1, 2, 19, 20, 21, 22, 23, 24, 25, 26]. Figure 2shows the schematics of the taper polishing method. A machined sample is cut at an angle β to the machined surface and then the cut surface of the sample is polished to remove possible damage induced by the cutting process. The polished sample is etched to remove the smearing layer of the polished surface and thus SSD is exposed in the sample. The SSD is then observed with a microscope and measured for damage dimensions. The depth H of SSD is calculated as in Eq. (1):

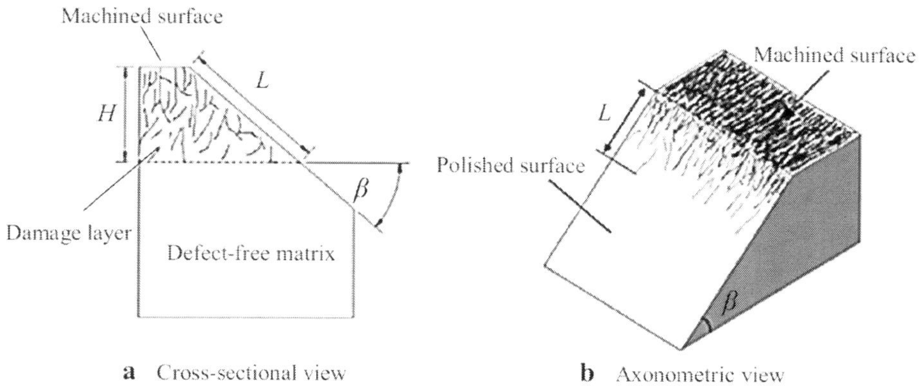

Figure 2: Schematics of the traditional taper polishing method [1].

$$H = L \cdot \sin\beta, \qquad (1)$$

Similar to the other traditional methods, the taper polishing method is direct, accurate, inexpensive, and easy to operate with satisfactory measurement results. However, it is a destructive method, and thus cutting and polishing the sample may introduce new damage to the sample, which adds errors to the measurement results. In addition, applying etching to a polished sample inevitably releases the strain energy at a crack tip, and extends the crack length. In this regard, the taper polishing method typically measures a crack with an extended length [27].

Dimpling

A novel method, named "ball dimpling", has been used to detect subsurface damage in machining of optical parts. The method was developed by the Optical Manufacturing Center of the University of Rochester [28, 29]. Similar to the traditional taper polishing method, the ball dimpling method polishes a sample part with a steel ball to make a dimple in order to expose the subsurface damage of the part. The depth of the dimple is large enough so as to cross the damage layer to the matrix. The depth of SSD is calculated by measuring the characteristic dimensions of the dimple. There are also some other forms of dimpling [30]. The ball dimpling method is simple in operation and low in cost. However, the scratches in a dimpling process may remain on the part surface, which can result in measurement errors. Additionally, the ball dimpling method is time consuming in SSD measurements for the parts made of hard materials due to a low dimpling efficiency [31].

MRF Polishing

With the development of precision machining technology, the magnetorheological finishing (MRF) polishing is used in the SSD measurements [11, 19, 31, 32, 33, 34, 35]. It has been reported that the shear effect plays an essential role in material removal with the MRF polishing, which ensures no extra damage introduced into the workpiece [28, 36, 37, 38]. The MRF polishing technology is to create a spot or a wedge on the surface of a part. The spot or the wedge crosses the damage layer to the damage-free matrix. The depth of SSD can be inferred by measuring the width of the spot or the wedge. The depth of SSD is usually tens of microns but the width of the spot is in the order of millimeters. The depth of damage is magnified hundreds of times in this way, which improves the accuracy of the measurement.

Only the local SSD information is detected with either the taper polishing or the MRF polishing method, the global information and spatial distribution of SSD are not provided. A sample has to be etched before observation with a microscope. The size and the distribution of SSD cannot be correctly assessed if a polished sample is over etched. Since sample etching is highly depended on the type and concentration of an etchant, the temperature and the etching time, it is rather difficult to precisely control an etching process. In addition, a subsurface damage may be overly measured because an etching process can release the energy at the crack tip so as to extend crack length.

Bonded-Interface Technique

The bonded-interface technique (BIT) was first proposed by Mulhearn et al. [39] to investigate the material deformation of metals in an indentation test. Figure 3 shows the schematics of BIT that is divided into multiple steps. Firstly polish the surfaces which need to be bonded together; secondly bond the polished surfaces, and machine the work surface; then remove the bonding glue, and observe SSD on the two polished interfaces with a microscope. BIT has a limited application to measuring subsurface damage in the parts made of ceramics, optical glasses and other hard and brittle materials by researchers [40, 41, 42, 43, 44] due to its false representation in the resulted stress-strain fields [45].

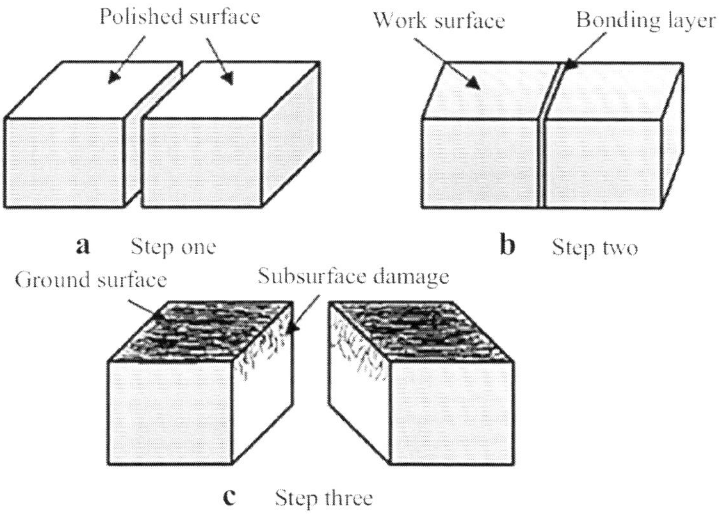

Figure 3: Schematics of bonded-interface technique [40].

One critical problem with this technique is that the sample has a bond layer, which changes the stress-strain fields nearby the interface when the sample is subjected to indentation at the interface. The Hertzian theory can no longer be applied to analyzing the stress-strain field in this case. Helbawi et al. [45] found that the maximum stress position shifted away from the bonded-interfaces in both finite element simulation and indentation experiments. In consequence, the maximum subsurface damage depth was away from the bonded-interfaces and was larger than the measured damage depth. However, the damage zone observed with BIT was with a higher damage density than that induced in the practical machining processes. The findings from Helbawi et al. [45] verified the limitation in BIT. Therefore, BIT cannot provide precise information on subsurface damage induced in the practical machining processes, and should not thus be recommended for quantitative detection of SSD in hard and brittle materials.

Cross-sectional Microscopy

Due to the amorphous layer induced in a machining process, it is not convincing to observe SSD from a machined surface. An additional cross-sectional inspection gives more precise information on SSD. Carr et al. [46] observed SSD of optical glasses from the cross section of a fracture part with a scanning electron microscope. Zhang et al. [14] presented a method to prepare a fractured cross section and procedures to observe SSD. However, in their method, a sample has to be split up. In addition, only the SSD in the fractured surface can be observed and measured, but their overall distribution and three-dimensional configurations remain unknown.

Transmission Electron Microscopy

The transmission electron microscopy (TEM) has been used to observe micro damage in a sample at a high resolution [47, 48, 49]. Subsurface microcracks, dislocation, amorphous and polycrystalline layers and other micro and nano defects can precisely be observed with TEM. The sample is often observed from both the horizontal and cross-sectional views to give a more comprehensive estimation on SSD. Although it provides more precise information, TEM requires the sample to be thin enough due to the high scattering on and poor penetration into the surface of the sample by electrons. Generally, a sample is first thinned to 20-50 μm, and then further to less than 1 μm by ion milling [2, 21, 49]. Therefore, the preparation procedure is complicated and time-consuming.

Chemical Etching

Chemical etching is to use a suitable etchant to remove the damage layer of a machined sample. The surface morphology and roughness after etching, as well as etching rate are used to evaluate subsurface damage. There are different inspection methods for estimating SSD using chemical etching. One method is based on the surface roughness of an etched sample [12, 35]. Wong et al. [50] measured the peak-to-valley (PV) value and root-mean-square deviation R_q of the surface profile with a detection probe at each etching stage. At the initial etching stages, a crack is exposed, which allows the tip of the probe to reach the inside of the crack, thus resulting in a high measurement value of surface roughness. With an increase in etching time, more and more cracks are exposed and intersect with each other to result in material removal. The number of subsurface cracks decreases after each etching stage, which leads to a decrease in surface roughness. The removal of the damage layer stabilizes surface roughness, which facilitates an accurate calculation of SSD depth.

There are constant- and differential-rate inspection methods based on etching rate. The constant-rate etching method controls the influence factors as constants, such as temperature, etchant concentration, and time. A sample is etched step by step, and etching rates are recorded at every step. Since damage introduces more contact surfaces with the etchant, the contact area between the etchant and the surface layer decreases as the damage layer is removed step by step. Therefore, etching rate is decreased correspondingly, shown in Figure 4. After the damage layer is removed, etching rate is stabilized, and SSD depth can thus be calculated according to the etching rate graph in Figure 4.

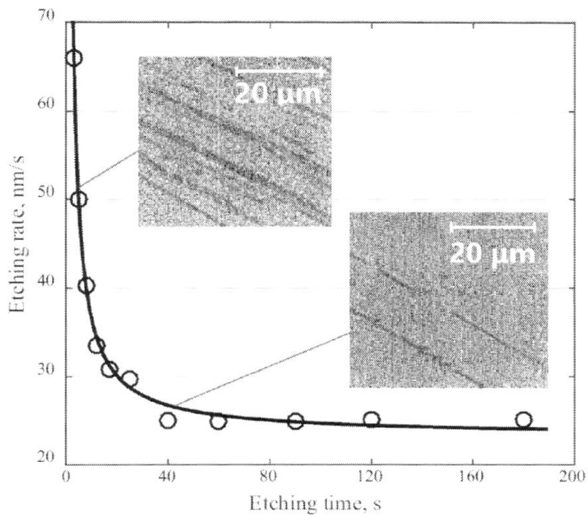

Figure 4: Etching rate vs. etching time for the silicon wafers ground by #2,000 diamond wheel [2].

The constant-rate etching method has been widely used to measure SSD in optical glasses, ceramics, and semiconductor materials [2, 22, 50, 51, 52]. The method records the absolute etching rate. However, the etching process is affected by many factors, such as etching time, etchant concentration, and temperature. It is difficult to simultaneously control all the influence factors.

In order to eliminate the influence of these factors and improve the measurement accuracy, Wang and Li et al. [19, 53, 54] proposed the differential-rate etching method. The method was to etch the machined and damage-free samples in the same environment simultaneously, and then to compute the differential-rate. SSD depth can be calculated based on the differential-rate.

As argued before, chemical etching promotes crack propagation, which results in larger measurement values of SSD compared to the actual ones. It has low repeatability and measurement efficiency. In addition, most etchant is poisonous, which is highly detrimental to human health and environmental protection.

The destructive methods for SSD are relatively reliable. However, there are still many disadvantages to be overcome in the future. A comparison of the destructive methods is presented in Table 1.

Table 1: Comparison of the destructive methods for SSD detection

Methods	**Advantages**	**Disadvantages**
Polishing techniques	Simple, mature, and easy to operate	1. Etching after polishing leading to crack propagation and thus measurement errors 2. Acquiring local SSD only; inability to detect the global distribution and three-dimensional configuration of SSD
Cross-sectional microscopy	Simple and observable with high precision	Acquiring local SSD only; inability to detect the global distribution and three-dimensional configuration of SSD
TEM microscopy	High resolution for micro/nano damage detection	1. Sophisticated procedures for sample preparation 2. Inability to detect SSD larger than tens of microns
Chemical etching	Easy to operate with low cost	1. Etching leading to crack propagation and thus measurement errors 2. Chemicals detrimental to human health and environmental protection

NON-DESTRUCTIVE METHODS

The destructive methods are time-consuming, which reduces production efficiency and increases cost. Non-destructive methods preserve the samples and detect SSD based on the knowledge of optics and acoustics, etc. In this section, four non-destructive methods are presented and discussed.

Predictive Methods

The predictive methods are mainly to predict SSD depth. There are two types of prediction based on process parameters and surface roughness (SR) of a machined part.

SSD depth can be empirically predicted with the size of abrasive grains during grinding or lapping, for which several researchers arrived at different results [40, 55, 56, 57, 58]. Lambropoulos et al. [59] proposed a relationship between the damage depth and the size of abrasive grains, expressed as Eq. (2),

$$0.3d^{0.68} < \delta < 2d^{0.85}, \qquad (2)$$

where d is abrasive grain size and δ is SSD depth.

Zhang et al. [17] built a predictive model based on material properties and grinding parameters to predict the depth of subsurface damage δ of ceramics induced in grinding, expressed as Eq. (3),

$$\delta = (200a_g)^{\frac{1}{\log(\lambda \frac{H}{K_C})}}, \qquad (3)$$

where a_g is the maximum cutting depth of abrasive grains, taking the form shown in Eq. (4); λ is constant, and $\lambda = 10^{-2}m^{1/2}$; m is the working condition coefficient, and 1/2<m<2/3; H and K_c are the hardness and the toughness of a workpiece material, respectively.

$$a_g = \sqrt{\frac{4v_w}{v_s r C}\sqrt{\frac{a_p}{d_s}}}, \qquad (4)$$

where v_w is the feed of a workpiece, v_s is the peripheral speed of a grinding wheel, ris the ratio of the width to the thickness of the non-deformed chips, a_p is the cutting depth of abrasive grains, d_s is the diameter of the wheel, and C is the grain density on the wheel surface.

The other predictive method uses the surface roughness information of a machined part for SSD depth prediction. The method is derived from fracture mechanics that is used to study cracking behavior. In studying lapping of glasses and ceramics with SiC abrasives, Aleinikov et al. [60] reported that the ratio of surface roughness and SSD depth was 3.9±0.2. However, the ratio was different due to different workpiece materials, machining processes and SSD measurement methods [31, 32, 33, 40, 58, 61, 62].

Based on the indentation fracture mechanics, Lambropoulos et al. [61] considered that the formation of SSD was due to the brittle fracture of a workpiece material, and proposed an explicit equation to predict SSD depth, as shown in Eq. (5),

$$\frac{\delta}{SR} = 2.326\alpha_K^{2/3}\left(\frac{E}{H}\right)^{(2-5m)/3}\frac{(\cot\psi)^{1/9}}{(\sin\psi)^{1/2}}\left(\frac{P}{K_c^4/H^3}\right)^{1/6}, \tag{5}$$

where E, H, and K_c are the elastic modulus, the hardness, and the toughness of the workpiece material, respectively. ψ is the semi-angle of the sharp cone of an abrasive grain; P is the force acting on the grain. α_K and m are numerical factors in ranges of 0.03–0.04 and 0.33–0.50, respectively. Force P determines the ratio, yet in practical machining, it varies with process parameters, which renders the model inconvenient to use.

Li et al. [11, 19] improved the model in Eq. (5), stated that the relationship between SSD and R_a should be non-linear, as expressed in Eq. (6). Damage depth is directly related to surface roughness by eliminating grain force P,

$$\delta = 3.08\alpha_K^{2/3}\frac{1}{(\sin\psi)^{2/3}}\frac{H^{2m}}{E^{2m-2/3}K_c^{2/3}}SR^{4/3}, \tag{6}$$

Lv et al. [9] built a model to predict damage in a BK7 glass sample machined by the rotation-ultrasonic machining and grinding. In their study, damage depth was found to be proportional to the second order function of surface roughness, namely, $\delta=\chi SR^2+l$, where constant l is related to the geometric shape of abrasives and to material properties. The constants in the model were derived from the experimental data fitting.

It is necessary to point out that these models reveal the relationship between SSD depth and surface roughness in terms of indentation fracture mechanics. Indentation is a quasi-static interaction between the tool and the workpiece. They ignored the effect of cutting speed on machining-induced damage. Li et al. [6] studied the effect of grinding wheel speed and workpiece feed rate on SSD of the BK7 glass, shown in Eq. (7),

$$\delta = \lambda\left(\frac{v_W \cdot L}{v_S}\right)^{\frac{4}{3}} \cdot \left[\arccos\left(2\sqrt{\frac{a}{d}}\right)\right]^{\frac{4}{3}} - \frac{v_W \cdot L \cdot \arccos\left(2\sqrt{\frac{a}{d}}\right)}{v_S \cdot \arccos\left(2\sqrt{\frac{a}{d}}\right)}, \quad (7)$$

where λ is related to the material properties of the workpiece and the geometric shape of the grains; v_w is the workpiece velocity; L is the spacing between the successive cutting edges; v_s, a, and are the speed, the depth of cut, and the diameter of the wheel. However, this model assumed a constant grit protrusion height, which does not conform to the practical grinding process.

Chen et al. [8] studied the influence of the wheel vibrations on SSD depth. They built a model that was fitted with a quadratic polynomial, i.e., $\delta=p_1SR^2+p_2SR+p_3$. Their study demonstrated that fitting accuracy was high with relatively high surface roughness, but deteriorated with surface roughness lower than 6 μm. In addition, the constants in the model were determined by the process parameters, especially grinding speed and the vibration frequency of the grinding wheel. Consequently, the model should be calibrated for different process parameters, which obstructs its practical applications.

Wang et al. [7] correlated SSD depth with cutting force in the rotary ultrasonic face milling of a K9 glass. They found that SSD depth was proportional to the exponent of cutting force, i.e., $\delta=\gamma F_c^{\chi}$. Due to the simplifications in deriving the model, the model may not be applicable to other materials and processes.

Although it is simple and easy to predict SSD depth by the predictive methods, the largest SSD depth, which is the most detrimental to the performance and the lifetime of a part, cannot precisely be predicted. Besides, because the prediction results mainly rely on surface roughness, subsurface cracks may propagate to a large depth beneath a relatively smooth surface [63]. Lakhdari et al. [30] found that SSD depth induced in ultrasonically assisted grinding was 35% less than the conventional grinding, even the surface quality is worse.

It should also be noted that many prediction models revealing the relationship between SSD and surface roughness were based on the quasi-static indentation fracture model proposed by Lawn et al. [64]. However, the situation in the practical processes is different from the indentation. High cutting speeds are often used in the practical machining processes, which leads to a high strain rate and may also cause high temperature in a machining area. The effects of temperature and strain rate on the material responses in machining should be considered in the predictive modeling of SSD.

Optical Coherent Tomography

The optical coherent tomography (OCT) is a new imaging detection technology. It was first applied to biology and medical diagnosis [65, 66]. The principle of OCT is based on the Michelson interferometry, as described in Figure 5. A light beam is emitted from a light source, then split into two beams by a beam splitter, one to the detection arm and the other to the reference arm. The light beam to the reference arm is reflected by a mirror, while the light beam to the detection arm irradiates the sample and is then reflected back. By moving the reference mirror, the two reflected light beams interfere with each other with the interference signal acquired by a detector. The signal is demodulated to receive the envelope function of the signal. The maximum of the interference function corresponds to the position of the mirror, which is used to infer the information on the sample.

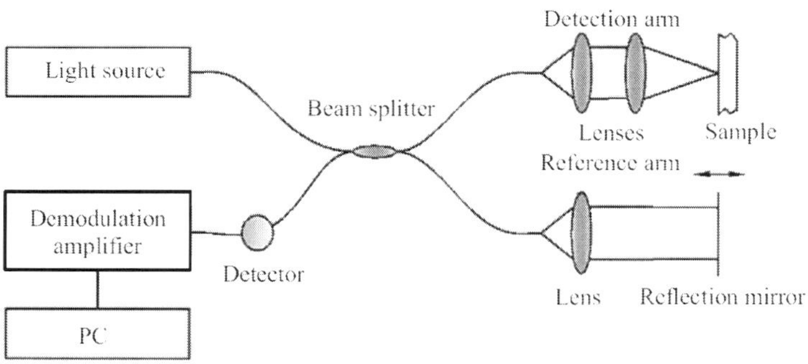

Figure 5: Schematic diagram of optical coherent tomography [57].

The resolution of OCT depends on the coherent length of light L_c, given in Eq. (8),

$$L_c = \frac{2c \cdot \ln 2}{\pi} \cdot \frac{1}{\Delta f} \approx 0.44 \frac{\lambda_0^2}{\Delta \lambda}, \quad (8)$$

where c and λ_0 are the propagation speed and the wavelength of the light, respectively; $\Delta\lambda$ is the full-width of the coherence function at half-maximum measured in wavelength units; and Δf is the spectral width of the source in the optical frequency domain. Because of the short coherence of the light, the interference between the detection and reference light beams takes place in the aplanatism condition at a longitude resolution of 10 μm.

Bashkansky et al. [67, 68, 69, 70, 71, 72] used OCT to detect SSD depth up to 140 μm in ceramics and obtained the three-dimensional characteristics of SSD. However, the images were blur because of the strong surface scattering, weak subsurface scattering and big tomography interval. Sergeeva et al. [57] used the short coherent interferometry to measure the artificial SSD in a glass, and reported that the depth and lateral resolution were 10 μm and 4 μm, respectively. While in a practical situation, cracks show more complex configurations, such as "chevron", "branch" and "fork" [73]. Some cracks are parallel to the surface, while others initiate at the middle of the existing cracks, with their propagation directions changing all the time.

Applicable to the materials that have strong light scattering property, OCT is usually used in the biology and medical diagnosis fields. For the semiconductor and optical materials, light scattered from SSD is too weak to have interference [74]. In addition, noises induced from surface scattering and reflection obstruct the processing of the targeted signals.

Laser Scattering

Various technologies have been proposed to detect SSD with laser scattering. Temple et al. [75, 76] used the total internal reflection microscopy (TIRM) to detect surface/subsurface damage of the transparent samples, and several other researchers applied TIRM to optical materials later [77, 78, 79, 80, 81]. Shown in Figure 6, a sample is put on a prism, with a layer of the matching fluid in between. Linearly polarized light passes through the prism and the matching fluid, and irradiates the sample. If the sample is damage-free, the total reflection takes place on the interface between the sample and air. Otherwise, the incident light is scattered by the damage within the sample. A portion of the scattered light passes through the sample to facilitate observations of the damage with a microscope. TIRM requires the sample with a low surface roughness and transparency to the incident light [40]. Additionally, TIRM is not suitable for quantitative measurements since it only provides qualitative results of SSD in a sample.

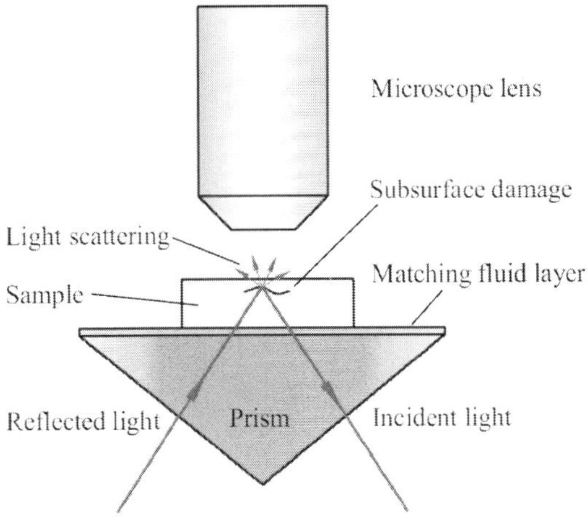

Figure 6: Schematic diagram of total internal reflection microscopy [75].

Confocal microscopy is also used recently for SSD detection. A schematic arrangement of the confocal microscopy is shown in Figure 7, in which light goes through an illumination pinhole, is reflected by a beam splitter, and is then focused by an objective lens on the sample. Light scattered from the sample is received by a light detector behind a detection pinhole (aperture) at the focal point of lens 2. The light offset from the focus is shut off by the detection pinhole plate. Therefore, only the light reflected from the focal point in the sample goes through the detection pinhole and is then received. The confocal microscopy provides higher resolution detection compared to the traditional microscopy, and has been applied to detecting SSD in ceramics [15, 82], optical parts [83] and semiconductor materials [5].

Winn et al. [82] used the confocal laser scanning microscopy (CLSM) to measure SSD in alumina. However, the maximum depth of measurement was limited within 20 μm in their study. Fine et al. [65] detected SSD in the polished and fine ground optical glass samples by CLSM with the resolution of 150 nm. Bertussi et al. [83] observed SSD in the fused silica optical components with the resolutions of 1 μm in depth and 0.26 μm in the scanning plane also by CLSM. Wang et al. [84] studied SSD by CLSM using a lens of a large numerical aperture and obtained the horizontal resolution of approximately 0.13 μm. However, CLSM requires that the samples under detection be with low surface roughness. Generally, CLSM is used for detection of a smooth surface as the light scattered from a rough surface interferes with the measurement.

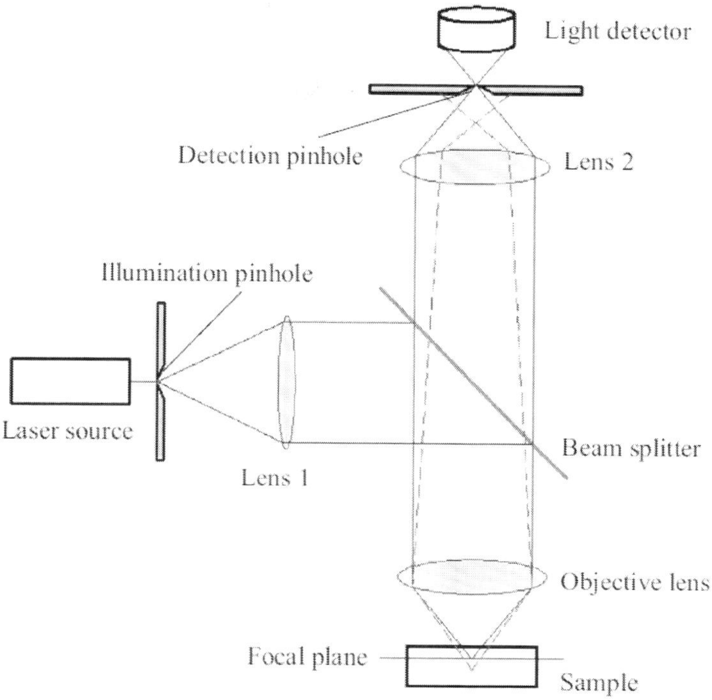

Figure 7: Schematic arrangement of confocal microscopy [83].

The Argonne National Laboratory and the Kansas State University jointly developed the cross-polarization confocal microscopy (CPCM) to measure SSD in the ground ceramics and silicon wafers [5, 15, 85, 86, 87, 88, 89, 90, 91, 92]. As shown in Figure 8, an incident light is linearly polarized and hits the surface of a sample. A portion of the light is reflected and scattered on the surface of the sample, while the other portion of the light penetrates into the sample. The surface scattering or reflecting light has the same polarization direction as the incident light, but the polarization direction of the transmission light is changed after being scattered by SSD, such as cracks and pores. The subsurface scattering light then refracts back to the air and combines with the surface scattering light, and is then filtered from the mixed lights with a polarized optical lens. The filtered light reflecting SSD is collected by a confocal microscopy system.

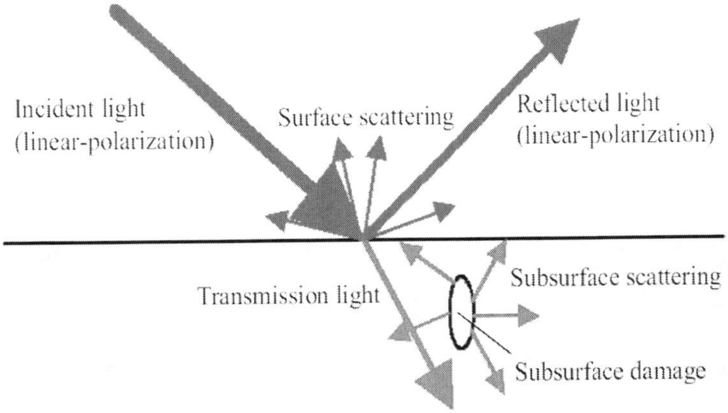

Figure 8: Schematic diagram of reflection, scattering and transmission of light on the surface of a sample [89].

The measurement system developed by Sun et al. [87] provides resolution higher than 1 μm in depth and enables to estimate the depth and extent of SSD in the sample of a rough surface. However, the system is limited in the lateral resolution. In a confocal microscopy, the system resolution is mainly depended on the diameter of the detection pinhole. The diameter of the pinhole is difficult to determine since it should match the wavelength of the incident light. In addition, the aberrations of the optical components and the errors in the adjustment of the light path should be also considered [74]. On the other hand, the refraction coefficient of the damage layer may be different from that of the damage-free matrix in a sample, and may vary along the depth because of the variation of SSD density. The effects of the variation of the refraction coefficient on the direction of light propagation are neglected in the previous studies [88, 90, 92].

In addition, the mechanism of the light scattering from SSD remains unknown. There are two light scattering mechanisms when an incident light encounters a scatter or non-homogeneity, i.e., Rayleigh scattering and Mie scattering. Rayleigh scattering predominates the scattering from a scatter which is much smaller than the wavelength. For the scatter similar to or larger than the wavelength, Mie scattering predominates. Both the two scattering mechanisms are applicable to the scattering from a general spherical scatter [93], whereas subsurface cracks are generally with the length of tens of microns and the width smaller than one micron in complex shapes. To our knowledge, the mechanism of the scattering from SSD is not fully interpreted. Therefore, further study and development are necessary.

Scanning Acoustic Microscopy

The scanning acoustic microscopy (SAM), as one of the ultrasonic detection methods, is a widely-used nondestructive technique for interior defect detection. A schematic of SAM is shown in Figure 9, in which an electrical radio frequency (RF) signal is sent to a piezoelectric transducer to transform into ultrasonic waves on the top surface of a sapphire delay rod. The waves propagate along the rod and are then focused into a sample through a coupling liquid with the concave end of the rod. While some waves reflect on the surface, the remaining waves penetrate into the sample and are scattered by the damage in the sample. The surface waves together with the subsurface scattering waves and noises are simultaneously received by a detector. The useful signals can be extracted from the received wave signals.

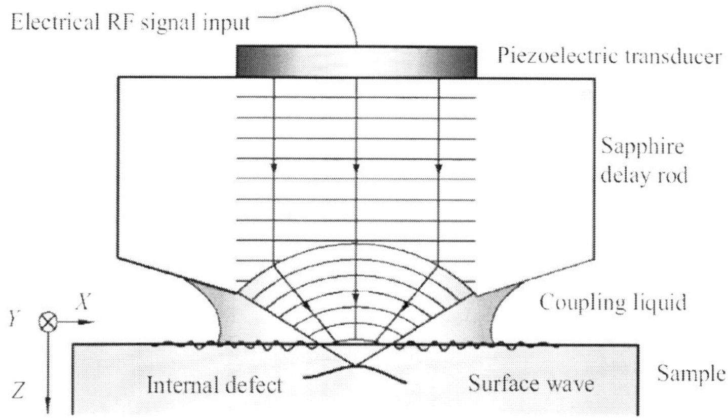

Figure 9: Schematic arrangement of scanning acoustic microscopy [97].

SAM has many advantages for the SSD detection. Both surface and subsurface defects can be detected at a high efficiency. However, SAM provides lateral resolution in the order of tens of microns [94, 95, 96, 97, 98], which is not applicable to the ultra-precision manufacturing. Generally, the signal acquired by SAM is combined with strong spurious signals which come from the surface waves and elastic waves in acoustic elements [96], making it difficult to distinguish the SSD information. In addition, SAM is normally applied to detecting the sample of a smooth surface. It may be difficult to apply to a sample with a rough surface due to the influence of the surface scattering waves.

The non-destructive methods preserve samples and save sample preparing time, which is in favor of improving overall manufacturing efficiency and reducing cost. However, there are still some limitations to overcome in the future. The methods are listed and compared in Table 2.

Table 2: Comparison among the non-destructive methods for SSD detection

Methods	**Advantages**	**Disadvantages**
Predictive methods	Simple, rapid and low cost	1. Applicable to predicting the statistic SSD depth rather than the maximum depth 2. Unable to predict the three-dimensional configuration and the distribution of SSD 3. Bad universality
OCT	Global scanning and applicable to three-dimensional configuration and spatial distribution of SSD	1. Inapplicable to detecting low light-scattering materials 2. Difficult to process images 3. Interfered by surface scattering
TIRM	Global scanning and in-process detection	1. Hard to measure SSD depth quantitatively 2. High requirement on surface condition, suitable for polished samples
CLSM	1. Global scanning and applicable to three-dimensional configuration and spatial distribution of SSD 2. High resolution	1. High requirement on surface condition, suitable for polished samples 2. Interfered by surface scattering
CPCM	1. Rapid global scanning 2. High resolution in depth direction 3. Low requirement on surface roughness	Low lateral resolution
SAM	1. High detection efficiency 2. High resolution in depth direction	1. Low lateral resolution 2. High requirement on surface condition

OTHER METHODS

Neauport et al. [35] added a little amount of barium into slurry during grinding, tracked the barium at different etching depths with the inductively coupled plasma-atomic emission spectrometry (ICP-AES), and inferred the depth of subsurface cracks. Similarly, "quantum dots" were used during lapping to detect the subsurface cracks [99, 100]. The quantum dots were fluorescent particles of the nanometer-size. The method was to reveal the depth of subsurface cracks by detecting the intensity of the fluorescence emitted from the quantum dots trapped in the subsurface cracks. However, since the width of a subsurface crack is normally less than one micron, the slurry is prevented from wetting the tip of the crack because of the surface tension of the slurry. Therefore, the measurement result of SSD is typically smaller than the real depth of the subsurface crack, which is also a problem with the dye penetration method [13, 101].

Podymova et al. [102, 103] measured the thickness of damage layer in the mono-crystalline silicon wafer with the laser optoacoustic method. The method was based on different mechanisms of laser-excited ultrasonic in the damaged layer and damage-free matrix, i.e., concentration-deformation mechanism in the matrix and thermal elastic deformation in the damaged layer. The thickness of the damaged layer could be inferred through the received ultrasonic signal with the method. However, the method is limited in detection depth to less than 10 μm and inapplicable to detecting the spatial distribution of SSD.

The X-ray diffraction (XRD) method is used to measure the machining-induced stress in the semiconductor materials [23, 85]. Generally, the diffraction spectrum is narrow and sharp if a sample does not have residual stresses; otherwise, the spectrum is broadened with diffraction peaks shifted. Residual stresses are determined based on a comparison between the measured and the standard diffraction spectra. However, since the X-ray penetration depth is limited in the most-used materials, XRD is only applicable to detecting the surface residual stresses.

The X-ray computed tomography (CT) is a powerful technique in revealing the three-dimensional microstructures of SSD. This technique is to take a number of 2D radiographic projections and then to reconstruct a 3D model without destroying a sample. The laboratory micro-focus computed tomography (micro CT), the synchrotron radiation computed tomography (SRCT), and the synchrotron radiation computed laminography (SRCL) are also utilized to detect micro defects with a high resolution of a few microns or higher [104, 105, 106]. The major drawback of CT is the radiation of X-ray which may hazard operator health.

The micro-Raman spectroscopy is usually used to measure residual stresses and phase transformation [107, 108, 109]. In this technique, light radiates a sample surface and is then scattered by the surface. Most of the incident light is elastically scattered, which means the scattered light has the same frequency as the incident light. However, the remaining of the incident light is non-elastically scattered, which means the scattered light has a different frequency than the incident light. This phenomenon is the inherent property of a material. Based on a Raman spectrum, residual strain, impurities, lattice distortion, and phase transformation of a sample can thus be estimated. Frequency shifts of a Raman spectrum indicates the information of the residual stresses and the phase transformation in the sample. This technique is simple and precise, but cannot detect subsurface cracks. Additionally, the technique is usually to detect a sample point-by-point rather than to scan the surface of the sample, which is time-consuming and less efficient.

Eddy current detection (ECD) is an efficient non-destructive method for detection of surface and subsurface defects in conductive materials. Eddy currents are loops of electrical current induced by an alternating magnetic field in a conductor due to Faraday's law of induction. In ECD, an ECD probe applies an excitation magnetic field to a part being detected to induce eddy currents in the part. The eddy currents create a magnetic field that reacts back on the excitation field, which results in a change in the impedance of the probe. Thus, a surface or subsurface defect can be detected based on the change in the impedance. ECD is widely applied to detecting subsurface defects [110] and residual stresses [111]. HE et al. proposed the induction thermography [112] and the eddy current pulse phase thermography [113] to detect SSD in steels and carbon fiber reinforced plastics. Practically, ECD has a limited application to the conductive and non-ferromagnetic materials, such as titanium alloys, aluminum alloys and carbon fiber reinforced composites.

For conductive ferromagnetic materials, such as maraging steels, the magnetic Barkhausen noise (MBN) detection technique can be used to evaluate residual stresses [114], impurities [115] and phase transformation [116] in the surface or subsurface of a part. An MBN signal represents discontinuous changes in the magnetization within a ferromagnetic material resulting from the change in an external magnetic field. The ECD and MBN techniques can only be applied to conductive and ferromagnetic materials, respectively. However, they are not suitable for non- or semi-conductive materials, such as ceramics, glass or silicon wafers.

DISCUSSION

Different hard and brittle materials have different properties that can result in different responses to a machining process. If the response of a material is known, it becomes possible to control the process so as to obtain the desirable surface integrity of the part made of the material. The response could be indicated by the spatial run and the distribution of the cracks in the machined part. In the process, the spatial run and the distribution of the induced subsurface cracks are influenced by many factors, including direction, magnitude and rate of the acting force and the anisotropy of the workpiece material. For example, in a monocrystalline silicon wafer, cracks initiate more easily in the {111} planes than other planes in the process since the {111} planes are the cleavage planes. The planes in which the cracks initiate are generally incline to the machined surface [2]. The crack behavior is complex and a propagating crack may branch into multiple cracks [117, 118, 119, 120]. Consequently, SSD presents complex configurations in the machined workpiece [73].

In a destructive method, only the SSD in the prepared section can be measured. The way to access the spatial run and the distribution is to prepare a set of sections, obtain the SSD in each section, and then reconstruct the spatial configurations. The predictive methods ignore the effect of the crystallographic orientation of a crystal on crack propagation in machining and consider that SSD and surface roughness are determined by median cracks and lateral cracks, respectively.

Tool wear is common in machining hard and brittle materials and significantly influences the SSD in a part [121, 122, 123]. In grinding, the SSD induced in machining generally increases with tool wear. It is unreasonable to choose the same parameters for subsequent processes due to the increased SSD from the previous process. Consequently, in-process detection provides real-time SSD information and helps optimize the subsequent processes. The destructive methods are difficult to integrate into a production line, while the non-destructive methods could achieve the in-process monitoring [40].

SUMMARIES AND OUTLOOK

Hard and brittle materials are hard to machine since SSD is easily induced in a machining process. Detecting SSD is a key link in an overall manufacturing chain, while the aforementioned methods applied to detecting SSD have associated limitations with respect to system resolution, detection depth and efficiency, as well as material type.

In this paper, the methods for detection of SSD in hard and brittle materials are reviewed. The destructive methods can provide observable, reliable, and quantitative information but are inapplicable to estimating the global information of SSD in a sample. Besides, the sample has to be destroyed, which is costly and time-consuming. The detection cost is acceptable in the mass production and non-flexible manufacturing processes. Otherwise, the non-destructive methods are more applicable and efficient. In the future, customized parts become more popular, which demands non-destructive methods for rapid estimation of the SSD induced in a part. In addition, the non-destructive methods are easier to integrate into a production line. It is anticipated that nondestructive methods are developed to meet the ever increasing demand on high quality manufacturing of the industrial parts made of hard and brittle materials.

AUTHORS' CONTRIBUTIONS

J-FY, QB and BZ were all in charge of the whole trial; J-FY wrote the manuscript; QB and BZ assisted with the revision. All authors have read and approved the final manuscript.

REFERENCES

1. S Gao. Fundamental research on silicon wafer thinning by ultra-precision grinding. Dalian University of Technology, 2013. (in Chinese)
2. Y X Zhang. Study on the surface layer damage of monocyrstalline silicon wafer induced by ultra-precision grinding. Dalian University of Technology, 2006. (in Chinese)
3. Z J Pei. A study on surface grinding of 300 mm silicon wafers. International Journal of Machine Tools and Manufacture, 2002, 42(3): 385–393.
4. R K Kang, Y X Zhang, D M Guo, et al. Study on the Surface and Subsurface Integrity of Ground Monocrystalline Silicon Wafers. Key Engineering Materials, 2005, 291–292: 425–432.
5. J M Zhang, J G Sun, Z J Pei. Application of Laser scattering on detection of subsurface damage in silicon wafers. ASME 2003 International Mechanical Engineering Congress and Exposition, Washington, DC, USA, November 15–21, 2003: 15–24.
6. H N Li, T B Yu, L D Zhu, et al. Evaluation of grinding-induced subsurface damage in optical glass BK7. Journal of Materials Processing Technology, 2016, 229: 785–794.
7. J J Wang, C L Zhang, P F Feng, et al. A model for prediction of subsurface damage in rotary ultrasonic face milling of optical K9 glass. The International Journal of Advanced Manufacturing Technology, 2016, 83(1–4): 347–355.
8. J B Chen, Q H Fang, P Li. Effect of grinding wheel spindle vibration on surface roughness and subsurface damage in brittle material grinding. International Journal of Machine Tools and Manufacture, 2015, 91: 12–23.
9. D X Lv, Y H Huang, Y J Tang, et al. Relationship between subsurface damage and surface roughness of glass BK7 in rotary ultrasonic machining and conventional grinding processes. The International Journal of Advanced Manufacturing Technology, 2013, 67(1–4): 613–622.
10. Z Q Yao, W B Gu, K M Li. Relationship between surface roughness and subsurface crack depth during grinding of optical glass BK7. Journal of Materials Processing Technology, 2012, 212(4): 969–976.
11. S Y Li, Z Wang, Y L Wu. Relationship between subsurface damage

and surface roughness of optical materials in grinding and lapping processes. Journal of Materials Processing Technology, 2008, 205(1–3): 34–41.

12. J Neauport, C Ambard, H Bercegol, et al. Optimizing fused silica polishing processes for 351nm high-power laser application. Proceedings of SPIE-The International Society for Optical Engineering, 2008, 7132: 51.
13. W Kanematsu. Visualization of subsurface damage in silicon nitride from grinding by a plasma etching and dye impregnation method. Journal of the American Ceramic Society, 2006, 89(8): 2564–2570.
14. B Zhang, T Histoshi, Y Masanori. Study on surface cracking of alumina scratched by single-point diamonds. Journal of Materials Science, 1988, 23(9): 3214–3224.
15. J G Sun, M Shirber, W A Ellingson. Laser-based optical scattering detection of surface and subsurface defects in machined Si3N4 components. John Wiley & Sons, Inc., 1997: 273–280.
16. N Azeggagh, L Joly-Pottuz, D Nélias, et al. Hertzian contact damage in silicon nitride ceramics with different porosity contents. Journal of the European Ceramic Society, 2015, 35(8): 2269–2276.
17. B Zhang, T D Howes. Subsurface evaluation of ground ceramics. CIRP Annals—Manufacturing Technology, 1995, 44(1): 263–266.
18. B Zhang, J Meng. Grinding damage in fine ceramics. Nanotechnology & Precision Engineering, 2003, 1(1): 48–56. (in Chinese)
19. Z Wang. Study on the detection and control techniques of subsurface damage in optical fabrication. National University of Defense Technology, 2008. (in Chinese)
20. H K Tonshoff, B Karpuschewski, M Hartmann, et al. Grinding-and-slicing technique as an advanced technology for silicon wafer slicing. Machining Science and Technology, 1997, 1(1): 33–47.
21. B Zhang, T D Howes. Material-removal mechanisms in grinding ceramics. CIRP Annals—Manufacturing Technology, 1994, 43(1): 305–308.
22. P P Hed, D F Edwards, J B Davis. Subsurface damage in optical materials: Origin, measurement and removal: Summary. Lawrence Livermore National Lab., CA (USA), 1988.
23. H K Tönshoff, W V Schmieden, I Inasaki, et al. Abrasive machining of silicon. CIRP Annals-Manufacturing Technology, 1990, 39(2): 621–

635.

24. X Tonnellier, P Shore, X Luo, et al. Wheel wear and surface/subsurface qualities when precision grinding optical materials. Proc. SPIE, 2006, 6273: 627308.
25. F W Huo. Study on the mechanism of ductile mode grinding of silicon wafers. Dalian University of Technology, 2006. (in Chinese)
26. Y X Zhang, D L Li, G Wei, et al. Experimental investigation on the detection technique for surface layer damage of machined silicon wafers. Rengong Jingti Xuebao/Journal of Synthetic Crystals, 2011, 40(2): 359–364.
27. Y Masanori, B Zhang, T Hitoshi. Observations of ceramic surface cracks by newly proposed methods. Yogyo-Kyokai-Shi, 1987, 95(10): 961–969. (in Japanese)
28. S D Jacobs, J C Lambropoulos, D Golini, et al. Use of magnetorheological finishing (MRF) to relieve residual stress and subsurface damage on lapped semiconductor silicon wafers. Proceedings of SPIE-The International Society for Optical Engineering, 2001, 4451: 286–294.
29. Y Y Zhou, P D Funkenbusch, D J Quesnel, et al. Effect of etching and imaging mode on the measurement of subsurface damage in microground optical-glasses. Journal of the American Ceramic Society, 1994, 77(12): 3277–3280.
30. F Lakhdari, D Bouzid, N Belkhir, et al. Surface and subsurface damage in Zerodur® glass ceramic during ultrasonic assisted grinding. The International Journal of Advanced Manufacturing Technology, 2017, 90(5–8): 1993–2000.
31. J A Randi, J C Lambropoulos, S D Jacobs. Subsurface damage in some single crystalline optical materials. Appl. Opt., 2005, 44(12): 2241–2249.
32. T Suratwala, L Wong, P Miller, et al. Sub-surface mechanical damage distributions during grinding of fused silica. Journal of Non-Crystalline Solids, 2006, 352(52–54): 5601–5617.
33. P E Miller, T I Suratwala, L L Wong, et al. The distribution of subsurface damage in fused silica. Proceedings of SPIE-The International Society for Optical Engineering, 2005: 599101–599101–25.
34. J A Menapace, P J Davis, W A Steele, et al. MRF applications: measurement of process-dependent subsurface damage in optical materials using the MRF wedge technique. Proceedings of SPIE-The

International Society for Optical Engineering, 2005: 599103–599103–11.

35. J Neauport, C Ambard, P Cormont, et al. Subsurface damage measurement of ground fused silica parts by HF etching techniques. Opt. Express, 2009, 17(22): 20448–20456.
36. A B Shorey, S D Jacobs, W I Kordonski, et al. Experiments and observations regarding the mechanisms of glass removal in magnetorheological finishing. Appl. Opt., 2001, 40(1): 20–33.
37. A B Shorey. Mechanisms of material removal in magnetorheological finishing (MRF) of glass. University of Rochester, 2000.
38. C Miao, S N Shafrir, J C Lambropoulos, et al. Shear stress in magnetorheological finishing for glasses. Appl. Opt., 2009, 48(13): 2585–2594.
39. T O Mulhearn. The deformation of metals by Vickers-type pyramidal indenters. Journal of the Mechanics and Physics of Solids, 1959, 7(2): 85–88.
40. J Wang, Y G Li, J H Han, et al. Evaluating subsurface damage in optical glasses. Journal of the European Optical Society-Rapid Publications, 2011, 6(11001).
41. S Van der Zwaag, J T Hagan, J E Field. Studies of contact damage in polycrystalline zinc sulphide. Journal of Materials Science, 1980, 15(12): 2965–2972.
42. H H Xu, S Jahanmir. Simple technique for observing subsurface damage in machining of ceramics. Journal of the American Ceramic Society, 1994, 77(5): 1388–1390.
43. F Guiberteau, N P Padture, B R Lawn. Effect of grain size on Hertzian contact damage in alumina. Journal of the American Ceramic Society, 1994, 77(7): 1825–1831.
44. D S Anderson, M E Frogner. A method for the evaluation of subsurface damage. Technical Digest of the Optical Fabrication and Testing Workshop. Optical Society of America, Washington DC, USA, 1985.
45. H Helbawi, L C Zhang, I Zarudi. Difference in subsurface damage in indented specimens with and without bonding layer. International Journal of Mechanical Sciences, 2001, 43(4): 1107–1121.
46. C W Carr, M J Matthews, J D Bude, et al. The effect of laser pulse duration on laser-induced damage in KDP and SiO2. Proc. SPIE, 2006: 6403.

47. I Zarudi, L Zhang. Subsurface damage in single-crystal silicon due to grinding and polishing. Journal of Materials Science Letters, 1996, 15(7): 586–587.

48. T Abe, Y Nakazato, M Daito, et al. The ductile mode grinding technology applied to silicon wafering process. Semiconductor Silicon/1994, Proceedings of the 7th International Symposium on Silicon Materials Science and Technology, The Electrochemical Society, Pennington, NJ, USA, 1994: 207–217.

49. H Z Wu, S G Roberts, G Möbus, et al. Subsurface damage analysis by TEM and 3D FIB crack mapping in alumina and alumina/5vol.%SiC nanocomposites. Acta Materialia, 2003, 51(1): 149–163.

50. L Wong, T Suratwala, M D Feit, et al. The effect of HF/NH4F etching on the morphology of surface fractures on fused silica. Journal of Non-Crystalline Solids, 2009, 355(13): 797–810.

51. B Ma, Z X Shen, Z Zhang, et al. Fabrication and detection technique of fused silica substrate with extremely low subsurface damage. High Power Laser & Particle Beams, 2010, 22(9): 2181–2185.

52. G Spierings. Wet chemical etching of silicate-glasses in hydrofluoric-acid based solutions. Journal of Materials Science, 1993, 28(23): 6261–6273.

53. G L Li. Theoretical and experimental research on the measurement of grinding subsurface damage for optical materials. National University of Defense Technology, 2006. (in Chinese)

54. Z Wang, Y L Wu, Y F Dai, et al. Rapid detection of subsurface damage of optical materials in lapping process and its influence regularity. Guangxue Jingmi Gongcheng/Optics & Precision Engineering, 2008, 16(1): 16–21.

55. R Sabia, H J Stevens, J R Varner. Pitting of a glass-ceramic during polishing with cerium oxide. Journal of Non-Crystalline Solids, 1999, 249(2–3): 123–130.

56. B Ma, Z X Shen, P F He, et al. Detection of subsurface defects of fused silica optics by confocal scattering microscopy. Chinese Optics Letters, 2010, 8(3): 296–299.

57. M Sergeeva, K Khrenikov, T Hellmuth, et al. Sub surface damage measurements based on short coherent interferometry. Journal of the European Optical Society-Rapid Publications, 2010, 5: 10003–1-5.

58. P P Hed, D F Edwards. Optical glass fabrication technology. 2:

Relationship between surface roughness and subsurface damage. Appl. Opt., 1987, 26(21): 4677–4680.

59. J C Lambropoulos. From abrasive size to subsurface damage in grinding. Convergence, 2000, 8: 1–3.
60. F K Aleinikov. The effect of certain physical and mechanical properties on the grinding of brittle materials. Soviet Physics-Technical Physics, 1957, 2(12): 2529–2538.
61. J C Lambropoulos, Y Li, P D Funkenbusch, et al. Noncontact estimate of grinding-induced subsurface damage. SPIE's International Symposium on Optical Science, Engineering, and Instrumentation. International Society for Optics and Photonics, 1999: 41–50
62. J Neauport, J Destribats, C Maunier, et al. Loose abrasive slurries for optical glass lapping. Appl. Opt., 2010, 49(30): 5736–5745.
63. K R Fine, R Garbe, T Gip, et al. Non-destructive real-time direct measurement of subsurface damage. Defense and Security. International Society for Optics and Photonics, Orlando, Florida, USA, March 28, 2005: 105–110.
64. B Lawn, R Wilshaw. Indentation fracture: principles and applications. Journal of Materials Science, 1975, 10(6): 1049–1081.
65. D Huang, E A Swanson, C P Lin, et al. Optical coherence tomography. Science, 1991, 254(5035): 1178–1181.
66. V V Tuchin, R K Wang, V V Tuchin. Optical coherence tomography: light scattering and imaging enhancement. In: Handbook of coherent-domain optical methods, Springer New York, 2013: 665–742.
67. M Bashkansky, M D Duncan, M Kahn, et al. Subsurface defect detection in ceramics by high-speed high-resolution optical coherent tomography. Optics Letters, 1997, 22(1): 61–63.
68. M Bashkansky, P R Battle, M Duncan, et al. Subsurface defect detection in ceramic materials using optical gating techniques. Review of Progress in Quantitative Nondestructive Evaluation, Springer, 1996: 1565–1572.
69. D O Thompson, D E Chimenti, M Bashkansky, et al. Subsurface defect detection in ceramic materials using optical gating techniques. Review of Progress in Quantitative Nondestructive Evaluation. Springer US, 1996: 1565–1572.
70. P R Battle, M Bashkansky, R Mahon, et al. Subsurface defect detection in ceramic materials using optical gating techniques. Optical

Engineering, 1996, 35(4): 1119–1123.

71. M Bashkansky, P R Battle, M D Duncan, et al. Subsurface defect detection in ceramics using an optical gated scatter reflectometer. Journal of the American Ceramic Society, 1996, 79(5): 1397–1400.
72. M Duncan, M Bashkansky, J Reintjes. Subsurface defect detection in materials using optical coherence tomography. Opt. Express, 1998, 2(13): 540–545.
73. Z J Pei, S R Billingsley, S Miura. Grinding induced subsurface cracks in silicon wafers. International Journal of Machine Tools and Manufacture, 1999, 39(7): 1103–1116.
74. C H Wang. Study on optical sub-surface damage evaluation technology. Xi'an: Xi'an Technological University, 2010. (in Chinese)
75. P A Temple. Total internal reflection microscopy: a surface inspection technique. Appl. Opt., 1981, 20(15): 2656–2664.
76. P A Temple. Examination of laser damage sites of transparent surfaces and films using total internal reflection microscopy. Natl. Bur. Std. Spec. Publ. 1980, 568: 333–341.
77. F Draheim, B Harnisch, T Weigel. Subsurface damage of optical components and the influence on scattering properties. Proceedings of SPIE-The International Society for Optical Engineering, 1994: 709–720.
78. C F Kranenberg, K C Jungling. Subsurface damage identification in optically transparent materials using a nondestructive method. Appl. Opt., 1994, 33(19): 4248–4253.
79. Z M Liao, S J Cohen, J R Taylor. Total internal reflection microscopy (TIRM) as a nondestructive subsurface damage assessment tool. Proc. SPIE, 1994: 43–53.
80. L M Sheehan, M Kozlowski, D W Camp. Application of total internal reflection microscopy for laser damage studies on fused silica. Proceedings of SPIE-The International Society for Optical Engineering, 1998, 3244(6): 282–295.
81. Y Deng, Q Xu, L Q Chai, et al. Total internal reflection microscopy: A subsurface defects identification technique in optically transparent components. High Power Laser & Particle Beams, 2009, 21(6): 835–840.
82. A J Winn, J A Yeomans. A study of microhardness indentation fracture in alumina using confocal scanning laser microscopy. Philosophical

Magazine A-Physics of Condensed Matter Structure Defects and Mechanical Properties, 1996, 74(5): 1253–1263.

83. B Bertussi, P Cormont, S Palmier, et al. Initiation of laser-induced damage sites in fused silica optical components. Opt. Express, 2009, 17(14): 11469–11479.

84. C H Wang. Optical subsurface damage evaluation using LSCT. Proc. SPIE, 2009, 7522: 75226K–75226K-7.

85. W K Lu, Z J Pei, J G Sun. Non-destructive evaluation methods for subsurface damage in silicon wafers: A literature review. International Journal of Machining and Machinability of Materials, 2007, 2(1): 125–142.

86. W K Lu, J G Sun, Z J. Pei Subsurface damage measurement in silicon wafers with cross-polarisation confocal microscopy. International Journal of Nanomanufacturing, 2006, 1(2): 272–282.

87. J Sun. Device and nondestructive method to determine subsurface micro-structure in dense materials. Argonne National Laboratory (ANL), Argonne, IL, 2006.

88. J M Zhang, J G Sun, Z J Pei. Subsurface damage measurement in silicon wafers by laser scattering. Trans. NAMRI/SME, 2002, 30: 535–542.

89. J G Sun, A Wolosewicz, M H Haselkorn, et al. Laser scattering detection and characterization of defects and machining damage in silicon nitride components. Office of Scientific & Technical Information Technical Reports, 1998.

90. J M Zhang, J G Sun. Quantitative assessment of subsurface damage depth in silicon wafers based on optical transmission properties. International Journal of Manufacturing Technology and Management, 2005, 7(5–6): 540–552.

91. J M Zhang, J G Sun, Z J Pei. Optical transmission properties of silicon wafers: theoretical analysis. ASME 2004 International Mechanical Engineering Congress and Exposition. American Society of Mechanical Engineers, Anaheim, California, USA, November 13–19, 2004: 17–24.

92. J M Zhang. Laser scattering techniques for subsurface damage measurements: system development, experimental investigation, and theoretical analysis. Kansas State University, 2006.

93. D W Hahn. Light scattering theory. Department of Mechanical and Aerospace Engineering, Florida, 2006.

94. A Y Sun, B F Ju. The acoustic micro integrated detection technique

for silicon wafer processing. Advanced Materials Research, 2012, 497: 151–155.

95. O Balogun, G D Cole, R Huber, et al. High-spatial-resolution subsurface imaging using a laser-based acoustic microscopy technique. IEEE Transactions on Ultrasonics, Ferroelectrics and Frequency Control, 2011, 58(1): 226–233.

96. Y V Korkh, A M Burkhanov, A B Rinkevich. Scanning acoustic microscope for visualization of microflaws in solids. Russian Journal of Nondestructive Testing, 2009, 45(10): 677–684.

97. C T Leondes, J Janting. Techniques in scanning acoustic microscopy for enhanced failure and material analysis of microsystems. Leondes C T, Springer US, 2006: 905–921.

98. D Rats, J V Stebut, F Augereau. High frequency scanning acoustic microscopy: a novel non-destructive surface analytical tool for assessment of coating-specific elastic moduli and tomographic study of subsurface defects. Thin Solid Films, 1999, s355–356(2): 347–352.

99. W Williams, B Mullany, W Parker, et al. Using quantum dots to evaluate subsurface damage depths and formation mechanisms in glass. CIRP Annals-Manufacturing Technology, 2010, 59(1): 569–572.

100. W B Williams, B A Mullany, W C Parker, et al. Using quantum dots to tag subsurface damage in lapped and polished glass samples. Applied Optics, 2009, 48(27): 5155–5163.

101. T Yonushonis. Manufacturing fluid including fluorescent dye penetrant and method for using to make components: U.S. Patent 6,677,584. 2004-1-13.

102. N Podymova, A Karabutov. Laser optoacoustic non-destructive method of thickness measurement of subsurface damaged layer in machined silicon wafers. Journal of Physics: Conference Series. IOP Publishing, 2010, 214(1): 012054.

103. A A KARABUTOV, N B PODYMOVA. Study on the subsurface damage depth in machined silicon wafers by the laser-ultrasonic method. Case Studies in Nondestructive Testing and Evaluation, 2014, 1: 7–12.

104. L Lavery, W Harris, J Gelb, et al. Recent advancements in 3D X-ray microscopes for additive manufacturing. Microscopy and Microanalysis, 2015, 21(S3): 131–132.

105. D J Bull, L Helfen, I Sinclair, et al. A comparison of multi-scale 3D

X-ray tomographic inspection techniques for assessing carbon fibre composite impact damage. Composites Science and Technology, 2013, 75: 55–61.

106. J Lambert, A R Chambers, I Sinclair, et al. 3D damage characterisation and the role of voids in the fatigue of wind turbine blade materials. Composites Science and Technology, 2012, 72(2): 337–343.

107. S Dutta, G Saxena, K Jindal, et al. Comparison of residual stress in deep boron diffused silicon (100), (110) and (111) wafers. Materials Letters, 2013, 100: 44–46.

108. J W Yan. Laser micro-Raman spectroscopy of single-point diamond machined silicon substrates. Journal of Applied Physics, 2004, 95(4): 2094–2101.

109. D W Ingrid. Raman spectroscopy: Chips and stress. Spectroscopy Europe, 2003, 15(2): 6–13.

110. W Du, Q Bai, Y B Wang, et al. Eddy current detection of subsurface defects for additive/subtractive hybrid manufacturing. The International Journal of Advanced Manufacturing Technology, 2017: 1–11.

111. M Morozov, G Y Tian, P J Withers. Noncontact evaluation of the dependency of electrical conductivity on stress for various Al alloys as a function of plastic deformation and annealing. Journal of Applied Physics, 2010, 108(2): 024909.

112. Y He, M Pan, F Luo. Defect characterization based on heat diffusion using induction thermography testing. Review of Scientific Instruments, 2012, 83(10): 104702.

113. Y He, M Pan, G Y Tian, et al. Eddy current pulsed phase thermography for subsurface defect quantitatively evaluation. Applied Physics Letters, 2013, 103(14): 144108.

114. P Vourna, A Ktena, P E Tsakiridis, et al. An accurate evaluation of the residual stress of welded electrical steels with magnetic Barkhausen noise. Measurement, 2015, 71: 31–45.

115. J A Perez-Benitez, J Capo-Sanchez, J Anglada-Rivera, et al. A study of plastic deformation around a defect using the magnetic Barkhausen noise in ASTM 36 steel. Ndt & e International, 2008, 41(1): 53–58.

116. A Stupakov, M Neslušan, O Perevertov. Detection of a milling-induced surface damage by the magnetic Barkhausen noise. Journal of Magnetism and Magnetic Materials, 2016, 410: 198–209.

117. K Ravi-Chandar, W G Knauss. An experimental investigation into dynamic fracture: I. Crack initiation and arrest. International Journal of Fracture, 1984, 25(4): 247–262.

118. K Ravi-Chandar, W G Knauss. An experimental investigation into dynamic fracture: II. Microstructural aspects. International Journal of Fracture, 1984, 26(1): 65–80.

119. K Ravi-Chandar, W G Knauss. An experimental investigation into dynamic fracture: III. On steady-state crack propagation and crack branching. International Journal of fracture, 1984, 26(2): 141–154.

120. K Ravi-Chandar, W G Knauss. An experimental investigation into dynamic fracture: IV. On the interaction of stress waves with propagating cracks. International Journal of Fracture, 1984, 26(3): 189–200.

121. J Y Shen, J Q Wang, B Jiang, et al. Study on wear of diamond wheel in ultrasonic vibration-assisted grinding ceramic. Wear, 2015, 332: 788–793.

122. Q Zhang, S To, Q Zhao, et al. Impact of material microstructure and diamond grit wear on surface finish in micro-grinding of RB-SiC/Si and WC/Co carbides. International Journal of Refractory Metals and Hard Materials, 2015, 51: 258–263.

123. K Fathima, A S Kumar, M Rahman, et al. A study on wear mechanism and wear reduction strategies in grinding wheels used for ELID grinding. Wear, 2003, 254(12): 1247–1255.

Chapter 9

Influence of Cutting Speed on Subsurface Damage Morphology and Distribution in Ground Fused Silica

Georg Schnurbusch [1], Ekkard Brinksmeier [2], and Oltmann Riemer [1]

[1]Laboratory for Precision Machining (LFM), University of Bremen, Badgasteiner Str. 2, 28359 Bremen, Germany

[2]Foundation Institute of Materials Science, Badgasteiner Str. 3, 28359 Bremen, Germany

ABSTRACT

In optical fabrication, brittle-hard materials are used for numerous applications. Especially for high-performance optics for laser or lithography applications, a complex and consistent production chain is necessary to account for the material properties. Particularly in pre-processing, e.g., for shaping optical components, brittle material behavior is dominant which leads to a rough surface layer with cracks that reach far below the surface. This so called subsurface damage (SSD) needs to be removed in subsequent

Citation: Schnurbusch, Georg; Brinksmeier, Ekkard; Riemer, Oltmann. 2017. "Influence of Cutting Speed on Subsurface Damage Morphology and Distribution in Ground Fused Silica." Inventions 2, no. 3: 15. https://doi.org/10.3390/inventions2030015

processes like polishing. Therefore, it is essential to know the extent of the SSD induced by shaping for an efficient design of precise corrective processes and for process improvement. Within this work the influence of cutting speed on SSD, in fused silica, induced by grinding has been investigated. To analyze the subsurface crack distribution and the maximum crack depth magnetorheological finishing has been appointed to polish a wedge into the ground surface. The depth profile of SSD was analyzed by image processing. For this purpose a coherent area of the polished wedge has been recorded by stitching microscopy. Taking the form deviation of the ground surface in to account to determine the actual depth beneath surface, the accuracy of the SSD-evaluation could be improved significantly. The experiments reveal a clear influence of the cutting speed on SSD, higher cutting speeds generate less SSD. Besides the influence on the maximum crack depth an influence on the crack length itself could be verified. Based on image analysis it was possible, to predict the maximum depth of cracks by means of crack length.

Keywords: optics manufacturing, precision grinding, subsurface damages, brittle hard materials

INTRODUCTION

Grinding is still an indispensable process step in the manufacturing of high precision optics, due to its high removal rates and well-controlled shaping. However, due to the prevailing material removal mechanisms, the grinding of brittle materials inevitably leads to microscopic cracks beneath the surface. This so called subsurface damage (SSD) influences the operability and lifetime of high-performance optics like semiconductor or high power laser applications [1,2]. The shaping process by grinding is followed by various corrective processes like polishing for smoothing the surface, improving the shape accuracy and removing SSD. The maximum crack depth at each process step decisively determines the extent of the polishing and corrective processes required and thus the costs of the overall optic manufacturing process [3]. According to this, a main issue in the production of high performance optics is to minimize the process-related SSD in grinding.

The undefined cutting edge and brittle material removal behavior make grinding a stochastic process. Single grains with high protrusions lead to a local overload of the normal force and cause the deepest cracks [4]. For fine-grained diamond grinding wheels the influencing variables grain protrusion and number of active grains cannot be determined directly, therefore it is nearly impossible to model the grinding process or to predict SSD. Thus, the process design for optics manufacturing still depends on empirical data. To date, the grain size serves as an orientation [5]. The stability of grinding processes as well as the accuracy in ascertaining the maximum depth of SSD is essential for increasing the efficiency in optics manufacturing. The same counts for the investigation of the influencing variables on the process-induced SSD. The accuracy of the evaluation is even more important for fine grinding where crack depths are only a few micrometers.

Nature and extent of subsurface damage in glasses makes quantitative assessment of damage difficult [6]. To date, it is still challenging to determine subsurface damage in optical glasses. Since the measurement of SSD with destructive methods is complex as well as cost-intensive and time-consuming, several attempts were undertaken to correlate process parameters and measurable output values, like surface roughness. The results of Lambropoulos et al. and Wang et al. show, however, a broad distribution in the results of these methods [7,8]. Even though there are numerous non-destructive approaches to determine the crack depths beneath the machined surface, the destructive methods still provide the most accurate results [7]. Over the past few years, the combination of magnetorheological finishing (MRF) and etching has proven to be the most appropriate strategy [7,8]. MRF provides a highly deterministic material removal process without generating new or propagating existing SSD [9]. The subsequent etching process serves to open the cracks on the polished surface to provide a better contrast in optical imaging. The damage free polishing of a wedge allows for a glance under the ground surface. Thus, not only the maximum crack depth can be determined, but also the density and shape of cracks can be observed at different depths. Resolution of this analysis obviously depends on the wedge gradient, as the area increases with decreasing wedge angles.

Simple investigations of SSD by MRF technique are limited to the determination of the maximum crack depth [10,11]. Miller et al. first analyzed the depth profile of the crack density by single micrographs along the wedge [4]. To enable analysis for a larger wedge area, adjacent microscope images were stitched by using a computer-based microscope system with a driven stage [8]. This method offers the great advantage, to analyze the recorded data by image processing. The SSD depth is calculated from the difference in height between the profiles of the ground surface and the wedge surface. In previous publications, a perfect initial surface was assumed, however, even fine grinding generates form deviations of a few micrometers. For rough cuts and pre-shaping this seems negligible, but in the case of fine grinding these deviations can be on the same scale as the SSD.

Within this work, the influence of cutting speeds on SSD for ground fused silica was investigated. To analyze the maximum crack depth and crack distribution MRF technique was used in conjunction with dry etching. Finally, for the first time, the initial surface profile was considered to enhance the accuracy to determine the actual crack depth.

MATERIALS AND METHODS

Fused silica samples with a diameter of 100 mm were ground on an ultraprecision 5-axis grinding machine (Nanotech 500 Freeform Generator, Moore Nanotechnology Systems). The experiments comprise a variation of cutting speeds between 30 and 62 m/s, in five subdivisions, with three repetitions (Table 1). For process monitoring, the grinding forces were measured by a multicomponent dynamometer (CompactDyn, Kistler, Winterthur, Switzerland) mounted beneath the hydrostatic grinding spindle. The grinding kinematics is shown in Figure 1.

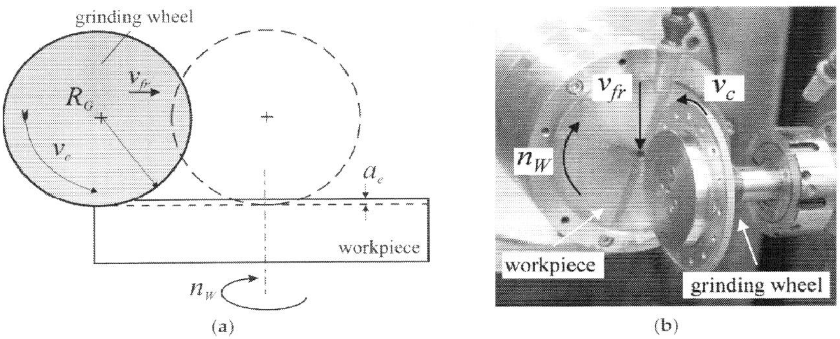

Figure 1: (a) Grinding kinematics—cross grinding with rotating workpiece; (b) Machine setup.

Table 1: Process parameters

Process Step	Cutting Speed v_c (m/s)	Depth of Cut a_e (μm)	Worktable Speed n_W (rpm)	Radial Feed v_{fr} (mm/min)
1st step: pre-processing	30	100 (4 × 25)	50	10
2nd step: SSD experiments	30–62	90 (4 × 20, 1 × 10)	50	10

A resin bonded diamond grinding wheel with a diameter of 100 mm, D20 (C70) was applied. All samples were ground with a constant rotary worktable speed $n_W = 50$ rpm and radial feed $v_{fr} = 10$ mm/s. Based on a fine grinding process, the material removal was conducted in four passes with 20 μm and one final pass with 10 μm depth of cut. In order to guarantee a uniform initial state, all samples were pre-ground to plane shape before running the actual experiments (pre-processing, $a_e = 100$ μm; four passes with 25 μm). Furthermore, the grinding wheel was dressed and balanced before each grinding experiment. This ensures that the measured SSD results from the actual process parameters.

RESULTS

Grinding Force and Surface Roughness

Based on the constant rotary worktable speed, the feed rate decreases continuously from edge to center. Since the remaining parameters are kept constant, this leads to a reduction in material removal rate and thus a reduction in chip thickness. Consequently, normal forces (Figure 2a) and roughness decrease from edge to center. In conventional grinding, it is generally assumed that higher cutting speeds lead to a lower chip thickness and thus also to lower grinding forces [12]. In the experiments presented in this work, however, the opposite is shown. Larger cutting speeds led to higher normal forces (Figure 2b). For brittle materials like fused silica, the removal mechanisms depend on the chip thickness. Due to a higher rate of ductile material removal, the forces increase with the cutting speed [13]. To make the results comparable and regarding the position of the maximum crack depth c_{max} at the MRF wedge, the presented values were obtained at a distance of 15 mm from workpiece center.

The surface roughness was measured by white light interferometry (WLI, Talysurf CCI HD, Taylor Hobson) with 20× magnification. Along circles of different radii six spots with an area of 840 × 840 μm² were measured with an angular distance of 60°. To exclude the waviness of the surface a long pass filter was set to 250 μm.

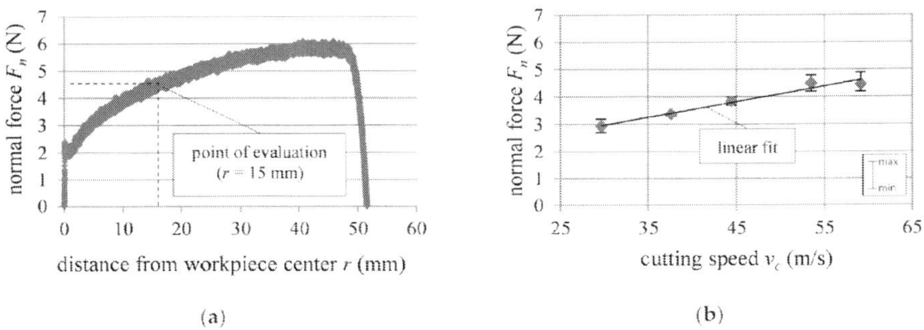

Figure 2: (a) Typical force curve (F_n) at v_c = 62 m/s) for constant worktable speed n_W; (b) Normal forces for different cutting speeds (distance from workpiece center 15 mm).

The results of the roughness measurement (e.g., in Figure 3) show an opposite trend towards normal force. This agrees well with the assumption that the proportion of ductile material behavior increases with higher cutting speeds.

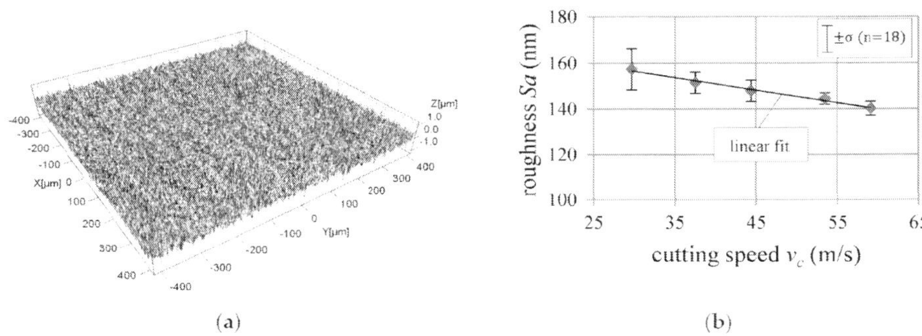

Figure 3: (a) White light interferometer (WLI) measurement (20× magnification): 2D-surface roughness (v_c = 30 m/s, Sa = 160 nm); (b) Surface roughness Sa for different cutting speeds (distance from workpiece center 15 mm) with linear fit.

Subsurface Damage Evaluation

For the evaluation of subsurface damage, the micrographs of the MRF wedge need to be merged with the depth profile. In order to maintain a high accuracy, the samples were provided with a fiducial mark by laser engraving after grinding. It is thus possible to align the samples for different measurements and to link the data of tactile measurements and microscope images. Figure 4 shows the wedge geometry and the entire evaluation procedure.

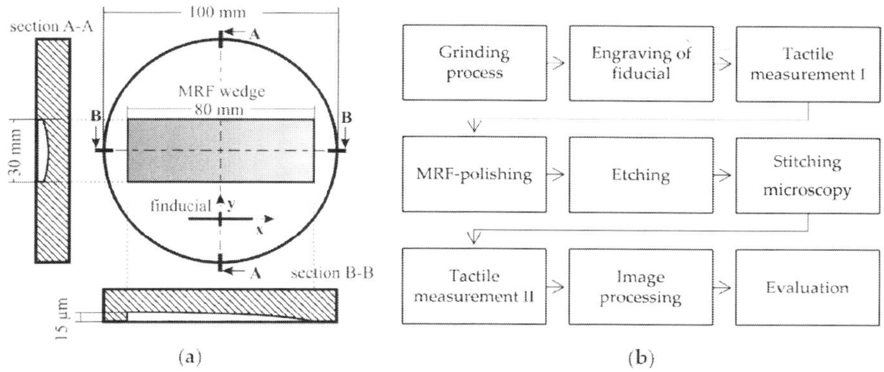

Figure 4: (a) Position and geometry of the magnetorheological finishing (MRF) wedge; (b) Sample processing chain.

After MRF preparation, the wedge surfaces were dry etched. Compared to wet etching the material removal by dry etching can be controlled with a sub-micron precision [14]. The micrographs of the wedge surface were taken with a ZEISS Axio Imager.Z2 Vario with 500× magnification. 750 micrographs were stitched together to an interrelated area of 40 × 1 mm^2. To minimize contamination from particles the samples were cleaned in an ultrasonic bath and microscopy was carried out under clean room conditions (ISO 7). In order to determine the actual depth of the wedge as accurately as possible, the distance between the ground surface (tactile measurement I) and the wedge surface (tactile measurement II) was calculated from the two tactile measurements.

Due to the form deviation, differences between the polished wedge surface and the actual depth profile of up to 2 µm (Figure 5) occur in the individual experiments. This is particularly relevant for fine grinding processes with low SSD. Within the experiments, the differences correspond to 30–40% of the measured maximum crack depth.

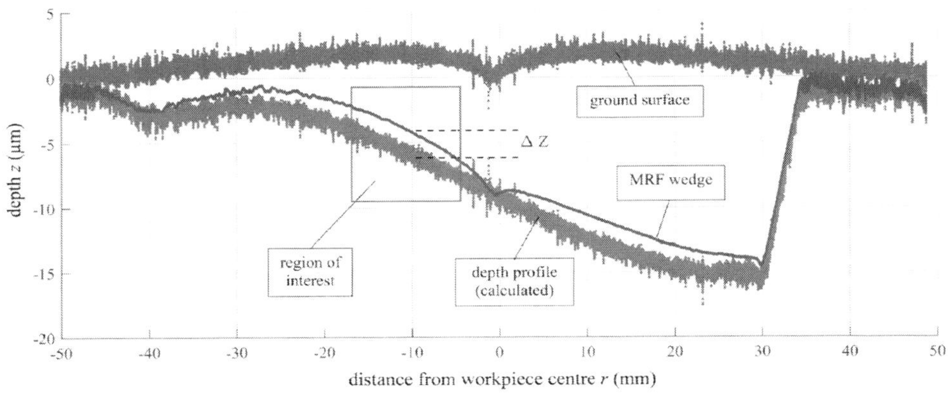

Figure 5: Tactile measurement of ground surface, MRF wedge and depth profile.

The number of cracks decreases with depth along the MRF wedge. With increasing depth of the wedge, more and more cracks are removed by polishing until the last fracture event can be determined as the maximum crack depth in the micrographs (Figure 6).

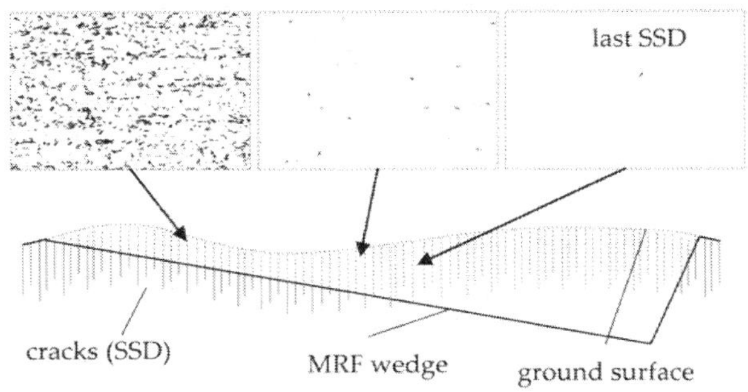

Figure 6: Sketch of subsurface damage (SSD) appearance along the MRF wedge.

To determine the maximum crack depth c_{max} the lateral position of the last defect on the micrograph of the polished and etched wedge surface is merged together with the corresponding information of the calculated depth profile (Figure 7).

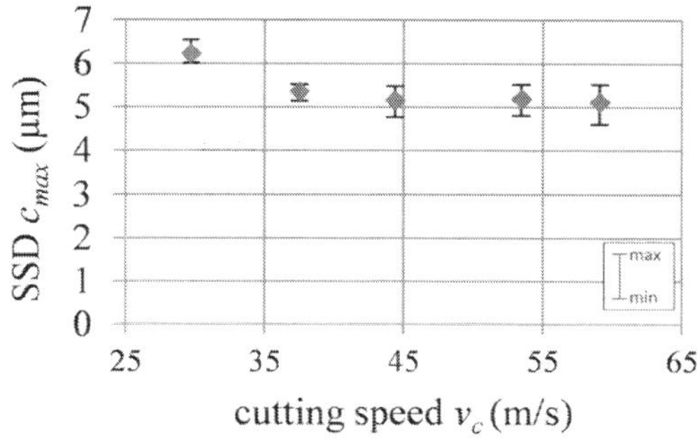

Figure 7: Maximum crack depth depending on the cutting speed.

SSD Analysis by Image Processing

Besides the evaluation of the maximum crack depth, the micrographs of the MRF wedge provide the opportunity to examine the crack distribution with increasing depth. Thus, a more extended examination on the influence of process parameters on grinding results is possible. The high number of cracks resulting from the size of the inspected wedge area requires a software-supported evaluation. Therefore the open source software, ImageJ, has been utilized. The software is a highly popular tool in biology in a wide range of applications and provides a segmentation algorithm. This enables the analysis of the position and the geometrically relevant data for each individual crack on the stitched microscope image. After binarization of the images, the cracks appear as uniform black lines on a white background. To specify the crack length, a best fit ellipse is applied to every single crack (Figure 8).

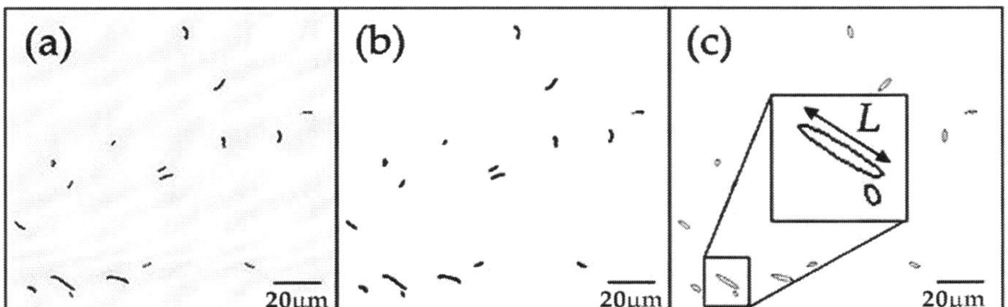

Figure 8: SSD-Image Processing (ImageJ): (**a**) raw microscopy data; (**b**) binary conversion; (**c**) determination of crack length L by best fit ellipse.

On average, 150,000 cracks were evaluated for each sample with this method. Figure 9 shows the ratio of the number of cracks of different lengths to the total crack number (relative frequency) for different cutting speeds.

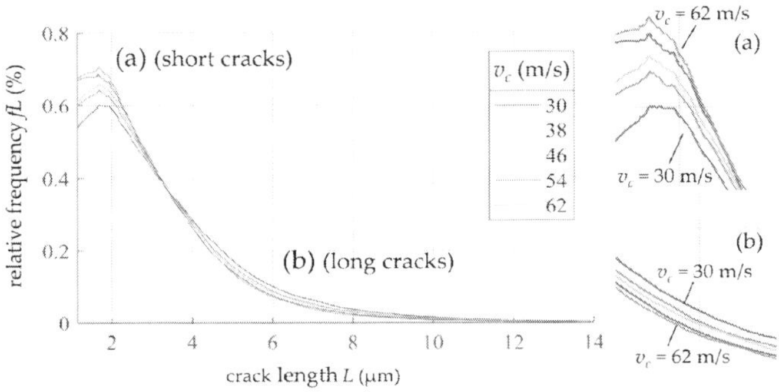

Figure 9: Relative frequencies of the crack length L (filtered data).

After smoothing by a Savitzky–Golay filter, it is easy to see that the distributions differ. As the cutting speed increases, the amount of shorter cracks (Figure 9a) increases and the amount of longer cracks decreases (Figure 9b).

DISCUSSION

Maximum Crack Depth

Typically, the crack density decreases along the wedge, i.e., with increasing depth, following a power law function [15]. At the scale of the surface roughness, the individual cracks intersect and Suratwala et al. describe this as the rubble zone. A few microns below the surface, the cracks can be detected individually. Within our experiments, with the given MRF wedge geometry, the lateral distance between the last cracks could be up to 500 μm. This underlines the importance of size and the slope of the polished wedge.

After an initial decrease in the maximum crack depth with an increasing cutting speed, a seemingly constant depth is obtained above 45 m/s (Figure 7). Besides chip thickness reduction, higher spindle speeds and resulting higher cutting speeds lead to stronger vibrations [16]. This may explain why there is no further improvement of the maximum SSD-depth at cutting speeds higher than 45 m/s. The single-grain load, as the main cause for SSD, seems to be confirmed. In theory, the influence of the material removal rate and thus process parameters such as cutting speed, feed rate and infeed can be easily assessed. In practice, however, the stochastic nature of the grinding wheel as well as other influences such as limited machine rigidity have influence on the maximum crack depth.

Miller et al. have shown that the addition of a small amount of larger grains to a fine slurry for polishing leads to a significant increase in the measured crack depth [4]. The normal force is distributed to a smaller number of particles than in a homogeneous particle distribution and the maximum individual grain loads are increased. In the grinding process, the effective grain size corresponds to the individual grain protrusion, which directly depends on the grain size.

The individually polished and etched wedge surfaces differ considerably, in crack distribution. Figure 10 shows samples with low and high maximum crack depths. The sample with low SSD shows a rather arbitrarily arranged crack pattern (Figure 10a), whereas in case of significantly larger SSD (Figure 10b), the ordered crack formation is particularly remarkable. Large protrusion of individual grains leads to higher individual loads. As a result, it is not possible to assess the process results by means of maximum SSD only in conjunction with the cutting speed e.g., vibrations or variation in grain protrusion.

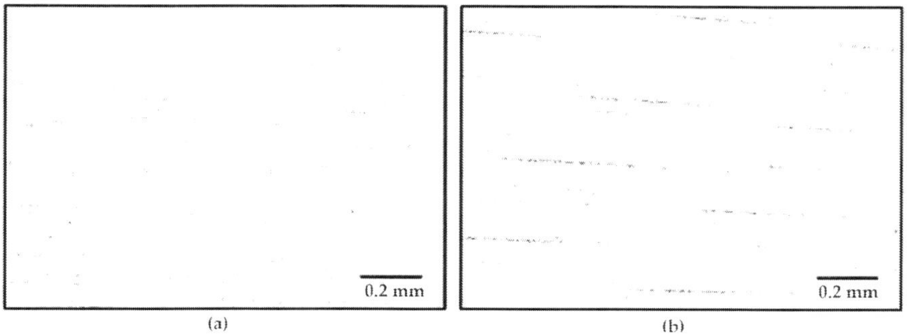

Figure 10: Micrograph of the MRF wedge surface after etching: (a) sample with low SSD (v_c = 54 m/s; c_{max} = 4.6 µm); (b) sample with larger SSD (v_c = 30 m/s; c_{max} = 6.8 µm).

As stated above, the maximum crack depth c_{max} results from one particular or only a few grains. It is thereby subject to a certain variation by the nature of the microstructure of the grinding wheel and the grinding process. Both, crack depth and crack length, depend directly on the grain load. The maximum crack depth c_{max} should therefore also lead to the maximum crack length L_{max}. Within the experiments L_{max} shows a linear decay with depth. Figure 11 shows L_{max} at intervals of 50 µm along the polished wedge. To avoid the influence of outliers by intersecting cracks on the linear fit, the double median crack length L_{Mdn} is set as cut-off criteria.

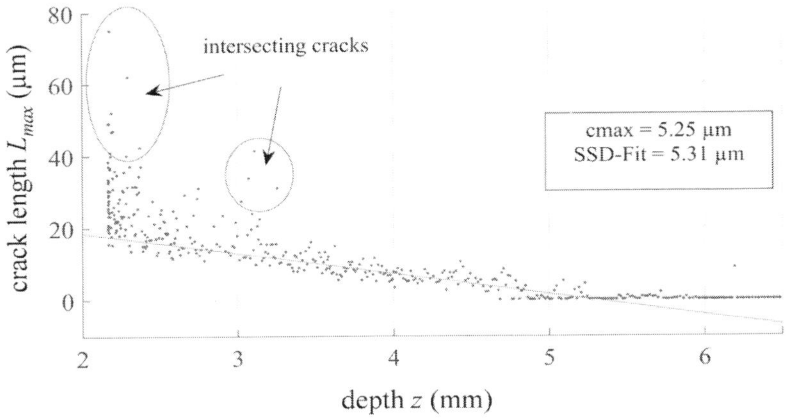

Figure 11: Maximum crack lengths L_{max} as a function of depth (v_c = 54 m/s) with linear fit.

The zero point of the fit provides a very good correlation with the measured maximum crack depths (Figure 12).

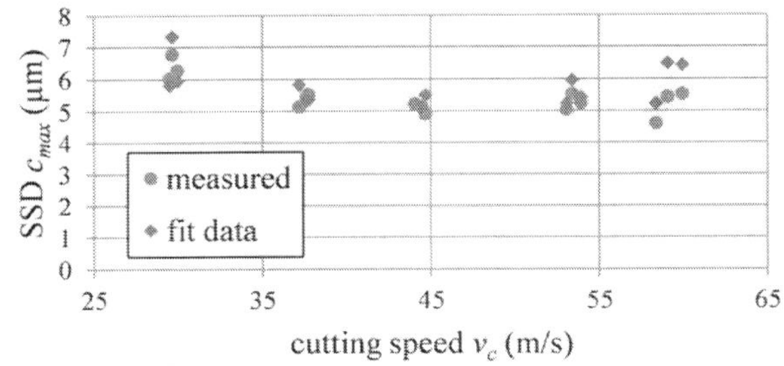

Figure 12: Measured and calculated maximum crack depths.

Fracture Mechanics

Determination of the exact surface topography of a grinding wheel is difficult, even for coarse-grained tools. Furthermore, the surface changes continuously by wear or is completely reshaped by dressing. Therefore, attempts are often made to simplify the grinding process by single grain scratch behavior. The basis for most of the considerations is the work of

Lawn, who investigated the cracking mechanisms of brittle materials under indentation [17]. His experiments showed the formation of median and lateral cracks as a function of loading and unloading on the workpiece surface. He showed that the maximum radial crack length c_r correlates with the load on the indenter, i.e., $P \sim c_r^{3/2}$[18].

$$c_r = \left(\frac{\chi_r P}{K_{Ic}}\right)^{2/3} \text{ with } \chi_r \sim E/H \tag{1}$$

With the indentation constant χ_r (0.046 for fused silica [19]), indenter load P, fracture toughness K_{Ic}, hardness H and Young's modulus E.

Due to the stochastic nature of the microtopography of grinding wheels, loads on a single grain can not be determined. Equation 1, nevertheless, reveals the maximum grain load to be calculated from the maximum radial crack length c_r, which corresponds to the maximum crack depth c_{max}.

$$P = \frac{K_{Ic}\, c_{max}^{3/2}}{\chi_r} \tag{2}$$

Using equation 2 and the measured maximum crack depths (Figure 7), the maximum individual grain loads $P$are about 0.2 N for the grinding wheel applied in this work. This corresponds to approximately 5% of the measured average normal force. In addition to the evaluation of the maximum crack depth, the wedge method also makes it possible to determine the crack length L on the polished surface. Based on the Hertzian indentation mechanics Miller et al. found a correlation between average crack length and max crack depth [4].

Assuming that the diamond grains are idealized spherical abrasives the circles of contact are defined by the Hertzian contact zone, with the contact circle radius a.

$$a = \left(\frac{4}{3}\frac{k}{E}Pr\right)^{1/3} \tag{3}$$

With the elastic mismatch factor k, the applied load P and the radius of the abrasive r. The factor k relates to the different Young's moduli and Poisson's ratio ν of the workpiece and indenter material, where the primes refer to the indenter.

$$k = \frac{9}{16}\left[\left(1 - \nu^2\right) + \frac{E}{E'}\left(1 - \nu'^2\right)\right] \tag{4}$$

In contrast to static indentation, a sliding indenter causes an asymmetry in the tensile stress field. Since the tensile stresses are relatively small near the front end of the indenter the likelihood of cone cracks increases in this region. This also predicts that the crack may not completely encircle the entire contact radius [20]. Miller et al. assumed that the crack propagates over only a quarter of the contact circumference.

$$L \approx \frac{2\pi a}{4} = \frac{\pi}{2}\left(\frac{4}{3}\frac{k}{E'}Pr\right)^{1/3} \tag{5}$$

Finally, the calculated crack length L_{calc} as a function of c_{max} according to Equations (2) and (5) shows appropriate accordance with the determined average crack length L_{Mdn} from the optical evaluation (Figure 13).

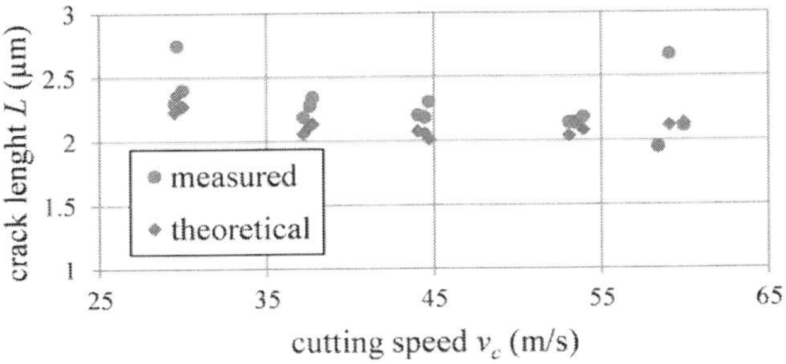

Figure 13: Measured and calculated crack length for different cutting speeds.

CONCLUSIONS

To date there is no alternative to the destructive techniques for the evaluation of SSD in optical glasses. By using modern stitching microscopes and image processing, the MRF wedge method offers the highest accuracy in the determination of crack distribution and maximum crack depth. Furthermore, a thorough investigation of the removal mechanisms of the grinding process of optical glasses is enabled.

Within this work subsurface damage (SSD) was directly determined by the MRF wedge technique for a series of fused silica ground with different cutting speeds from 30 to 62 m/s. To improve the accuracy of the SSD evaluation method wet etching was replaced by dry etching. Furthermore, the form deviation of the ground surfaces was taken into account to determine the actual crack depth. Thus, a significant influence of the cutting speed on SSD for a fine grinding process, as it is found in process chains of optical fabrication, could be demonstrated. With stitching microscopy and image processing the gapless evaluation of the density of cracks and the distributions of crack lengths along the wedge surface has been implemented. Thus, a dependency of the crack length distribution on the cutting speed could be confirmed. Based on the model of Miller [4], the relation between calculated crack length L_{calc} as a function of c_{max} and median crack length L_{Mdn} could be demonstrated for SSD of only a few microns. Furthermore, a relationship between the maximum crack length as a function of depth and the maximum crack depth could be found. This and the different crack patterns of the SSD samples supports the thesis that the maximum crack depth is caused by a few individual grains that also produce the largest crack length.

AUTHOR CONTRIBUTIONS

Georg Schnurbusch, Ekkard Brinksmeier and Oltmann Riemer designed the experiments; Schnurbusch performed the experiments and materials investigation; Georg Schnurbusch and Oltmann Riemer analyzed the data. Ekkard Brinksmeier contributed the experimental infrastructure; Georg Schnurbusch, Ekkard Brinksmeier and Oltmann Riemer discussed the results and wrote the paper.

REFERENCES

1. Hed, P.P.; Edwards, D.F.; Davis, J.B. Subsurface damage in optical materials: Origin, measurement & removal. In Proceedings of the Optical fabrication and testing workshop, Santa Clara, CA, USA, 2 November 1988.
2. Neauport, J.; Ambard, C.; Bercegol, H.; Cahuc, O.; Champreux, J.P.; Charles, J.L.; Cormont, P.; Darbois, N.; Darnis, P.; Destribats, J.; et al. Optimizing fused silica polishing processes for 351 nm high power laser application. In *Laser-Inducted Damage in Optical Materials, Proceedings of the SPIE, Volume 7132, Boulder, CO, USA, 22 September 2008*; Exarhos, G.J., Ristau, D., Eds.; SPIE: Bellingham, WA, USA, 2008.
3. Comley, P.; Morantz, P.; Shore, P.; Tonnellier, X. Grinding metre-scale mirror segments for the E-ELT ground based telescope. *CIRP Annals-Manufacturing Technology* 2011, *60*, 379–382.
4. Miller, P.E.; Suratwala, T.I.; Wong, L.L.; Feit, M.D.; Menapace, J.A.; Davis, P.J.; Steele, R.A. The distribution of subsurface damage in fused silica. In *Laser-Inducted Damage in Optical Materials, Proceedings of SPIE Volume 5991, Boulder, CO, USA, 19 September 2005*; Exarhos, G.J., Guenther, A.H., Eds.; SPIE: Bellingham, WA, USA, 2006.
5. Lambropoulos, J.C. From Abrasive Size to Subsurface Damage in Grinding. *TOPS* 2000, *42*, 17–18.
6. Lucca, D.A.; Brinksmeier, E.; Goch, G. Progress in Assessing Surface and Subsurface Integrity. *CIRP Ann.*1998, *47*, 669–693.
7. Wang, L.; Li, Y.; Han, J.; Xu, Q.; Guo, Y. Evaluating subsurface damage in optical glasses. *JEOS:RP 6* 2011, *6*, 1–16.
8. Menapace, J.A.; Davis, P.J.; Steele, W.A.; Wong, L.L.; Suratwala, T.I.; Miller, P.E. MRF Applications: Measurement of process-dependent subsurface damage in optical materials using the MRF wedge technique. In *Laser-Inducted Damage in Optical Materials, Proceedings of SPIE Volume 5991, Boulder, CO, USA, 19 September 2005*; Exarhos, G.J., Guenther, A.H., Eds.; SPIE: Bellingham, WA, USA, 2006.
9. Menapace, J.A.; Davis, P.J.; Steele, W.A.; Wong, L.L.; Suratwala, T.I.; Miller, P.E. Utilization of magnetorheological finishing as a diagnostic tool for investigating the three-dimensional structure of fractures in fused silica. In *Laser-Inducted Damage in Optical Materials, Proceedings of SPIE Volume 5991, Boulder, CO, USA, 19 September*

2005; Exarhos, G.J., Guenther, A.H., Eds.; SPIE: Bellingham, WA, USA, 2006.

10. Randi, J.A.; Lambropoulos, J.C.; Jacobs, S.D. Subsurface damage in some single crystalline optical materials. *Appl. Opt.* 2005, *44*, 2241–2249.
11. Li, Y.; Zhenga, N.; Li, H.; Houa, J.; Lei, X.; Chena, X.; Yuana, Z.; Guoa, Z.; Wanga, J.; Guob, Y.; Xua, Q. Morphology and distribution of subsurface damage in optical fused silica parts: Bound-abrasive grinding. *Appl. Surf. Sci.* 2011, *257*, 2066–2073.
12. Klocke, F. *Manufacturing Processes 2: Grinding, Honing, Lapping*; Springer: Heidelberg, Germany, 2009; pp. 251–269.
13. Wang, H.; Subhash, G.; Chandra, A. Characteristics of single-grit rotating scratch with a conical tool on pure titanium. *Wear* 2001, *249*, 566–581.
14. Bollinger, L.D.; Gallatin, G.M.; Samuels, J.; Steinberg, G.; Zarowin, C.B. Rapid, non-contact optical figuring of aspheric surfaces with Plasma Assisted Chemical Etching (PACE). In *Advanced Optical Manufacturing and Testing, Proceedings of SPIE 1333, San Diego, CA, USA, 1 July 1990*; Sanger, G.M., Reid, P.B., Eds.; SPIE: Bellingham, WA, USA, 1990.
15. Suratwala, T.I.; Wong, L.L.; Miller, P.E.; Feit, M.D.; Menapace, J.A.; Steele, R.A.; Davis, P.J.; Walmer, D. Subsurface mechanical damage distributions during grinding of fused silica. *J. Non.-Cryst. Solids* 2006, *352*, 5601–5617.
16. Huang, H.; Yin, L. High speed grinding performance and material removal mechanism of silicon nitride. *Initiat. Precis. Eng. Begin. Millenn.* 2002, 416–420.
17. Lawn, B.R.; Swain, M.V. Microfracture beneath point indentations in brittle solids. *J. Mater. Sci.* 1975, *10*, 113–122.
18. Lawn, B.R.; Evans, A.G.; Marshall, D.B. Elastic/plastic indentation damage in ceramics: The mediad/radial crack system. *J. Am. Ceram. Soc.* 1980, *63*, 574–581.
19. Anstis, G.R.; Chantikul, P.; Lawn, B.R.; Marshall, D.B. A critical evaluation of indentation techniques for measuring fracture toughness: I, direct crack measurements. *J. Am. Ceram. Soc.* 1981, *64*, 533–538.
20. Lawn, B.R. Partial cone crack formation in a brittle material loaded with a sliding spherical indenter. *Proc. R Soc. Lond. Ser. A* 1967, *299*, 307–316.

Chapter 10

Subsurface Damage in Polishing Process of Silicon Carbide Ceramic

Yan Gu, Wenhui Zhu, Jieqiong Lin, Mingming Lu , and Mingshuo Kang

School of Mechatronic Engineering, Changchun University of Technology, Changchun 130012, China

ABSTRACT

Subsurface damage (SSD) in the polishing process of silicon carbide (SiC) ceramic presents one of the most significant challenges for practical applications. In this study, the theoretical models of SSD depth are established on the basis of the material removal mechanism and indentation fracture mechanics in the SiC ceramic polishing process. In addition, the three-dimensional (3D) models of single grit polishing are also developed by using the finite element simulation; thereby, the dynamic effects of different

Citation: Gu, Yan; Zhu, Wenhui; Lin, Jieqiong; Lu, Mingming; Kang, Mingshuo. 2018. "Subsurface Damage in Polishing Process of Silicon Carbide Ceramic." Materials 11, no. 4: 506. https://doi.org/10.3390/ma11040506

process parameters on SSD depth are analyzed. The results demonstrate that the material removal was mainly in brittle mode when the cutting depth was larger than the critical depth of the brittle material. The SSD depth increased as the polishing depth and abrasive grain size increased, and decreased with respect to the increase in polishing speed. The experimental results suggested a good agreement with the theoretical simulation results in terms of SSD depth as a function of polishing depth, spindle speed, and abrasive grain size. This study provides a mechanistic insight into the dependence of SSD on key operational parameters in the polishing process of SiC ceramic.

Keywords: SiC ceramic, subsurface damage, polishing process, finite element simulation

INTRODUCTION

A new generation of space-to-ground optical information collection systems requires high ground resolution and a large coverage width, so as to make the optical systems constantly highly lightweight, as well as have a large diameter and an off-axis. This also inspires the development of optical remote sensing camera key components that are made of a high-quality light optical reflector material [1]. Silicon carbide (SiC) ceramic represents a kind of ideal space mirror material by virtue of its high hardness and strength, excellent chemical stability, and good wear resistance [2,3]. However, there remain challenges associated with the inherent properties of SiC, such as its high brittleness and low fracture toughness, which may lead to subsurface microcracks, dislocation, deformation, and residual stresses. This may heavily degrade the mechanical strength and fatigue properties, as well as its adoption of surface integrity [4]. Computer-controlled optical surfacing has been widely used in precision machining of hard and brittle materials. However, the precision polishing process of silicon carbide ceramic is limited in many aspects; its process is simple, and there may be polishing subsurface damage (SSD).

At present, it is common practice to investigate the polishing mechanism by establishing the models of material removal depth based on the Preston equation according to $dh=kpvdt$ [5,6]. However, the constant k (i.e., factor influence coefficient) is an empirical parameter that needs correction depending on the results of time and labor-intensive experiments. In addition, it is also noticed that the Preston equation is a macroscopic model, i.e., it cannot reflect the effect of the lack of abrasive characteristics on the depth of removal. From the microscopic point of view, a number of studies described the polishing process, focusing on the complicated interaction between abrasive grains and the workpiece. For example, Wang et al. established equations describing the relationship between pressure and cutting depth, taking into account abrasive particle size as an important parameter of removing depth; they also gave mathematical models for linear removal strength [7]. Qi et al. considered the impact of the characteristics of abrasive grains on removal depth and linear removal strength when establishing the microscopic models of material removal depth during polishing [8]. These studies, from the perspective of abrasion, will be helpful for better understanding the effect of material removal depth on SSD.

Recently, many theoretical analyses, numerical simulation methods, and experimental methods have been developed for investigating the material removal mechanism and SSD in the process of brittle material processing. In theory, Wang et al. applied the indentation fracture mechanics of brittle materials to establish the predictive models of cutting force and the SSD depth of K9 optical glass during ultrasonic rotary surface machining [9]. Liu et al. reported the mathematical relationship between the depth of subsurface cracks and the processing parameters during the silicon wafer cutting process, according to the principle of indentation fracture mechanics [10]. Esmaeilzare et al. used a cup diamond wheel in the grinding process of the specimens made of Zerodur® glass–ceramic to investigate the influences of grinding parameters on the SSD and surface roughness (SR), and established a statistical model for the prediction of SR and SSD depth [11]. Chen et al. also proposed a new model to analyze the relationship between SR and SSD depth on the basis of grinding kinematics analysis and the indentation fracture mechanics of brittle materials that took the wheel spindle vibration into account [12]. Recently, the simulation technology of hard and brittle materials in precision machining has been applied more and more in the research of SSD. For example, Zhu et al. utilized single-grit simulation to investigate the initiation and propagation of individual cracks under controllable maximum undeformed chip thickness in SiC grinding [13]. Liu et al. conducted single-grit engagement simulations to investigate the effects of the high speed on the surface/subsurface cracks and grinding forces [14]. Komanduri et al. carried out a molecular dynamics (MD) simulation of nanometric cutting with tools of different edge radii relative to the depth of cut, and investigated the variations of cutting forces, thrust forces, the force ratio, the specific energy, and the subsurface deformation with the tool geometry and depths of cut; their results found that they all have a significant influence [15]. Dai et al. employed three-dimensional molecular dynamics (MD) simulations to investigate the effect of tool geometry on subsurface damage and material removal in nanometric cutting single-crystal silicon [16]. Li et al. also used MD simulations to investigate the effects of grinding speed on grinding damage and grinding surface integrity through the analysis of chips, dislocation motions, and phase transition in a single-crystal silicon grinding process [17]. In experimental testing, Lucca et al. employed a variety of techniques to assess the surface alterations that occur as a result of processing, and they presented examples of the processing of metals, ceramics, and glasses in order to assess the nature and extent of surface alterations [18]. Agarwal et al. studied the grinding characteristics, surface integrity, and material removal mechanisms in diamond wheel-ground SiC surfaces, and the study combined surface roughness (SR), surface damage, and subsurface damage with grinding parameters, which provided valuable insights for the material removal mechanism and the dependence of grinding damage on grinding conditions [19]. Jiang et al. carried out a series of K9 glass precision grinding tests, and studied the influence of grinding parameters on SSD. The experimental results show that the material removal rate was closely related to SSD [20]. Blaineau et al. obtained the relationship between SSD depth and grinding force by comparing experimental results with discrete element method simulation results [21]. Wang et al. used the method of theoretical analysis, numerical simulation, and experimental testing to analyze the SSD during the high-speed grinding of brittle materials [22]. However, theoretical studies on the SSD of hard brittle materials in the polishing process have been reported with limited success, which is likely because the majority of existing studies rarely take systematic theoretical models for SSD and three-dimensional (3D) dynamic analysis into consideration. Thus, it will be highly desirable to develop the methods based on finite element simulation for more accurately predicting and controlling the SSD depth during the polishing of hard and brittle materials.

The main purpose of this study is to establish the theoretical models of SSD by taking into account the removal mechanism for the SiC ceramic polishing process. In order to obtain the relationship between SSD and processing parameters (polishing depth, tool speed, and abrasive grain size) on a quantitative basis, the 3D finite element simulation is performed to simulate the process of single-grit polishing SiC ceramic. A series of experiments were carried out to validate the theoretical and simulation results.

THEORETICAL ANALYSIS

SSD Models

The contact polishing of hard and brittle materials can be simplified in an abstract way into a single-grit scratch process [23], which can be schematically illustrated in Figure 1. Figure 2 shows the lateral cracks in association with material removal and surface formation. The median cracks causes SSD, and thus, the degradation of strength of the material [24]. cm is the median crack length at the bottom of the plastic zone under peak load or critical conditions in the scratch process; cl is the transverse crack depth from the bottom of the plastic deformation zone; hi is the depth of penetration of the abrasive particles and the workpiece; and α is abrasive grain apex angle.

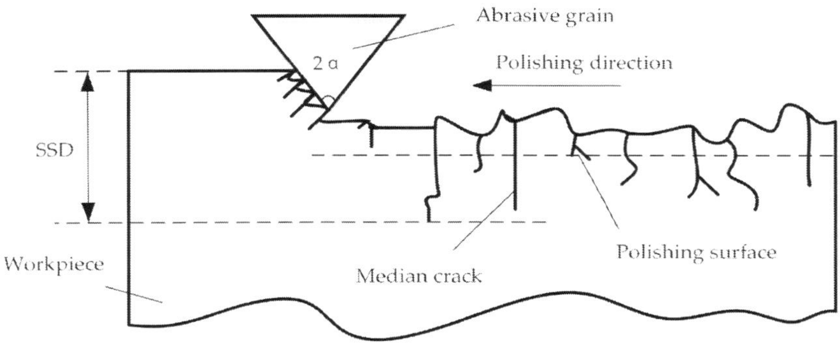

Figure 1: Scratch process of single abrasive grain.

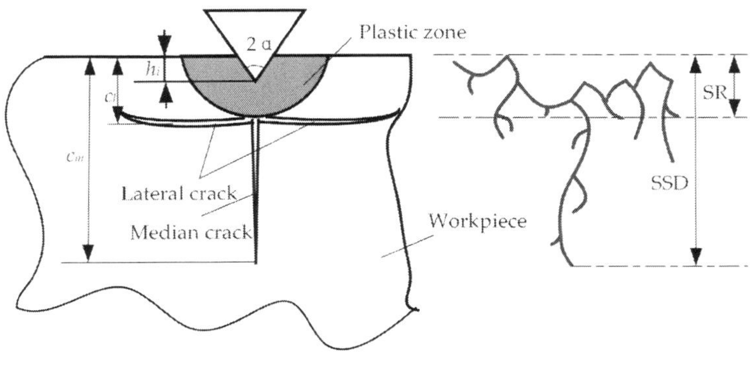

Figure 2: Subsurface damage model.

The median crack can be classified as typical SSD, as it is perpendicular to the polishing surface, and extends to the machined plane. According to the indentation fracture mechanics, the depth of median crack c_m can be correlated to material properties, the geometry of the abrasive grain, and penetration depth [25] according to:

$$c_m = 0.206\frac{(E \cdot H_s)^{1/3}}{(K_c \cdot \beta)^{2/3}}(\cot\alpha)^{4/9}(\tan\alpha)^{4/3} \cdot (h_i)^{4/3}, \quad (1)$$

where H_s (GPa) is the scratch hardness, K_c (3.5 MPa·m$^{(1/2)}$) is the fracture toughness, and β (0.363) is the material parameter determined by elastic recovery.

The SSD depth can be derived from the geometric relationship between the abrasive and the workpiece, which is shown in Figure 2 as:

$$\text{SSD} = \max(c_m), \quad (2)$$

Models of Single-Grit Penetration Depth

The Height Distribution of ABRASIVE Protrusion

Polishing is a complex material removal process involving rubbing, scratching, plowing, and cutting, which is necessarily analyzed by using probability statistics. The abrasive grain sizes are determined in terms of the value of grit (generally indicated by M) [26]. The maximum diameter d_{max} and minimum diameter d_{min} of the abrasive grains can be determined when the grit is known. The mean diameter d_m is defined by $d_m=(d_{max}+d_{min})/2$. The number of the abrasive grains per unit area can be calculated as:

$$N = \frac{6V_g}{\pi {d_m}^2}, \quad (3)$$

where V_g (%) is the grain ratio, and the structure number S stands for the volume ratio of the grains in the whole element volume. The relationship between S and the volume ratio is $V_g=2(31-S)$, for instance: V_g is 62 and 44, when S = 0 and S = 9, respectively [27].

Massive experiments have shown that both d_{max} and d_m are very close to the maximum and the average heights of the abrasive protrusion, respectively [7]. Hence, the distribution of the abrasive grain protrusion heights of the polishing tool surface accords well with Gaussian distribution as:

$$f(h) = \frac{1}{\sqrt{2\pi}\sigma} e^{\frac{-(h-u)^2}{2\sigma^2}}, \quad (4)$$

where h is the height of the abrasive protrusion, the mean value $u=d_m$, and the standard deviation $\sigma=(d_{max}-d_m)/3$. Values of d_m, d_{max}, and σ for different grain sizes are shown in Table 1.

Table 1: Values of d_m, d_{max} and σ for different grit designations

M (Mesh Size)	dmax (μm)	dm (μm)	σ (μm)
180	84.4	47.3	12.4
320	47.5	21.1	8.8
400	38.0	15.5	7.5
600	25.3	8.8	5.5
800	19.0	5.9	4.4
1000	15.2	4.3	3.6
1500	10.1	2.4	2.6
2000	7.6	1.6	2.0
4000	3.8	0.6	1.1
6000	2.5	0.35	0.7
10,000	1.5	0.17	0.4

Figure 3 illustrates a schematic diagram of the grain protrusion heights distribution in the abrasive tool surface. The origin of the coordinate system is fixed on the horizontal position of d_m, and h_0 is the vertical distance from the origin 0 and workpiece surface. It is based on the geometric relationship $h_0=3\sigma-m$, where m is the indentation depth $m\in[0,6\sigma]$. Considering that $\int_{-3\sigma}^{3\sigma} f(h)dh=0.9973$, the probability density function of the abrasive grain protrusion heights distribution can be written as:

$$f(h) = \begin{cases} \frac{1}{\sqrt{2\pi}\sigma} e^{\frac{-h^2}{2\sigma^2}} & |h| \le 3\sigma \\ 0 & |h| > 3\sigma \end{cases}, \quad (5)$$

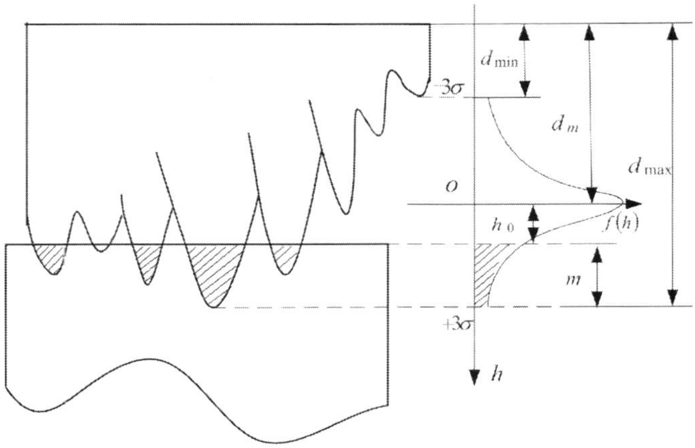

Figure 3: Schematic illustration of the abrasive grain heights distribution.

The Number of Effective Grains

Through the above analysis, we can conclude that not all of the grains are involved in the polishing process. The material removal process will happen only when the protrusion height is bigger than h_0. From Equation (5), the particle distribution probability of the abrasive grain height higher than h_0 in unit area is obtained:

$$P\{h_0 < h < 3\sigma\} = \int_{h_0}^{3\sigma} f(h)\mathrm{d}h, \tag{6}$$

Combining Equations (3) and (6) gives the number of effective grains in the following form:

$$N_r = \frac{6V_g}{\pi d_m{}^2}\int_{h_0}^{3\sigma} f(h)\mathrm{d}h, \tag{7}$$

The Depth of a Single Abrasive Grain

The contact deformation between diamond abrasive grains and the SiC workpiece can be classified as the plastic deformation [28]. According to the theory of elastic mechanics, the plastic contact force between a single abrasive grain and workpiece can be described as:

$$F_a = 2h_i{}^2 H_w, \tag{8}$$

where H_w is the hardness of the workpiece, and h_i is the penetration depth of the abrasive particles.

According to the principle of mechanical equilibrium, the contact force between the polishing tool and the workpiece (the contact force between the effective abrasive particles and the workpiece) is equal to the polishing force:

$$F = N_r F_a = 2{h_i}^2 H_w \frac{6V_g}{\pi {d_m}^2} \int_{h_0}^{3\sigma} f(h)\mathrm{d}h, \tag{9}$$

Therefore, the penetration of a single abrasive grain into the workpiece is obtained as:

$$h_i = \sqrt{\frac{F}{2H_w \frac{6V_g}{\pi {d_m}^2} \int_{h_0}^{3\sigma} f(h)\mathrm{d}h}} = \sqrt{\frac{\pi F {d_m}^2}{12 H_w V_g \int_{h_0}^{3\sigma} f(h)\mathrm{d}h}}, \tag{10}$$

Substituting Equation (10) for Equation (2) gives that:

$$\begin{aligned} SSD &= \max\left[0.206\frac{(E\cdot H_s)^{1/3}}{(K_c\cdot\beta)^{2/3}}(\cot\alpha)^{4/9}(\tan\alpha)^{4/3}\cdot\left(\frac{\pi F {d_m}^2}{12H_w \int_{h_0}^{3\sigma} f(h)\mathrm{d}h}\right)^{2/3}\right] \\ &= \max\left[\lambda\cdot\left(\frac{\pi F {d_m}^2}{12H_w V_g \int_{h_0}^{3\sigma} f(h)\mathrm{d}h}\right)^{2/3}\right] \end{aligned}, \tag{11}$$

where λ is the proportional coefficient related to material properties and head geometry, whose value can be calculated by:

$$\lambda = 0.206\frac{(E\cdot H_s)^{1/3}}{(K_c\cdot\beta)^{2/3}}(\cot\alpha)^{4/9}(\tan\alpha)^{4/3}, \tag{12}$$

The material characteristic parameters of SiC ceramics are shown in Table 2.

Table 2: The physical parameters of silicon carbide (SiC) material

Material Properties Parameters	Value
Density ρ (kg/m³)	3215
Elastic modulus E (GPa)	454
Poisson's ratio	0.25
Yield strength σ (MPa)	620
Specific heat [J/(kg·K)]	526.3

Conductivity [W/(m·K)]	180
Hardness (GPa)	29.4

Analysis for Dynamic Parameters

Equation (9) shows that the depth of a single abrasive cut into the workpiece is related to the particle size and contact force of the abrasive. Figure 4 schematically illustrates the diagram of the motion between the polishing head and the workpiece. The force as a function of velocity at any point on the polishing head can be calculated by a kinematic geometric relationship. As shown in Figure 4, the tangential unit vector can be obtained at the radius r as:

$$\hat{\tau} = -\sin\theta\,\vec{i} + \cos\theta\,\vec{j}, \qquad (13)$$

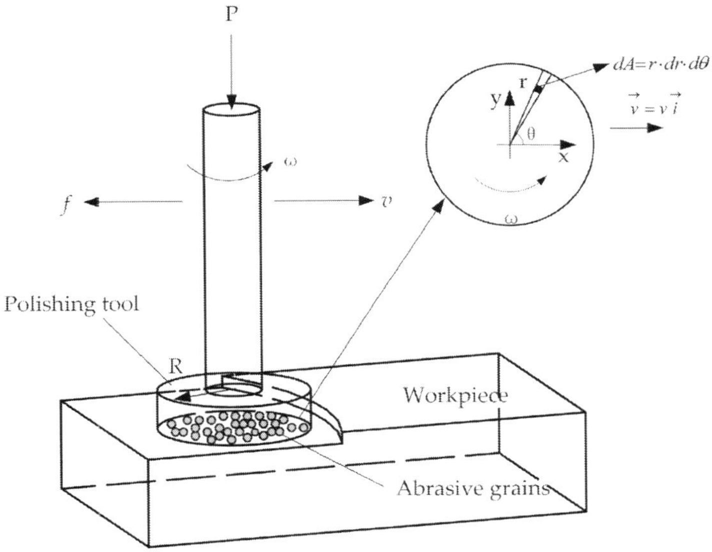

Figure 4: Schematic diagram of the motion between the polishing tool and the workpiece.

The linear velocity $\vec{v}$ at r can be written as:

$$\vec{v_r} = \omega r \cdot \hat{\tau} = -\omega r \sin\theta\,\vec{i} + \omega r \cos\theta\,\vec{j}, \qquad (14)$$

Since the feed rate of the polishing head is $\vec{v} = v\vec{i}$, the absolute rate $\vec{v_a}$ at radius r should be given as:

$$\vec{v_a} = \vec{v} + \vec{v_r} = (v - \omega r \sin\theta)\vec{i} + \omega r \cos\theta \vec{j}, \tag{15}$$

Therefore, the unit vector at radius r can be given as:

$$\hat{v}_a = \frac{\vec{v_a}}{v_a} = \frac{1}{\sqrt{v^2 + \omega^2 r^2 - 2v\omega r \sin\theta}}\left[(v - \omega r \sin\theta)\vec{i} + \omega r \cos\theta \vec{j}\right], \tag{16}$$

The normal stress on the disk surface is σ=P/A=P/(πR²). The frictional force at dA is df=μσdA, and the vector form is $d\vec{f} = -\hat{v}_a \cdot df$. Thus, the friction force of the disc can be expressed as:

$$\vec{f} = \int d\vec{f} = -\iint \frac{(v - \omega r \sin\theta)\vec{i} + \omega r \cos\theta \vec{j}}{\sqrt{v^2 + \omega^2 r^2 - 2v\omega r \sin\theta}} \cdot \frac{\mu P}{\pi R^2} \cdot r \cdot dr \cdot d\theta, \tag{17}$$

Clearly visible is the resultant force of zero on the disk in the y-direction. In order to simplify Equation (17), we consider $1/\sqrt{v^2 + \omega^2 r^2 - 2v\omega r \sin\theta}$ for McLaughlin expansion at v=0:

$$\begin{aligned} \frac{1}{\sqrt{v^2+\omega^2r^2-2v\omega r\sin\theta}} &= \frac{1}{\sqrt{v^2+\omega^2r^2-2v\omega r\sin\theta}}\Big|_{v=0} - \frac{\omega r\sin\theta - v}{(v^2+\omega^2r^2-2v\omega r\sin\theta)^{3/2}}\Big|_{v=0} \cdot v + (v) \\ &= \frac{1}{\omega r} + \frac{\sin\theta}{\omega^2 r^2} v + (v) \end{aligned}, \tag{18}$$

where o(v) is the infinitesimal of higher order.

Substituting Equation (18) for Equation (17) gives the friction of the disk as:

$$\vec{f} = -\frac{\mu P v}{\omega R}\vec{i}, \tag{19}$$

As can be seen from the geometric relations in Figure 4, the force between the polishing head and the workpiece can be written as a function of friction and pressure as:

$$F = \sqrt{f^2 + P^2} = P \cdot \sqrt{1 + \frac{\mu^2 v^2}{\omega^2 R^2}}, \tag{20}$$

Substituting Equation (20) to Equation (11) gives:

$$SSD = \max\left[\lambda \cdot \left(\frac{\pi P d_m^2}{12 H_w V_g \int_{h_0}^{3\sigma} f(h)dh}\right)^{2/3} \cdot \left(1 + \frac{\mu^2 v^2}{\omega^2 R^2}\right)^{1/3}\right], \tag{21}$$

Analysis for SSD Depth Models

In brittle material processing, if the material is removed to the critical value of brittle fracture, the brittle mode (transverse crack or intermediate crack) occurs. The critical cutting thickness hc can be written as [29]:

$$h_c = 0.15\left(\frac{E}{H_w}\right)\left(\frac{K_c}{H_w}\right)^2, \tag{22}$$

For $h_i > h_c$, the fragile mode takes place in the SiC polishing process, and thus, the value of h_c can be calculated as 0.0328 μm. The h_i value can be estimated as 0.5878 μm according to Equation (10), where P=10N and d_m=8.8 μm. In this case, the material removal is in the brittle mode, i.e., $h_i > h_c$.

In the brittle mode polishing process, Equation (21) can be used for initially determining the processing parameters. The formula gives the relationship between the parameters of the wheel speed ω, the feed speed v, the apex angle of abrasive grain α, and the applied load F and SSD depth. With the increasing of feed speed, particle angle and load, the SSD depth increases; however, with the increasing of polishing speed, the SSD depth decreases. In addition, there is a certain proportion of the external load on the surface of the workpiece in the polishing process, and the polishing depth in the experiment.

NUMERICAL SIMULATION

Constitutive Models

The polishing process of the SiC ceramic produces a new surface as a result of brittle fracture. Thus, it creates an easy way to fabricate subsurface microcracks, dislocations, deformation, and residual stress damage. Correspondingly, it is also necessary to establish a fracture failure model to predict subsurface damage. In this study, the JH-2 constitutive model was used to describe the fracture damage of ceramic polishing [14,30] as:

$$\sigma = \left(1 + C\ln\dot{\varepsilon}\right)\sigma_{HEL}\left[A\left(\frac{P+T}{P_{HEL}}\right)^N - D\left(A\left(\frac{P+T}{P_{HEL}}\right)^N - B\left(\frac{P}{P_{HEL}}\right)^M\right)\right], \tag{23}$$

where σ (GPa) is the actual equivalent stress of SiC, σ_{HFL} (GPa) is the equivalent stress at Hugoniot, P is the hydrostatic pressure, and P_{HEL} (GPa) is the hydrostatic pressure at Hugoniot. A, B, C, M, and N denote the predetermined parameters of the material. D is the material damage coefficient, and T (GPa) is the maximum hydrostatic tensile strength.

The hydrostatic pressure of the intact material can be given by [29]:

$$P = \begin{cases} K_1\mu + K_2\mu^2 + K_3\mu^3, & if\ \mu \geq 0 \\ K_1\mu, & if\ \mu \leq 0 \end{cases}, \tag{24}$$

where $\mu=\rho/\rho_0-1$, and K_1, K_2, and K_3 are the constants.

When the brittle material begins to damage (D > 0), there will be a volume expansion that equals the increase in ΔP. Thus, the hydrostatic pressure to produce damage can be written as [30]:

$$P = K_1\mu + K_2\mu^2 + K_3\mu^3 + \Delta P, \tag{25}$$

Based on the perspective of energy loss, the pressure increment is given by [13]:

$$\Delta P_{t+\Delta t} = -K_1\mu_{t+\Delta t} + \sqrt{(K_1\mu_{t+\Delta t} + \Delta P_t)^2 + 2\beta K_1 \Delta U}, \tag{26}$$

where β (0≤β≤1) is the conversion factor of elasticity loss energy and hydrostatic pressure potential, and ΔU(GPa) is the internal energy increment.

Geometrical and physical separation constitute the two main criteria for the separation of the debris and the workpiece. The separation criterion of the JH-2 constitutive equation belongs to the physical separation, which is suitable for characterizing the failure of brittle materials. The material failure parameter ω is defined as [31]:

$$\omega = \sum \frac{\Delta\bar{\varepsilon}^{pl}}{\bar{\varepsilon}_f^{pl}(P)}, \tag{27}$$

where $\Delta\bar{\varepsilon}^{pl}$ is the increment of the equivalent plastic strain, and $\bar{\varepsilon}_f^{pl}(P)$ is the failure strain under the pressure P, that is [31]:

$$\bar{\varepsilon}_f^{pl}(P) = D_1(P^* + T^*)^{D_2}, \bar{\varepsilon}_{f,\min}^{pl} \leq \bar{\varepsilon}_f^{pl} \leq \bar{\varepsilon}_{f,\max}^{pl}, \tag{28}$$

where D1 and D2 are the model constants, and D=ω when the JH-2 model is employed. Supposing that the brittle material undergoes failure (D > 1); then, the failure element is removed from the mesh to achieve debris separation.

Table 3 gives the typical constants for the constitutive and failure models, which are determined through a series of experiments.

Table 3: The model constants of SiC [13]

Constitutive Model	ρ_0(kg/m^3)	G(GPa)	A	N	B	M	C
	3215	193	0.96	0.65	0.35	1.0	0.009
	σ^i_{max}(GPa)	σ^f_{max}(GPa)	HEL(GPa)	P_{HEL}(GPa)	β	$\dot{\varepsilon}_0$	T(GPa)
	12.2	1.3	11.7	5.13	1.0	1.0	0.75
	β	K_1(GPa)	K_2(GPa)	K_3(GPa)			
	1.0	220	361	0			
Failure Model	D_1	D_2	$\bar{\varepsilon}^{pl}_{f,max}$	$\bar{\varepsilon}^{pl}_{f,min}$	FS	*Damage*	
	0.48	0.48	1.2	0.0	0.2	0	

Simulation Methods

In this study, 3D finite element models are developed using the finite element software ABAQUS/Explicit (Abaqus Inc., Providence, RI, USA). The JH-2 ceramic model is imported by custom materials, which can be used for discussing the relationship among the process parameters and stress distribution, surface morphology, surface, and SSD. In the realistic polishing process, the material removal is produced by the interaction of a large number of random particles distributed on the polishing head and the workpiece. In order to simplify the analysis, the single-grit simulation is used to investigate the subsurface cracks during SiC polishing.

A 3D model is utilized to visualize the initiation and propagation of transverse and median cracks (Figure 5). The size of the workpiece is set 30 μm × 10 μm× 20 μm, which is discretized with a mesh consisting of C3D8R elements. To maximize the calculation efficiency, the model is triturated, and the element is divided densely in the cutting path but has low density in the other regions. The grid encryption area is larger than the contact area between the abrasive particles and the workpiece, so as to prevent the abrasive particles and larger grid contact simulation deviation. The bottom of workpiece is fully constrained, and the boundary condition of abrasive speed is given. The tangential and normal friction on the abrasive–workpiece interface should be considered, which give the friction coefficient of 0.3. In dynamic analysis, a small number of units with small size control the steady time increment. To make the computation more time and cost-effective, the method of mass scaling is adopted. The scaling factor is set as 30. In the simulation process, we only consider the machining effect whilst neglecting the chemical action, the temperature effect of the polishing fluid, and the coupling effect between the abrasive particles. A diamond polishing tool with a diameter of 10 mm is used in the experiment, and the polishing speeds are represented by the maximum linear velocity of the polishing tool. The polishing simulation parameters are shown in Table 4.

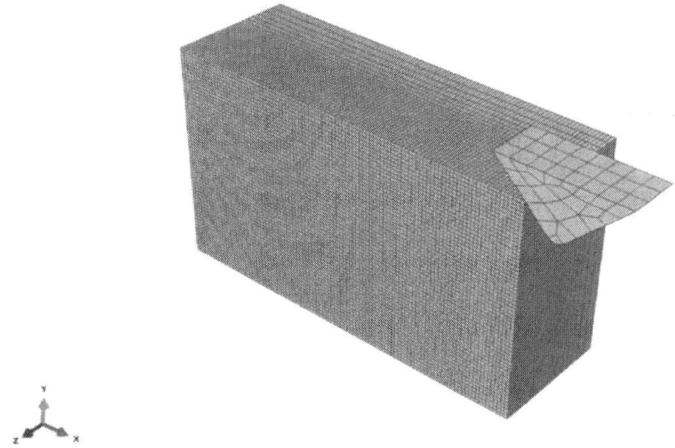

Figure 5: The geometric structure of the three-dimensional (3D) finite element model.

Table 4: Parameters for numerical simulation

Polishing Parameters	Value
Polishing depth a_p (μm)	1, 2, 3, 4
Polishing speed V_s (mm/s)	523, 733, 1151, 1364
Abrasive grain size h (μm)	5, 6, 7, 8

Simulation Results and Discussion

Analysis for Single-Grit Polishing Process

Brittle materials are mainly used to remove materials under the action of fracture and friction. Figure 6 shows the brittleness removal process of SiC ceramic polishing for the polishing depth a_p = 1 μm, the polishing speed V_s = 1151 mm/s, the apex angle of abrasive grain $\alpha = 90$ °, as well as the height of the abrasive grain (h = 7 μm). In the initial stage, both the front and bottom of the abrasive particles start to produce tiny cracks, indicating the removal of the workpiece in a brittle mode. As shown in Figure 6b, as the abrasive moves in the cutting direction, the cutting effect eliminates the crack at the front of the abrasive grain, but is unable to eliminate that below the abrasive grains. The initiation cracks begin to gather and expand on the subsurface, forming a median crack. In this way, the SSD tends to be deepened. As shown in Figure 6c, with further cutting of the abrasive grain, a transverse crack is produced near the median crack. Owing to the squeezing and shearing effect of the abrasive particles, the stress concentration occurs on the tip and the side of the crack. The abrasive particles continue to move forward (Figure 6d), and the transverse cracks expand in the direction of the stress concentration. The stress concentration on the side of the median crack also germinates the transverse crack. When the crack is extended to a certain length, it is connected under the surface of the workpiece, resulting in the brittle removal. Figure 6e–h demonstrates that the crack undergoes three stages under the action of abrasive cutting—i.e., initiation, expansion, and cracking—to form a median crack and a transverse crack. The cracks are interconnected with the processed subsurface to form the machined surface. Meanwhile, the SSD is caused by the median crack, which is unconnected with the lateral crack.

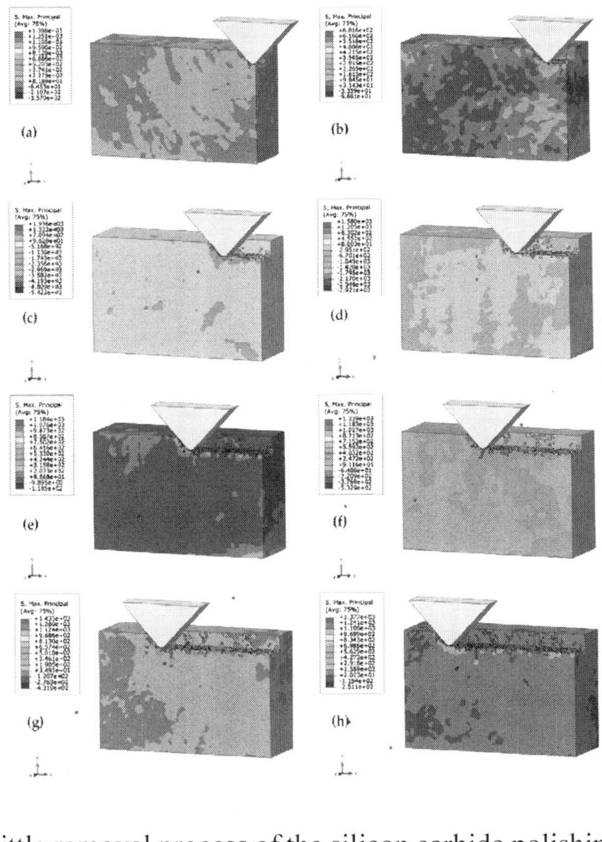

Figure 6: Brittle removal process of the silicon carbide polishing: (a) t = 2.63 × 10^{-6} s; (b) t = 5.25 × 10^{-6} s; (c) t = 7.88 × 10^{-6} s; (d) t = 1.04 × 10^{-5} s; (e) t = 1.31 × 10^{-5} s; (f) t = 1.58 × 10^{-5} s; (g) t = 1.86 ×10^{-5} s; (h) t = 2.13 × 10^{-5} s.

Effect of Polishing Depth on SSD Depth

Figure 7a–d shows the subsurface topographies of the machined workpiece at stepwise polishing depths of 1 μm, 2 μm, 3 μm, and 4 μm (α = 90 °, V_s = 1361 mm/s, h = 7 μm). For a_p = 1 μm, there is no observation of obvious subsurface cracks in the workpiece. a_p = 2 μm further increased the contact area of the abrasive grain and the workpiece. The edge cracks resulted from the median crack, whereas the edge damage was caused by the cutting action. Figure 7c,d illustrates the obvious damage caused by median cracks at a_p = 3 μm and 4 μm, while the lateral cracks caused the collapse under the effect of the cutting force. Taken together, the increase in polishing depth not only degraded the smoothness of the workpiece surface, it also destructed the subsurface and strength of the materials.

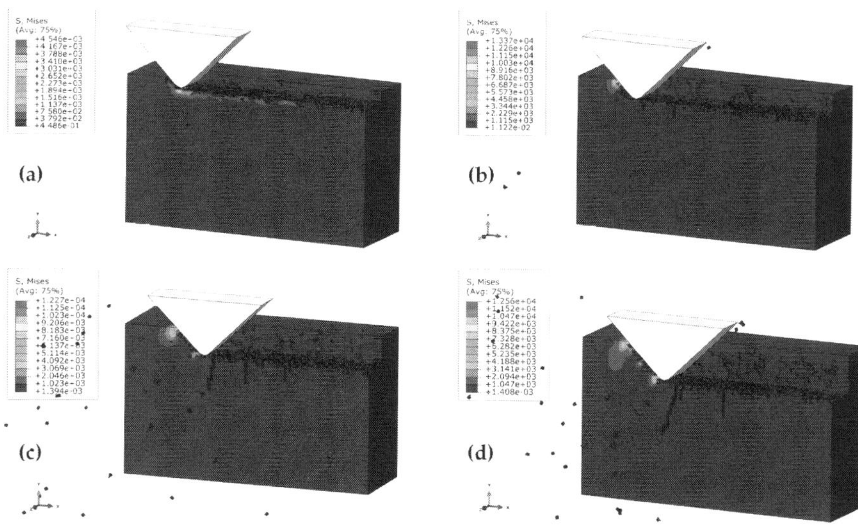

Figure 7: Effect of polishing depth on the depth of subsurface damage (SSD) at the polishing speed of 1361 mm/s, and the height of the abrasive grain of h = 7 μm: (a) a_p = 1 μm; (b) a_p = 2 μm; (c) a_p = 3 μm; (d) a_p = 4 μm.

Effect of Polishing Speed on SSD Depth

Figure 8 illustrates the subsurface topographies of the machined workpiece at various polishing speeds (523–1361 mm/s) at the conditions of a_p = 2 μm, α = 90°, and h = 7 μm. For polishing speed V_s = 523 mm/s, there has been obvious subsurface cracks, and the edge of the workpiece is severely damaged. In this case, the SSD depth is approximately 4.9 μm. When the polishing speed was increased to 733 mm/s, the edge breakage tended to be weakened (SSD depth of about 3.7 μm). Further increasing the polishing speed to 1151–1361 mm/s led to the disappearance of edge breakage, and a resultant decrease in SSD. In other words, the quality of the subsurface can be improved at a relatively higher polishing speed. Herein, the polishing speed can be optimized to be 1151 mm/s, based on the above conditions.

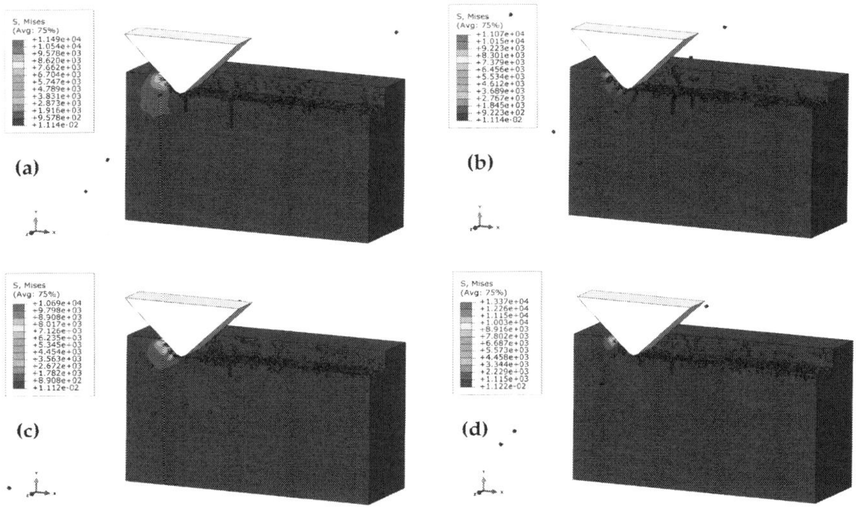

Figure 8: Effect of polishing speed on SSD depth at the polishing depth of 2 μm, and the height of the abrasive grain h = 7 μm: (a) V_s = 523 mm/s; (b) V_s = 733 mm/s; (c) V_s = 1151 mm/s; (d) V_s = 1361 mm/s.

Effect of Abrasive Grain Size on SSD Depth

Figure 9 shows the morphologies of subsurface cracks at the abrasive height h = 5–8 μm. It is found that there is no obvious subsurface crack on the surface of the workpiece when the abrasive heights h = 5 μm, and the polishing surface is relatively smooth. There is a depth of damage caused by tiny cracks on the subsurface, and the presence of stress concentration reduced the material strength. From Figure 9a–d, it can be seen that the increase of the abrasive height will cause the propagation at the subsurface crack, and even aggravate the damage of the edge of the workpiece. Therefore, in actual processing, the small size of the abrasive can be selected to reduce the SSD.

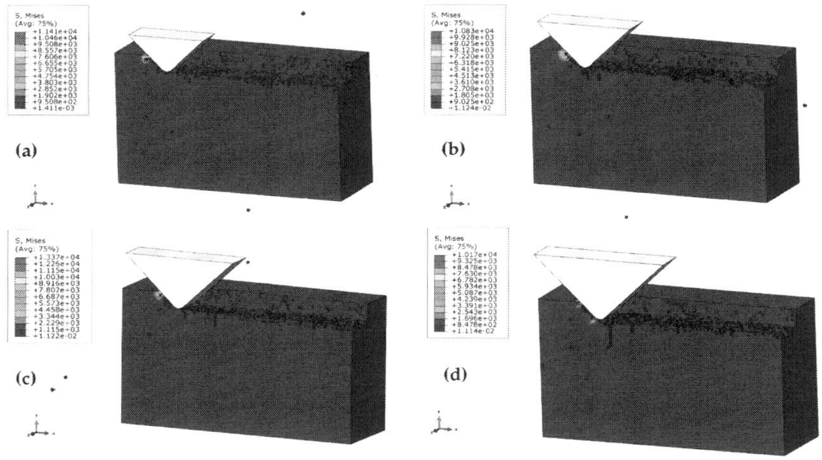

Figure 9: Effect of abrasive grain size on SSD depth at the polishing depth of 2 μm and V_s = 1361 mm/s: (a) h = 5 μm; (b) h = 6 μm; (c) h = 7 μm; (d) h = 8 μm.

VALIDATION OF EXPERIMENTAL RESULTS

Experimental Preparation

We performed the polishing experiments by using an in-house five-axis linkage computer-controlled precision polishing machine equipped with a diamond polishing tool (diameter of 10 mm) and polishing paste with different sizes of diamond grits (Figure 10). The rotating speed of the spindle can reach 3000 rpm, accounting for the maximum linear velocity of 1570 mm/s. The samples used here are the square-shaped pressureless sintered SiC ceramic, which is sized 10 mm × 10 mm × 2 mm.

Figure 10: Optical photograph of experimental setups.

We studied the dependence of SSD on three parameters, i.e., polishing depth (a_p = 1 μm, 2 μm, 3 μm, and 4 μm), spindle speed (V_s = 1000 rpm, 1400 rpm, 2200 rpm, and 2600 rpm), and abrasive grain size (W0.5, W1.5, W2.5, and W3.5), in the polishing process. The SSD depth was determined through the observation of a cross-section microscopy. Following all of the measurements, the SiC samples were chemically eroded by HF solution (1 vol. H_2O dest, 1 vol. HCl 38%, and 1 vol. HF 40%) to enable the visibility of subsurface cracks. After cleaning by ultrasonication several times, the cross-section images of the specimen were observed with a scanning electron microscope (SEM) (JSM-6700F, JEOL, Tokyo, Japan). Thereby, the micrographs of 10 different parts under each parameter were selected for the identification of the maximum depth of cracks (maximum depth of SSD) on a quantitative basis.

Experimental Results and Discussion

Figure 11a–d show the typical SEM images of subsurface cracks beneath the ground surface at a_p = 3 μm, V_s= 2200 r/min, and W2.5. Due to the coexistence of Si and the SiC phase involved in the material, the stress state would become very complex after the interaction between grains and the specimen, and thus, the direction of the cracks propagation would be random. After the propagation of these cracks, a large number of dimples were generated near the surface of the workpiece. Some cracks extended to the free surface to remove the material. As shown in Figure 11a, these pits were smooth, and other parts of the cracks continued to expand downward at the bottom of the pits. At some relatively weak points, the cracks extended to deeper locations, where the cracks were scattered. Although the interior of the material was not absolutely uniform, there were some inherent defects. In addition, the individual cracks were also found on the sites far away from the surface of the material. These should originate from the inherent defects produced within the material that were extended and driven by the polishing force (Figure 11d).

Substituting the process parameters in the experiment into Equation (21), the predicted values of the maximum SSD depth are obtained. As shown in the Table 5, the predicted and measured values of maximum SSD depth are listed. Figure 12 demonstrates the variations of maximum SSD depth with polishing depth a_p, spindle speed V_s, and an abrasive grain size that includes measured and predicted values.

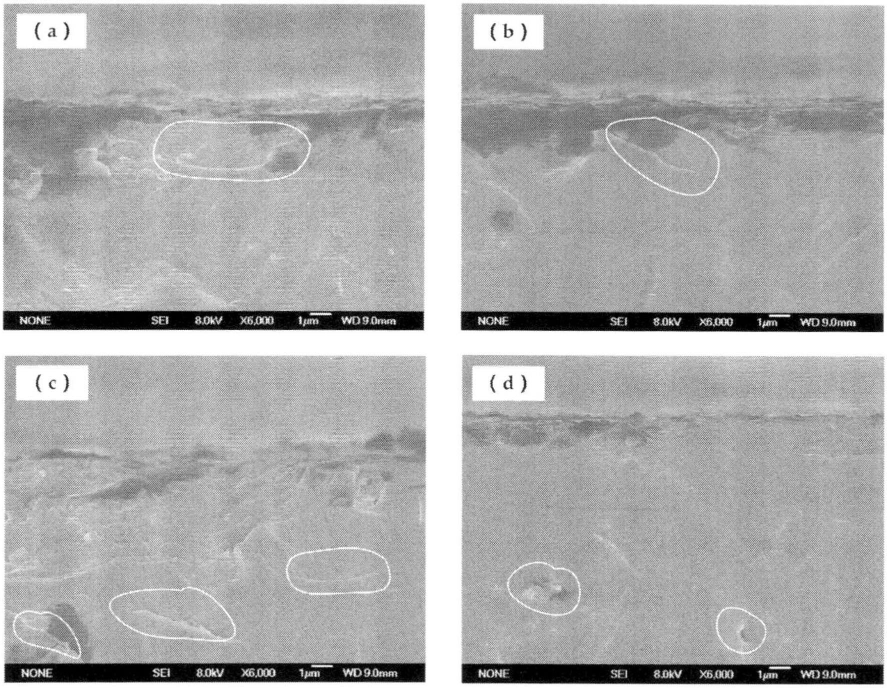

Figure 11: SEM images of SSD morphology of SiC for a_p = 3 μm, V_s = 2200 r/min, and W2.5: (a) near the finished surface; (b) bottom of the pit; (c) extended surface; (d) far away from the surface.

The SSD depth was positively correlated with an increase in the polishing depth. The most likely reason should be that the polishing depth a_p and polishing force F of the single grit increased as the polishing depth increased. This result appeared to be in good consistence with Equation (12), i.e., increasing the polishing force eventually led to an increase in the depth of the SSD. Thereafter, the relationship between the spindle speed and the SSD depth could be obtained by a stepwise change of spindle speed. Increasing the spindle speed decreased the normal force of the single grit, which gave rise to the formation of debris, and the decrease in the plastic deformation of surface layer and the average SSD depth. However, on the other hand, the upper limitation of the spindle speed due to the spindle performance and strength of the polishing tool necessitated the consideration of the relationship between the spindle speed and the polishing tool. This determined the optimum polishing speed for improving the machining quality and processing efficiency. It is clearly visible in Figure 12c that the increased abrasive grain size worsened the SSD depth. When the abrasive grain size was increased, the actual cutting depth of the single grit was increased, which enhanced the normal contact force of the single grit and the SSD depth.

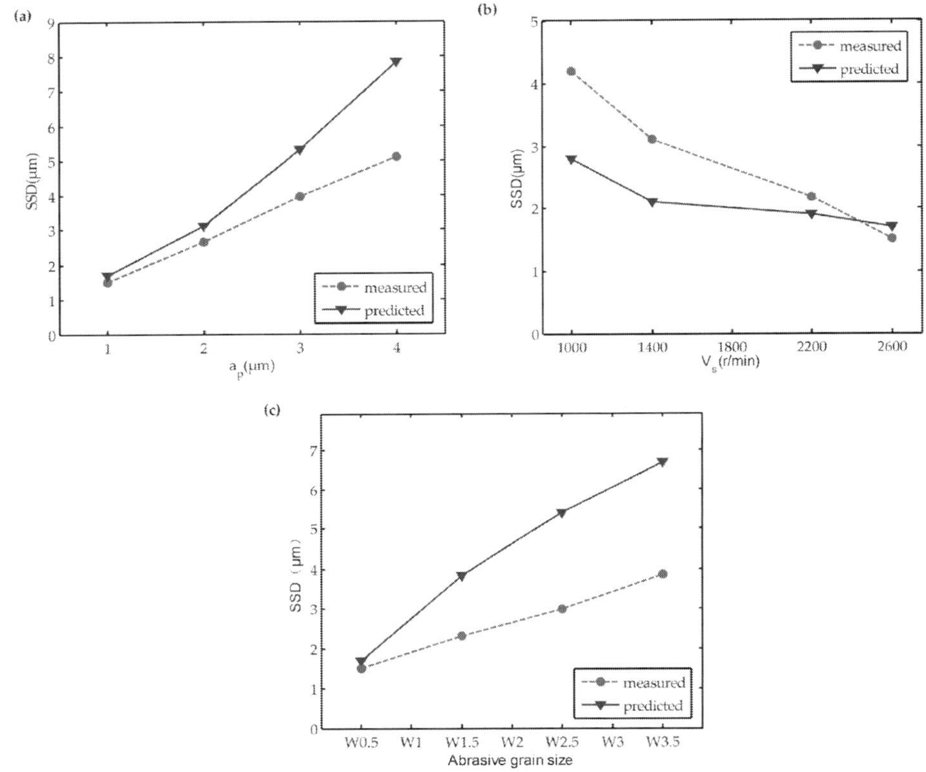

Figure 12: Dependence of SSD depth on polishing parameters: (a) polishing depth a_p; (b) spindle speed V_s; (c) abrasive grain size.

Table 5: The predicted and measured values of maximum SSD depth

Test	Polishing Depth a_p (μm)	Polishing Speed V_s (mm/s)	Abrasive Grain Size	Predicted SSD (μm)	Measured SSD (μm)
1	1	2600	W 0.5	1.697	1.515
2	2	2600	W 0.5	3.110	2.231
3	3	2600	W 0.5	5.340	3.961
4	4	2600	W 0.5	7.834	5.113
5	1	1000	W 0.5	2.804	4.212
6	1	1400	W 0.5	2.102	3.116

7	1	2200	W 0.5	1.898	2.176
9	1	2600	W 1.5	3.844	2.325
10	1	2600	W 2.5	5.398	2.987
11	1	2600	W 3.5	6.697	3.857

The above results and analysis clearly suggested a good agreement between the theoretical simulation and the experiments with respect to SSD depth as a function of the polishing depth, spindle speed, and abrasive grain size. However, the theoretical value was also found to be different from the experimental value. The most likely reason should be: (1) the theoretical equations are built in the static process; however, the real polishing process is a dynamic and complex situation; (2) the depth that was calculated by the theoretical equations failed to consider the problem of elastic and plastic deformation; (3) the etching time was barely controlled, which induced an etching error in the experimental tests.

CONCLUSIONS

Based on the above results, the kinematic characteristics of the polishing process and the fracture characteristics of the brittle materials are analyzed and discussed. The theoretical models were established for the description of SSD. The SSD of SiC ceramic was simulated and predicted by the finite element method. The effect of different parameters on SSD of the SiC ceramic workpiece was also analyzed, which provided a reasonable guidance for the subsequent polishing process. According to the discussion above, main conclusions can be drawn as follows:

- On the basis of the dynamic relationship between the abrasive particles and the workpiece in the theoretical analysis, when hi>hc, the material removal is mainly brittle fracture. The SSD depth increases as the abrasive grain angle and load (proportional to the polishing depth) increases, and decreases as the speed increases.
- In the simulation, the formation and propagation of cracks are presented. The simulation results indicate that the polishing depth, abrasive grain size, and polishing speed have the most significant effect on SSD, respectively. By increasing polishing depth and abrasive grain size, the SSD depth increases, and an increasing polishing speed results in a decrease in SSD depth.
- The polishing experiments under different processing parameters

are carried out. The experimental results validate the theoretical and simulation results.

ACKNOWLEDGMENTS

This work was supported by Research on Key Technology of Manufacturing of Information-enabled for complex light machine parts Complex Optical Machine Functional Parts, National Ministry of Science and Technology International Cooperation Project, number: 2016YFE0105100. Meanwhile, this work was supported by the Micro-Nano and Ultra-Precision Key Laboratory of Jilin Province (grant no. 20140622008JC), and Science and Technology Development Projects of Jilin Province (grant nos. 20180623034TC).

AUTHOR CONTRIBUTIONS

Y.G. and W.Z. designed the experiments, analyzed the results and wrote the paper; M.K. performed the simulation work; J.L. and M.L. contributed experimental reagents and materials, and provided support for scanning electron microscope experiments.

REFERENCES

1. Zhang, Y.; Zhang, J.; Han, J.; He, X.; Yao, W. Large-scale fabrication of lightweight Si/SiC ceramic composite optical mirror. Mater. Lett. 2004, 58, 1204–1208.
2. Ealey, M.A.; Weaver, G.Q. Developmental history and trends for reaction bonded silicon carbide mirrors. In Proceedings of the SPIE's 1996 International Symposium on Optical Science, Engineering, and Instrumentation, Denver, CO, USA, 4–9 August 1996; pp. 66–72.
3. Zhang, Z.; Yang, X.; Zheng, L.; Xue, D. High-performance grinding of a 2-m scale silicon carbide mirror blank for the space-based telescope. Int. J. Adv. Manuf. Technol. 2017, 89, 463–473.
4. Agarwal, S.; Rao, P.V. Grinding characteristics, material removal and damage formation mechanisms in high removal rate grinding of silicon carbide. Int. J. Mach. Tools Manuf. 2010, 50, 1077–1087.
5. Yang, M.Y.; Lee, H.C. Local material removal mechanism considering curvature effect in the polishing process of the small aspherical lens die. J. Mater. Process. Technol. 2001, 116, 298–304.
6. Brinksmeier, E.; Riemer, O.; Gessenharter, A. Finishing of structured surfaces by abrasive polishing. Precis. Eng. 2006, 30, 325–336.
7. Wang, G.; Wang, Y.; Xu, Z. Modeling and analysis of the material removal depth for stone polishing. J. Mater. Process. Technol. 2009, 209, 2453–2463.
8. Qi, J.; Zhang, D.; Li, S.; Chen, B. A micro-model of the material removal depth for the polishing process. Int. J. Adv. Manuf. Technol. 2016, 86, 2759–2770.
9. Wang, J.; Zhang, C.; Feng, P.; Zhang, J. A model for prediction of subsurface damage in rotary ultrasonic face milling of optical K9 glass. Int. J. Adv. Manuf. Technol. 2016, 83, 347–355.
10. Liu, T.; Ge, P.; Bi, W.; Gao, Y. Subsurface crack damage in silicon wafers induced by resin bonded diamond wire sawing. Mater. Sci. Semicond. Process. 2017, 57, 147–156.
11. Esmaeilzare, A.; Rahimi, A.; Rezaei, S.M. Investigation of subsurface damages and surface roughness in grinding process of Zerodur® glass–ceramic. Appl. Surf. Sci. 2014, 313, 67–75.
12. Chen, J.; Fang, Q.; Li, P. Effect of grinding wheel spindle vibration on surface roughness and subsurface damage in brittle material grinding.

Int. J. Mach. Tools Manuf. 2015, 91, 12–23.

13. Zhu, D.; Yan, S.; Li, B. Single-grit modeling and simulation of crack initiation and propagation in SiC grinding using maximum undeformed chip thickness. Comput. Mater. Sci. 2014, 92, 13–21.
14. Liu, Y.; Li, B.; Wu, C.; Zheng, Y. Simulation-based evaluation of surface micro-cracks and fracture toughness in high-speed grinding of silicon carbide ceramics. Int. J. Adv. Manuf. Technol. 2016, 86, 799–808.
15. Komanduri, R.; Chandrasekaran, N.; Raff, L.M. Effect of tool geometry in nanometric cutting: A molecular dynamics simulation approach. Wear 1998, 219, 84–97.
16. Dai, H.; Chen, G.; Zhou, C.; Fang, Q.; Fei, X. A numerical study of ultraprecision machining of monocrystalline silicon with laser nano-structured diamond tools by atomistic simulation. Appl. Surf. Sci.2017, 393, 405–416.
17. Li, J.; Fang, Q.; Zhang, L.; Liu, Y. Subsurface damage mechanism of high speed grinding process in single crystal silicon revealed by atomistic simulations. Appl. Surf. Sci. 2015, 324, 464–474.
18. Lucca, D.A.; Brinksmeier, E.; Goch, G. Progress in assessing surface and subsurface integrity. CIRP Ann.1998, 47, 669–693.
19. Agarwal, S.; Rao, P.V. Experimental investigation of surface/subsurface damage formation and material removal mechanisms in SiC grinding. Int. J. Mach. Tools Manuf. 2008, 48, 698–710.
20. Jiang, C.; Xu, J.; Wang, C. Experimental investigation of subsurface damage of optical glass in precision grinding using a brittle material removal fraction. Int. J. Adv. Manuf. Technol. 2017, 90, 725–730.
21. Blaineau, P.; André, D.; Laheurte, R.; Darnis, P.; Darbois, N.; Cahuc, O.; Neauport, J. Subsurface mechanical damage during bound abrasive grinding of fused silica glass. Appl. Surf. Sci. 2015, 353, 764–773.
22. Wang, C.; Fang, Q.; Chen, J.; Liu, Y.; Jin, T. Subsurface damage in high-speed grinding of brittle materials considering kinematic characteristics of the grinding process. Int. J. Adv. Manuf. Technol. 2016, 83, 937–948.
23. Jing, X.; Maiti, S.; Subhash, G. A New Analytical Model for Estimation of Scratch-Induced Damage in Brittle Solids. J. Am. Ceram. Soc. 2007, 90, 885–892.
24. Lambropoulos, J.C.; Jacobs, S.D.; Ruckman, J. Material removal

mechanisms from grinding to polishing. Ceram. Trans. 1999, 102, 113–128.

25. Gu, W.; Yao, Z.; Li, K. Evaluation of subsurface crack depth during scratch test for optical glass BK7. Proc. Inst. Mech. Eng. Part C J. Mech. Eng. Sci. 2011, 225, 2767–2774.
26. Xi, F.; Zhou, D. Modeling surface roughness in the stone polishing process. Int. J. Mach. Tools Manuf. 2005, 45, 365–372.
27. Doman, D.A.; Warkentin, A.; Bauer, R. A survey of recent grinding wheel topography models. Int. J. Mach. Tools Manuf. 2006, 46, 343–352.
28. Zhao, Y.; Chang, L. A micro-contact and wear model for chemical–mechanical polishing of silicon wafers. Wear 2002, 252, 220–226.
29. Bifano, T.G.; Dow, T.A.; Scattergood, R.O. Ductile-regime grinding: A new technology for machining brittle materials. J. Eng. Ind. 1991, 113, 184–189.
30. Holmquist, T.J.; Johnson, G.R. Characterization and evaluation of silicon carbide for high-velocity impact. J. Appl. Phys. 2005, 97, 093502.
31. Johnson, G.R.; Holmquist, T.J. An improved computational constitutive model for brittle materials. In Proceedings of the AIP Conference, Colorado Springs, CO, USA, 28 June–2 July 1993; pp. 981–984.

SECTION IV: SURFACE ROUGHNESS

Chapter 11

Surface Roughness Evaluation Based on Acoustic Emission Signals in Robot Assisted Polishing

Beatriz De Agustina [1], Marta María Marín [1], Roberto Teti [2], and Eva María Rubio [1]

[1]Department of Manufacturing Engineering, Industrial Engineering School, National University of Distance Education (UNED), C/Juan del Rosal, 12, E28040-Madrid, Spain

[2]Department of Materials and Production Engineering, University of Naples Federico II Piazzale Tecchio, 80, Naples 80125, Italy

ABSTRACT

The polishing process is the most common technology used in applications where a high level of surface quality is demanded. The automation of polishing processes is especially difficult due to the high level of skill and dexterity that is required. Much of this difficulty arises because of the lack of reliable data on the effect of the polishing parameters on the resulting

Citation: De Agustina, Beatriz; Marín, Marta M.; Teti, Roberto; Rubio, Eva M. 2014. "Surface Roughness Evaluation Based on Acoustic Emission Signals in Robot Assisted Polishing." Sensors 14, no. 11: 21514-21522. https://doi.org/10.3390/s141121514

surface roughness. An experimental study was developed to evaluate the surface roughness obtained during Robot Assisted Polishing processes by the analysis of acoustic emission signals in the frequency domain. The aim is to find out a trend of a feature or features calculated from the acoustic emission signals detected along the process. Such an evaluation was made with the objective of collecting valuable information for the establishment of the end point detection of polishing process. As a main conclusion, it can be affirmed that acoustic emission (AE) signals can be considered useful to monitor the polishing process state.

Keywords: robot assisted polishing, surface roughness, acoustic emission signals, contact force, monitoring

INTRODUCTION

The increasing demand for improved surface integrity of manufactured components with high production rate has acted as a driving force in the development of automated production methods. For this reason, the monitoring of surface quality obtained by a manufacturing process has been extensively studied and improved in the last years. In fact, a great effort has been made to evaluate and control on line the surface roughness obtained in finishing machining processes by the analysis of signals detected by different sensors such as dynamometer, acoustic emission, accelerometer, current/power among others [1–7]. This is an important challenge to avoid valuable product rejects and refinishing [8,9].

Among these processes, polishing is the most common technology used in those applications in which high surface qualities, in terms of low roughness level, minimized subsurface damage and high form accuracies, are demanded. It is an essential step in optics manufacturing and in mold finishing operations [4].

The automation of polishing is especially difficult due to the high level of skill and dexterity that is required. Much of this difficulty arises due to a lack of data in relation with the effect of the polishing parameters on the resulting surface roughness, although some experimental works have been published in this area [10,11]. In general, these studies built up a specific database with experimental data and working knowledge that includes a set of polishing conditions such a tool type, size and grit, feed rate, number of passes and cutting fluids used [12–14].

Until now, regarding the monitoring of polishing processes, there are few experimental studies carried out [12,15–21].

In some of them, sensors have been integrated during the automated polishing processes. Among them, the use of acoustic emission sensors has allowed the achievement of valuable results for the estimation of the process state [12,17–21].

In this work, an experimental study was developed to evaluate the surface roughness obtained during robot assisted polishing processes by the analysis of acoustic emission (AE) signals. The aim is to find out a trend of a feature or features calculated from the AE signals obtained along the process. Such an evaluation was made with the objective of collecting valuable information for the establishment of the end point detection of polishing process; this is when the required surface roughness is reached or it is necessary to replace the tool and select a finer one and/or change other polishing condition, to carry on with the following stage of polishing.

For this propose, dry polishing turning tests were carried out on a Robot Assisted Polishing machine (STRECON NanoRAP 200) performed during five polishing sessions, at different polishing contact force, using the same polishing stone tool. Along the tests, AE signals were acquired and roughness surface measurements were taken at the end of each polishing session.

EXPERIMENTAL PROCEDURE

The experimental procedure taken in this study is shown on the following points:

Polishing Tests

For the development of the tests, it is proposed to employ a Robot Assisted Polishing (RAP) system. This is a polishing machine equipped with a robot arm in which interchangeable polishing tools are mounted. A control module is integrated in the robot arm that allows the setting of, not only the conventional polishing parameters such a cutting speed and feed rate, also the contact force between the tool and workpiece, and the tool pulsation. This movement increases the cutting efficiency and improves the steadiness of the machining process.

In this study, dry polishing turning tests were carried out on a Robot Assisted Polishing machine (STRECON NanoRAP 200) performed during five polishing sessions, at different polishing contact force. Experimental tests were taken on a bar of alloy steel (UNS G52986) with a length of 75 mm and diameter of 30 mm (Figure 1). This machine operates similarly to a turning lathe. In this case, the operation performed was a cylindrical turning in which three movements are combined: the rotation of the workpiece, the feed movement of the tool and a third movement; the tool has also a pulsation movement on the feed direction with a frequency of 500 pulses/min on a length of 1 mm.

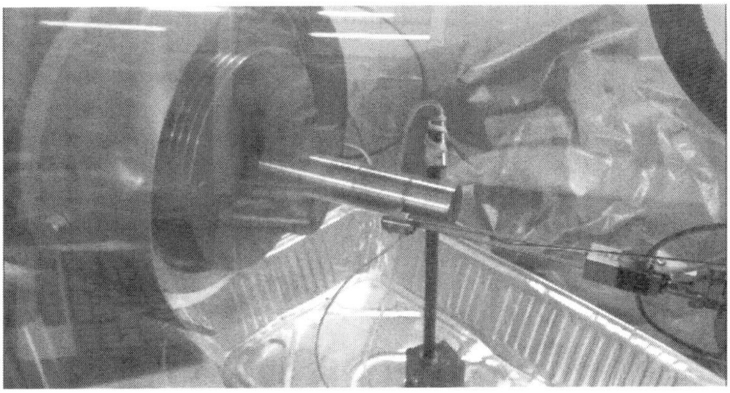

Figure 1: Robot Assisted Polishing machine.

The abrasive tool selected was the polishing stone with a grit number of 800 that corresponds to a grain size of approximately 11 μm from Gesswein (MP800) according to the Comercial Standard CS271-65 [22].

Each session consists of 60 polishing passes. The Table 1 summarizes polishing conditions used in each session.

Table 1: Polishing conditions at the five sessions performed

Polishing Session	Contact Force (N)	Spindle Speed (rpm)	Feed Rate (mm/s)	Pulsation Tool (Pulses/min)	Pulsation Stroke (mm)
1	17.64	300	5	500	1
2	9.8				
3	9.8				
4	9.8				
5	17.64				

Acquisition of AE Signal during Polishing Tests

An AE sensor was mounted on the tool holder as close as possible to the polishing stone to minimize signal loss and achieve a good signal-to-noise ratio. Signals samples were taken with an interval time between acquisitions of, approximately, 0.93 s. That is, along the five polishing sessions, a total number of 4650 acquisitions. Sample rate was set up for 1 MHz.

Surface Roughness Measurements

At the end of the each polishing session, five surface roughness measurements were taken on the workpiece (along one of its generatrix) by the use the roughness tester MahrSurf XD1 equipped with a 2 μm radius tip. For the measurements, it was applied a cut-off and an evaluation length of 0.25 and 1.25 mm, respectively.

In the measurement process, data (xi, zi) of the surface geometry workpiece were obtained. The arithmetical average roughness, *Ra*, was selected as a parameter to analyze. According to ISO 4288 (ISO 4288, 1996) [23] standards, this parameter is defined as the arithmetical average of the absolute values of the deviations of the roughness profile, *R*, and it is expressed mathematically by means of the Equation (1):

$$Ra = \frac{1}{l_m} \int_0^{l_m} |z(x)| dx \qquad (1)$$

Analysis of AE Data

Row AE signal data arrays, with 131,072 values each of them, are stored and analyzed by Matlab software. Different features were calculated to determine the process state, in terms of surface roughness, along the polishing time.

From AE data, an analysis in the frequency domain was carried out by using the Fast Fourier Transform (FFT) algorithm. This is an efficient algorithm that allows the representation of a digital signal in the frequency domain using less number of numerical operations compared to Digital Fast Fourier (DFT) with a substantial computational power saving [24].

Firstly, FFT power spectral graphics at different polishing time were plotted. The objective is to select a range or ranges of frequencies where the amplitudes of certain peaks shown in the FFT power spectral plots decrease with the time process, as it is pointed out in previous studies in which robot assisted polishing tests were carried out [18,24]. Once the range of frequencies was selected, maximum amplitude was calculated from all the AE acquisitions registered to achieve a possible trend and a correlation with the surface roughness.

RESULTS

As a first approach, FFT power spectral plots were obtained at different polishing time corresponding to the acquisitions taken, approximately, at the middle of each polishing session.

All of them show several peaks that appear around the same range of frequencies as it can be seen in Figure 2. This is the FFT power spectral plot at 0400 acquisition during the session 1. As it was mentioned, the amplitude of some of those peaks decreases as process progresses, whereas magnitudes on other frequencies do not show a variation over time.

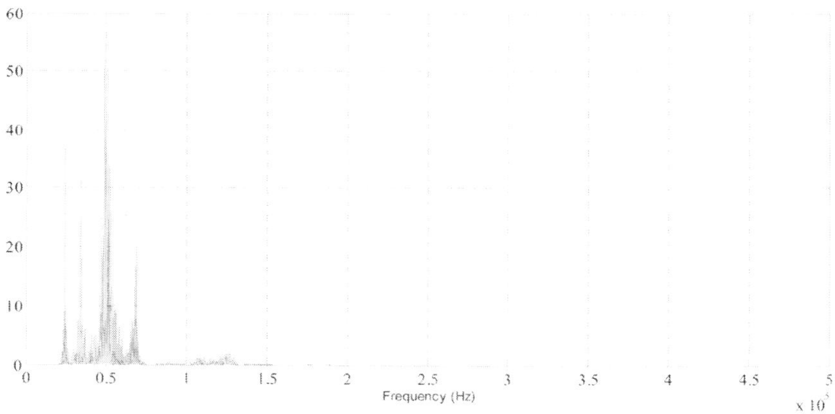

Figure 2: Fast Fourier Transform (FFT) power spectral plot at 0400 acquisition (session 1).

In addition, it is important to remark that when a higher magnitude of contact force was applied, the amplitudes of these peaks increased. This direct relationship, between the selected AE signal parameter and the contact force, was observed at the last session of polishing (session 5).

Further analysis was made to evaluate the whole polishing process; this is the total acquisitions taken during the five sessions. According to the peaks shown in the FFT power spectral representations, different ranges of frequencies were selected to determine the maximum amplitude of the different peak reached. Among the ranges of frequencies analyzed (ranges that contain peaks at 24, 34 50, 69 and 125 kHz) the selected range that is shown in this study, is the range between 60,000 and 100,000 Hz, so a clearer trend was found. Similar results are presented in the study developed by Pilný and partners [18], in which three main peaks were distinguished at 20, 45 and 150 kHz.

Maximum amplitudes calculated from each acquisition along the five polishing sessions are plotted in the following Figure 3.

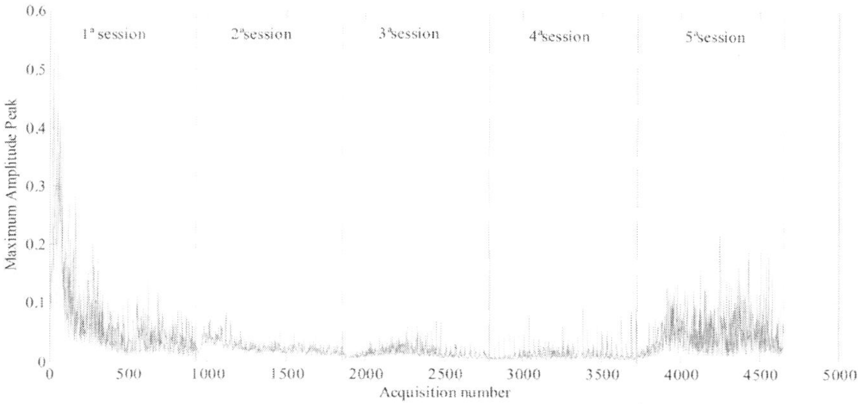

Figure 3: Maximum amplitudes of FFT power spectral of acoustic emission (AE) signal (60,000–100,000 Hz).

As it can be observed, as the polishing process progresses, the amplitudes tend to decrease, apart from the ones obtained at the fifth session, for the reason that was previously pointed out.

Besides, at the end of each polishing session, five surface roughness measurements were taken on the workpiece, around the same location.

Finally, to compare the results from the analysis of AE signals in the frequency domain and the obtained surface roughness, the averages of maximum amplitudes of FFT power spectral were plotted in Figure 4 and the five *Ra* measurements for each polishing season were plotted by means of box and whisker diagrams (Figure 5).

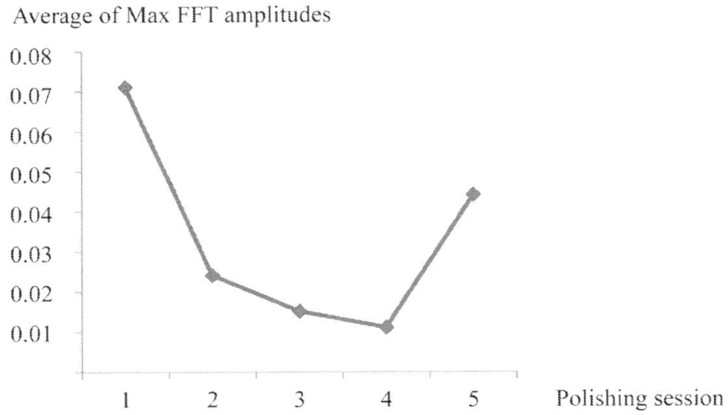

Figure 4: Plot of the average of maximum amplitudes FFT power spectral of AE signal *vs.* polishing session.

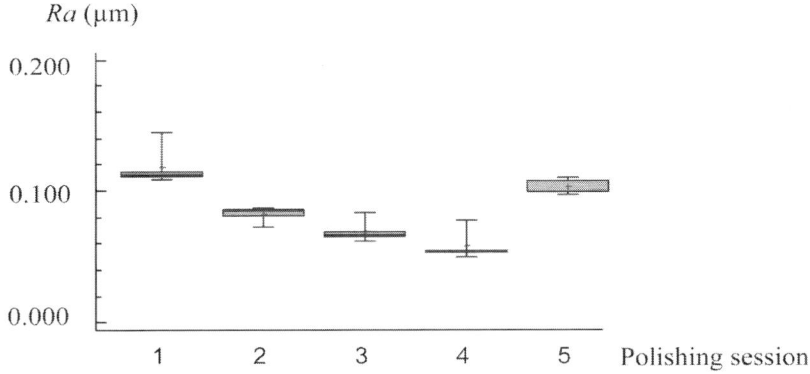

Figure 5: *Ra* measurements at the end of each season.

First of all, it is important to clarify with respect the Figure 5, that the values of the surface roughness measured on the workpiece before polishing is included and plotted at the polishing session number 0.

As shown in Figures 4 and 5, the feature extracted of AE signals and the surface roughness follow the same trend, obtaining the lowest value of *Ra* (0.059 µm) at the fourth polishing session. This is when it would be approached the end detection of the process. Afterwards, at the fifth polishing session, with the increase of the contact force applied up to 17.64 N, a larger value of *Ra* was obtained. Nevertheless, at first stage of polishing process (session 1) with the use of a higher magnitude of contact force, at the same cutting condition, the surface roughness improved from 0.142 to

0.116 μm. This shows that it is necessary to reduce such control parameters for the last polishing stages.

CONCLUSIONS

As a first approach, in this study, an evaluation of the surface roughness obtained by Robot Assisted Polishing with different contact force applied between the tool and the workpiece, was developed. Concretely, dry polishing turning tests were carried out on a bar of alloy (UNS G52986) by using a tool with a grit number of 800. An analysis of the AE signal in frequency domain detected during the process was made with the objective of finding a feature or features from AE signal sensitive to the variations of the surface roughness measured along five polishing sessions. The main conclusions extracted from the results obtained in this study can be summarized as follows:

- For the polishing conditions employed in this study, the surface roughness and the analyzed feature calculated from the frequency domain of AE signal (maximum amplitude of the FFT power spectral between the frequencies 60,000 and 100,000 Hz) during polishing process follow the same trend. Therefore, AE signal can be considered useful for the monitoring of polishing process state.
- The end point detection of the process was determined at the end of the polishing session 4. That corresponds to the polishing stage when the most improved surface roughness has been reached.
- The contact force between the tool and the workpiece is a control parameter of the process that should be reduced at the last polishing stages.
- Finally, it is important to indicate that from the results obtained in this study, it is possible to collect valuable data to use together with other features extracted from others sensors to further develop monitoring polishing systems.

ACKNOWLEDGMENTS

The authors thank to the Research Group of the UNED "Industrial Production and Manufacturing Engineering (IPME)" for the given support during the development of this work, and the funding of: the European project (FP7 IFaCOM reference: 285489), the Spanish Ministry of Science and Innovation (Project DPI2011-27135) and the Industrial Engineering School-UNED (Project REF2013-ICF03).

AUTHOR CONTRIBUTIONS

Teti and Rubio, has participated in the idea and the planning of the different stages of the experimental procedure in this research paper, as well as in the analysis of results and final conclusions reached. Marín, Rubio and Agustina have participated in the analysis of the results and final conclusions extracted.

REFERENCES

1. Eversona, C.E.; Cheraghi, S.H. The application of acoustic emission for precision drilling. *J. Mach. Tools Manuf.* 1999, *39*, 371–387.
2. Dongfeng, S.; Gindy, N. Development of an online machining process monitoring system: Application in hard turning. *Sens. Actuators* 2007, *135*, 405–414.
3. Karayel, D. Prediction and control of surface roughness in CNC lathe using artificial neural network. *J. Mater. Process. Technol.* 2009, *209*, 3125–3137.
4. Klocke, F.; Dambon, O.; Schneider, U.; Zunke, R.; Waechter, D. Computer-based monitoring of the polishing processes using LabView. *J. Mater. Process. Technol.* 2009, *209*, 6039–6047.
5. Marinescu, I.; Axinte, D. A time–frequency acoustic emission-based monitoring technique to identify workpiece surface malfunctions in milling with multiple teeth cutting simultaneously. *Int. J. Mach. Tools Manuf.* 2009, *49*, 53–65.
6. Teti, R.; Jemielniak, K.; O'Donnell, G.; Dornfeld, D. Advanced monitoring of machining operations. *CIRP Ann. Manuf. Technol.* 2010, *59*, 717–739.
7. Núñez, P.J.; Simao, J.; Arenas, J.M.; de la Cruz, C. Surface roughness characterization using cutting force analysis, regression and neural network prediction models. *Mater. Sci. Forum* 2006, *256*, 211–216.
8. Guo, Y.B.; Ammula, S.C. Real-time acoustic emission monitoring for surface damage in hard machining, damage in hard machining. *Int. J. Mach. Tools Manuf.* 2005, *45*, 1622–1627.
9. Azouzi, R.; Guillot, M. On-line prediction of surface finish and dimensional deviation in turning using neural network based sensor fusion. *Int. J. Mach. Tools Manuf.* 1997, *37*, 1201–1217.
10. Wu, B.H.; Wang, J.J. A neuro-fuzzy approachto generating mold/die polishing sequences. *J. Mater. Process. Technol.* 2009, *209*, 3241–3250.
11. Huissoon, J.P.; Ismail, F.; Jafari, A.; Bedi, S. Automated polishing of die steel surfaces. *Int. J. Adv. Manuf. Technol.* 2002, *19*, 285–290.
12. Ahn, J.H.; Shen, Y.F.; Kim, H.Y.; Jeong, H.D.; Cho, K.K. Development of a sensor information integrated expert system for optimizing die polishing. *Robot. Comput. Integr. Manuf.* 2001, *17*, 269–276.

13. Brinksmeier, E.; Riemer, O.; Gessenharter, A.; Autschbach, L. Polishing of structured molds. *CIRP Ann. Manuf. Technol.* 2004, *53*, 247–250.
14. Takaya, Y.; Tachika, H.; Hayashi, T.; Kokubo, K.; Suzuki, K. Performance of water-soluble fullerenol as novel functional molecular abrasive grain for polishing nanosurfaces. *CIRP Ann. Manuf. Technol.* 2009, *58*, 495–496.
15. Kim, S.Y.; Park, C.J.; Seo, Y.J. Signal analysis of the end point detection method based on motor current. *Microelectron. Eng.* 2003, *66*, 472–479.
16. Lee, D.E.; Hwang, I.; Valente, C.M.O.; Oliveira, J.F.G.; Dornfeld, D.A. Precision manufacturing process monitoring with acoustic emission. *Int. J. Adv. Manuf. Technol.* 2006, *46*, 176–188.
17. Ahn, J.H.; Lee, M.C.; Jeong, H.D.; Kim, S.R.; Cho, K.K. Intelligently automated polishing for high quality surface formation of sculptured die. *J. Mater. Process. Technol.* 2002, *133–134*, 339–344.
18. Chang, Y.P.; Hashimura, M.; Dornfeld, D.A. An investigation of the AE Signals in the lapping process. *CIRP Ann. Manuf. Technol.* 1996, *45*, 331–334.
19. Jeong, H.; Kim, H.; Lee, S.; Dornfeld, D. Multi-Sensor monitoring system in Chemical Mechanical Planarization (CMP) for correlations with crocess issues. *CIRP Ann. Manuf. Technol.* 2006, *55*, 325–328.
20. Pilný, L.; Bissacco, G.; de Chiffre, L.; Ramsing, J. Acoustic Emission based In-process Monitoring in Robot Assisted Polishing. Proceedings of the 11th International Symposium of Measurement Technology and Intelligent Instruments, Aachen, Germany, 1–3 July 2013.
21. Dunegan, H. *An Acoustic Emission Technique for Measuring Surface Roughness; the DECI Report*; Dunegan Engineering Consultants Inc.: Midland, TX, USA, 1998.
22. U.S. Department of Commerce. Grading of abrasive grain for grinding wheels Commercial Standard CS 271-65. 1965.
23. *ISO 4288:1996 Geometrical Product Specifications (gps). Surface Texture: Profile Method. Rules and Procedures for the Assessment of Surface Texture*; ISO: Genève, Switzerland, 1996.
24. Lazarev, R. Monitoring and Control of Fine Abrasive Polishing Processes. Ph.D Thesis, University of Southern Denmark, Sønderborg, Denmark, 2012.

Chapter 12

Interfacial Surface Roughness Determination by Coherence Scanning Interferometry using Noise Compensation

Hirokazu Yoshino[1,2], John Michael Walls[1], and Roger Smith[1]

[1] Loughborough University, Leicestershire LE11 3TU, UK

[2] Taylor Hobson Ltd., Leicestershire LE4 9JD, UK

ABSTRACT

The capability of coherence scanning interferometry has been extended recently to include the determination of the interfacial surface roughness between a thin film and a substrate when the surface perturbations are less than ~10 nm~10 nm in magnitude. The technique relies on introducing a first-order approximation to the helical complex field (HCF) function. This approximation of the HCF function enables a least-squares optimization to be carried out in every pixel of the scanned area to determine the heights

Citation: Hirokazu Yoshino, John Michael Walls, and Roger Smith, "Interfacial surface roughness determination by coherence scanning interferometry using noise compensation," Appl. Opt. 56, 4757-4765 (2017). https://doi.org/10.1364/AO.56.004757

of the substrate and/or the film layers in a multilayer stack. The method is fast but its implementation assumes that the noise variance in the frequency domain is statistically the same over the scanned area of the sample. This results in reconstructed surfaces that contain statistical fluctuations. In this paper we present an alternative least-squares optimization method, which takes into account the distribution of the noise variance-covariance in the frequency domain. The method is tested using results from a simulator and these show a significant improvement in the quality of the reconstructed surfaces.

INTRODUCTION

Surface metrology of form, flatness, roughness, and smoothness are important for quality assurance in many types of manufacturing. It is a particular issue for optical components or optical coatings where control of features in the nanometer or sub-nanometer range is required. Stylus-based surface profilometry is the conventional technique used to provide two-dimensional surface metrology. However, this technique can cause modification of the surface under measurement. *Coherence scanning interferometry* (CSI), previously referred to as*scanning white light interferometry*, is a well established, non-contact method that provides true three-dimensional measurements [1]. Optical microscopy provides lateral images without height information. *Scattermeters* measure the proportion of specular to diffuse reflection to calculate the root mean square roughness [2]. In comparison, CSI measures absolute heights at each pixel in the field of view with sub-nanometer vertical resolution [3]. This allows all the various surface roughness parameters measured using stylus profilometry to be computed using CSI [4].

CSI measures surface topography by locating and connecting peak positions in the interference pattern, referred to as an *interferogram*, at each pixel over the scanned area to reconstruct the measured surface. One of the prerequisites for the test surface is that the surface should have identical amplitude reflection coefficients over the field of view; otherwise, a phase shift on reflection occurs, which results in an erroneous vertical profile. Even if the refractive index is unchanged over the measurement area, problems can occur in the CSI measurement for transparent/semi-transparent thin films with thicknesses of $<\sim 1.5$ μm$<\sim 1.5$ μm. This is because the interferogram has multiple peaks corresponding to the interfacial surfaces in the thin film assembly. The peaks may be superimposed depending on the thin film structure. In the case where the films are >~1.5 μm>~1.5 μm, it is possible to detect and separate the peaks to reproduce the interfacial topographies [5–7]. However, this is not the case with thin films of a few hundred nanometers in thickness. A Fourier transform of the interferogram can be performed for investigations in this thickness regime. Subtle changes in phase and amplitude are compared with those synthesized mathematically [8–11].

One of the methods defined in the frequency domain approach uses a theory based on the *helical complex field* (HCF) function [12,13] and its extensions [14,15]. *Interfacial surface roughness* (ISR) has been determined using this method by introducing a first-order approximation to the HCF function, which enables fast real-time computation [3,16]. However, the method can introduce spurious surface roughness, as shown in a previous study [15].

This paper presents a methodology to reduce the spurious roughness which can occur with the existing ISR method and improves its computation stability for a wide variety of samples with no additional hardware, no changes in measurement procedure, and little extra computational effort.

THEORY

Helical Complex Field Function

The determined and synthesized HCF functions, which have a reflection coefficient averaged over the incident angle of the light of amplitude $\overline{r}$ [15], are expressed as follows:

$$\mathrm{HCF}^{d}(\nu;\mathbf{d}) = \overline{r}_{\mathrm{ref}}(\nu)\cdot\frac{\mathcal{F}[I]_{SB+}}{\mathcal{F}[I_{\mathrm{ref}}]_{SB+}},$$

$$\mathrm{HCF}^{s}(\nu,\mathbf{d}) = \overline{r}(\nu,\mathbf{d})\cdot\exp(j4\pi\nu\Delta z_{\mathrm{HCF}}\cos\overline{\theta}), \qquad (1)$$

where the thin film assembly consists of a substrate represented by its subscript sub and the thickness of the L film layers expressed by $\mathbf{d} = [d_{\mathrm{sub}}, d_1, \ldots, d_L]^{\mathrm{T}}$. Note that d_{sub} is not used in Eq. (1) but Δd_{sub} is considered in the following discussion. The determined HCF function, HCF^d, is given by the positive sideband of the Fourier transform of the actual interference signal I obtained from a test sample divided by that of a known reference material I_{ref}. Whereas the synthesized HCF function, HCFs, is derived from a mathematical model of the test thin film structure

[12]. Note that given any signal s, the positive sideband of the Fourier transform of the signal is termed $\mathcal{F}[s]_{SB+}$. The averaged incident angle of the CSI instrument is denoted by $\overline{\theta}$ and v represents the light frequency. The unknown parameter Δz_{HCF} is associated with the surface height [15].

Assuming that the thin films are completely flat over the field of view, the set of film thicknesses is numerically determined to be $\mathbf{d} = \hat{\mathbf{d}}$ together with Δz_{HCF} by minimizing the squared error between the two functions in Eq. (1). Normally the interference signals over M test sample pixels and M_{ref} reference sample pixels, typically a few hundred, are averaged to have each signal with less noise; i.e., $I = 1/M\sum_i I^i$ is effectively used in Eq. (1) and so is $I_{ref} = 1/M_{ref}\sum_i I^i_{ref}$.

First-Order Approximation to the Synthesized HCF Function

Let ϵ_{px} be the noise induced in the interference signal at any pixel I_{px} with its interfacial surface perturbation Δd, then the determined and the synthesized HCF functions are, respectively

$$\mathrm{HCF}^d_{\mathrm{px}} = \overline{r}_{\mathrm{ref}}(\nu) \cdot \frac{\mathcal{F}[I_{\mathrm{px}} + \epsilon_{\mathrm{px}}]_{SB+}}{\mathcal{F}[I_{\mathrm{ref}}]_{SB+}},$$

$$\mathrm{HCF}^s_{\mathrm{px}}(\nu; \hat{\mathbf{d}} + \Delta\mathbf{d}) \approx \overline{\mathrm{HCF}^d} + j4\pi\nu \cos\overline{\theta} \cdot \overline{\mathrm{HCF}^d} \times \left\{ \Delta d_{\mathrm{sub}} + \sum_{l=1}^{L} G_l(\nu; \hat{\mathbf{d}}) \Delta d_l \right\},$$

where

$$G_l(\nu; \hat{\mathbf{d}}) = 1 + \frac{1}{4\pi\nu \cos\overline{\theta}} \frac{\partial \chi(\hat{\mathbf{d}})}{\partial d_l}, \quad \arg(\overline{r}) = \chi,$$

$$\overline{\mathrm{HCF}^d} = \overline{r}_{\mathrm{ref}}(\nu) \cdot \frac{\mathcal{F}[E[I_{\mathrm{px}} + \epsilon_{\mathrm{px}}]]_{SB+}}{\mathcal{F}[I_{\mathrm{ref}}]_{SB+}}. \tag{2}$$

Note that the condition $E[I_{\mathrm{px}} + \epsilon_{\mathrm{px}}] = E[I_{\mathrm{px}}]$ holds. It follows that $\overline{\mathrm{HCF}^d}$ is smooth over the wavelength range of interest, as shown in Fig. 1(a). For the computations based on Eq. (2), the expressions for the HCF functions need to be re-written in a spectrally discrete manner $(\boldsymbol{\nu} = [\nu_1, \nu_2, \ldots, \nu_m]^{\mathrm{T}})$ as follows:

$$\mathbf{HCF^d_{px}} = [\mathrm{HCF}^d_{\mathrm{px}}(\nu_1), \mathrm{HCF}^d_{\mathrm{px}}(\nu_2), \ldots, \mathrm{HCF}^d_{\mathrm{px}}(\nu_m)]^{\mathrm{T}},$$

$$\mathbf{HCF^s_{px}} \approx \overline{\mathbf{HCF^d}} + \mathrm{Diag}[\overline{\mathbf{HCF^d}}]\mathbf{G}\Delta\mathbf{d}. \tag{3}$$

Accordingly the expression is re-written as a linear inverse problem with noise ε_o in the frequency domain:

$$\mathbf{HCF}^{\mathbf{d}}_{\mathbf{px}} \simeq \overline{\mathbf{HCF}^{\mathbf{d}}} + \mathrm{Diag}[\overline{\mathbf{HCF}^{\mathbf{d}}}]\mathbf{G}\Delta\mathbf{d} + \boldsymbol{\varepsilon}_o,$$

where

$$\mathbf{G} = j4\pi\cos\overline{\theta}\begin{bmatrix} \nu_1 & \nu_1\cdot G_1(\nu_1) & \cdots & \nu_1\cdot G_L(\nu_1) \\ \nu_2 & \nu_2\cdot G_1(\nu_2) & \ddots & \vdots \\ \vdots & \vdots & \ddots & \vdots \\ \nu_m & \nu_m\cdot G_1(\nu_m) & \cdots & \nu_m\cdot G_L(\nu_m) \end{bmatrix},$$

$$\mathrm{Diag}[\overline{\mathbf{HCF}^{\mathbf{d}}}] = \begin{bmatrix} \overline{\mathrm{HCF}^d(\nu_1)} & 0 & \cdots & 0 \\ 0 & \overline{\mathrm{HCF}^d(\nu_2)} & \ddots & \vdots \\ \vdots & \ddots & \ddots & 0 \\ 0 & \cdots & 0 & \overline{\mathrm{HCF}^d(\nu_m)} \end{bmatrix},$$

$$\overline{\mathbf{HCF}^{\mathbf{d}}} = [\overline{\mathrm{HCF}^d(\nu_1)}, \overline{\mathrm{HCF}^d(\nu_2)}, \ldots, \overline{\mathrm{HCF}^d(\nu_m)}]^{\mathrm{T}},$$

$$\boldsymbol{\varepsilon}_o = [\varepsilon_{o1}, \varepsilon_{o2}, \ldots, \varepsilon_{om}]^{\mathrm{T}} \sim N(\boldsymbol{\varepsilon}_o|\mathbf{0}, \sigma_o^2\mathbf{I}). \quad (4)$$

Note that the noise ε_{oi} existing in the frequency domain is assumed to follow a normal distribution $N(\varepsilon_{oi}|0, \sigma_o^2)$ with the mean at zero and variance σ_o for $i \in \mathbb{N}$. Given a random vector a, the operators $E[\mathbf{a}]$ and $E[(\mathbf{a} - E[\mathbf{a}])(\mathbf{a} - E[\mathbf{a}])^{\mathrm{T}}]$ are understood to be the ensemble and the variance-covariance matrix of the random vector respectively. The variance-covariance matrix of the noise ε_o is assumed to be $\sigma_o^2\mathbf{I}$, where I is the identity matrix.

Interfacial Surface Roughness Determination by the Present ISR Method

Using the discrete expressions introduced in the previous section, a merit function $J_{\mathrm{px}} = \|\mathbf{HCF}^{\mathbf{d}}_{\mathbf{px}} - \mathbf{HCF}^{\mathbf{s}}_{\mathbf{px}}\|^2$ is minimized for every pixel with respect to Δd such that $\boldsymbol{\varepsilon}_o \sim N(\boldsymbol{\varepsilon}_o|\mathbf{0}, \sigma_o^2\mathbf{I})$. The solution $\hat{\Delta\mathbf{d}}$ is equivalent to that established by the maximum likelihood estimation under the assumption that each element of the noise ε_o is stochastically independent and has the same variance, i.e., the relation $E[(\boldsymbol{\varepsilon}_o - E[\boldsymbol{\varepsilon}_o])(\boldsymbol{\varepsilon}_o - E[\boldsymbol{\varepsilon}_o])^{\mathrm{T}}] = \sigma_o^2\mathbf{I}$ holds. Then, the solution of the linear inverse problem in Eq. (4) is given analytically by [15,17,18]

$$\hat{\Delta\mathbf{d}} = (\mathbf{G}^{\mathrm{T}}\mathbf{G})^{-1}\mathbf{G}^{T}\mathbf{u}, \quad (5)$$

where

$$\mathbf{u} = \{\mathrm{Diag}[\overline{\mathbf{HCF}^{\mathbf{d}}}]\}^{-1}[\mathbf{HCF}^{\mathbf{d}}_{\mathbf{px}} - \overline{\mathbf{HCF}^{\mathbf{d}}}].$$

As expressed in Eq. (5), the vector u is an observed signal while Δd is an unknown original signal to be estimated. The problem can be re-expressed as

$$\mathbf{u} = \mathbf{G}\Delta\mathbf{d} + \boldsymbol{\varepsilon}, \quad \text{where } \boldsymbol{\varepsilon} = \{\mathrm{Diag}[\overline{\mathbf{HCF^d}}]\}^{-1}\boldsymbol{\varepsilon}_o, \tag{6}$$

where the noise ε and its variance σ^2 have been re-defined for simplification. Accordingly, the problem and the merit function are simplified such that min $J^{\dagger}_{\mathrm{px}} = \|\mathbf{u} - \mathbf{G}\Delta\mathbf{d}\|^2$ subject to $\boldsymbol{\varepsilon} \sim N(\boldsymbol{\varepsilon}|\mathbf{0}, \sigma^2\mathbf{I})$.

ISR Methodology with Noise Compensation

Although the existing ISR method [15,16,19] has been used to determine the surface topography of a layer buried under a transparent thin oxide film, the method can induce spurious roughness caused by system and environmental noise in the signal. This is illustrated in Fig. 1 for data taken from a 520 nm SiO_2 thin film. If the optimization process averages over the full scanned area as in Fig. 1(a), the result is a smoothly varying HCF function. However, if the HCF function is determined locally, as shown in Figs. 1(b) and 1(c), there is a different functionality

Consider that we have a variance-covariance matrix Σ for the noise, then the optimal solution given by the least-squares error method in Eq. (5) is not valid. This is because the probability distribution of an observed signal u now follows $N(\mathbf{u}|\mathbf{G}\Delta\mathbf{d}, \Sigma)$. It follows that we need to modify the merit function to deal with this different probability distribution, otherwise the least-squares method would lead to an erroneous solution or become unstable. Without the modification, the ISR method measures a higher interfacial surface roughness.

ISR with Noise Compensation (ISR-NC) Determination

Let p(u) be the probability density function (PDF) of the observed signal u, then using the assumption that the spectral noise ε follows the normal distribution $N(\boldsymbol{\varepsilon}|\mathbf{0}, \Sigma)$, the PDF also is a normal distribution. Therefore, the PDF and its log likelihood function L(Δd) are expressed as follows [18]:

$$p(\mathbf{u}) = \frac{1}{(2\pi)^{\frac{m}{2}}|\Sigma|^{\frac{1}{2}}} \exp\left[-\frac{1}{2}(\mathbf{u} - \mathbf{G}\Delta\mathbf{d})^{\mathrm{T}}\Sigma^{-1}(\mathbf{u} - \mathbf{G}\Delta\mathbf{d})\right],$$

$$L(\Delta\mathbf{d}) = \log p(\mathbf{u}) = -\frac{1}{2}(\mathbf{u} - \mathbf{G}\Delta\mathbf{d})^{\mathrm{T}}\Sigma^{-1}(\mathbf{u} - \mathbf{G}\Delta\mathbf{d}) + C, \tag{7}$$

where C is a constant independent of u and Δd. Maximization of the log-likelihood function with respect to Δd is equivalent to minimization of the following merit function $J^{\ddagger}_{\mathrm{px}}$:

$$\underset{\Delta \mathbf{d}}{\text{minimize}}\, J_{\text{px}}^{\ddagger} = (\mathbf{u} - \mathbf{G}\Delta\mathbf{d})^{T}\Sigma^{-1}(\mathbf{u} - \mathbf{G}\Delta\mathbf{d}),$$
$$\text{subject to } \boldsymbol{\varepsilon} \sim N(\boldsymbol{\varepsilon}|\mathbf{0}, \Sigma). \quad (8)$$

As with the existing ISR method, the optimal solution for this linear inverse problem $\hat{\Delta}\mathbf{d}$ is obtained analytically [18] with the variance-covariance matrix as follows [17,18,20]:

$$\hat{\Delta}\mathbf{d} = (\mathbf{G}^{T}\Sigma^{-1}\mathbf{G})^{-1}\mathbf{G}^{T}\Sigma^{-1}\mathbf{u}. \quad (9)$$

In statistical signal processing, this method is often referred to as pre-whitening [18]. One of the benefits of this method is that the multi-correlation (covariance) of the noise is also considered when minimizing the merit function. This means that the ISR with noise compensation (ISR-NC) method puts more importance on the wavelength domains with smaller noise variance when determining an optimal solution. The way in which the variance-covariance matrix is calculated is described in the following section.

Determination of the Variance-Covariance Matrix of the Noise

The variance-covariance matrix of the noise Σ is determined from the reference measurement with a known material. First the reference global interferogram I_{ref} is computed by averaging over the interferogram at the i-th pixel I^{i}_{ref} resulting in $\overline{\mathbf{HCF}^{\mathbf{d}}_{\text{ref}}}$. The HCF function at the i-th pixel on the reference sample $\mathbf{HCF}^{\mathbf{d,i}}_{\text{ref}}$ is also obtained.

Defining $\varepsilon_i = \mathbf{HCF}^{\mathbf{d,i}}_{\text{ref}} - \overline{\mathbf{HCF}^{\mathbf{d}}_{\text{ref}}}$ as the noise at the i-th pixel its variance-covariance matrix Σ_o can be computed to match the expression in Eq. (9) as shown in Eq. (10). Note that Σ_o is determined by computing the sample variance and covariance over an area corresponding to M_{ref} pixels (normally ~from 20 × 20 to 50 × 50):

$$\Sigma_o = E[(\mathbf{HCF}^{\mathbf{d}}_{\text{ref}} - \overline{\mathbf{HCF}^{\mathbf{d}}_{\text{ref}}})(\mathbf{HCF}^{\mathbf{d}}_{\text{ref}} - \overline{\mathbf{HCF}^{\mathbf{d}}_{\text{ref}}})^{T}]$$
$$\simeq \frac{1}{M_{\text{ref}}}\sum_{i=1}^{M_{\text{ref}}}(\mathbf{HCF}^{\mathbf{d,i}}_{\text{ref}} - \overline{\mathbf{HCF}^{\mathbf{d}}_{\text{ref}}})(\mathbf{HCF}^{\mathbf{d,i}}_{\text{ref}} - \overline{\mathbf{HCF}^{\mathbf{d}}_{\text{ref}}})^{T},$$
$$\therefore \Sigma \simeq \text{Diag}[\overline{\mathbf{HCF}^{\mathbf{d}}}]^{-1}\Sigma_o\{\text{Diag}[\overline{\mathbf{HCF}^{\mathbf{d}}}]^{-1}\}^{T}, \quad (10)$$

where

$$\overline{\text{HCF}^{d}_{\text{ref}}}(\nu) = \bar{r}_{\text{ref}}(\nu) \cdot \frac{\mathcal{F}[I_{\text{ref}}]_{SB+}}{\mathcal{F}[I_{\text{ref}}]_{SB+}} = \bar{r}_{\text{ref}}(\nu),$$

$$\text{HCF}^{d,i}_{\text{ref}}(\nu) = \bar{r}_{\text{ref}}(\nu) \cdot \frac{\mathcal{F}[I^{i}_{\text{ref}}]_{SB+}}{\mathcal{F}[I_{\text{ref}}]_{SB+}} + \varepsilon_i,$$

$$I_{\text{ref}} \simeq \frac{1}{M_{\text{ref}}}\sum_{i=1}^{M_{\text{ref}}} I^{i}_{\text{ref}}.$$

Note that no additional process is required to obtain Σ since the measurement of a reference sample is a prerequisite for the existing HCF-based techniques [13–15].

Figures 2 and 3 show the actual variance-covariance matrix of the noise and its variances (diagonal elements) obtained from a flat silicon surface used as a reference. It is clear that the noise variance is larger as the wavelength approaches its limits for both the real and imaginary parts. As expected, the noise variance is not constant over the spectral range of interest.

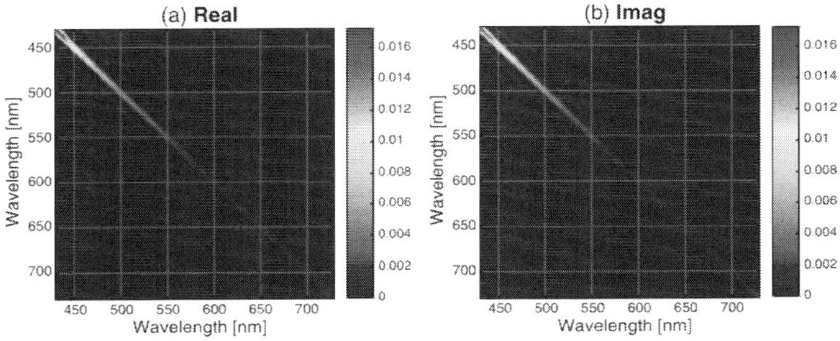

Figure 2: Noise variance-covariance matrix Σ_o from a silicon reference sample with $M_{ref} = 21 \times 21$ pixels (actual CSI measurement): (a) real part, (b) imaginary part (color available online).

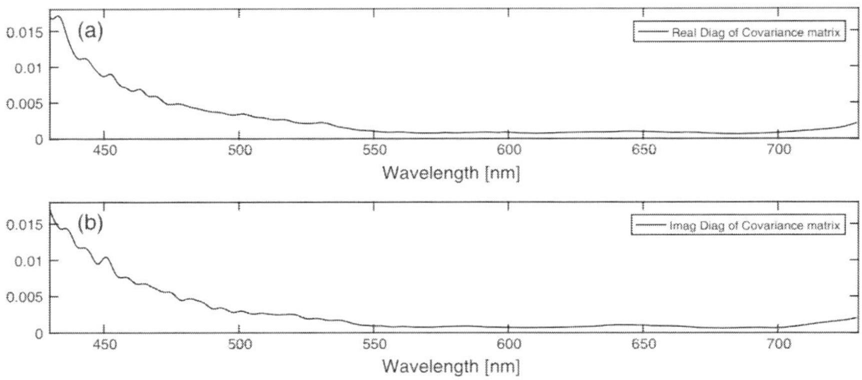

Figure 3: Noise variance in the frequency domain (actual CSI measurement). The diagonal element of the (a) real and (b) imaginary parts of the noise variance-covariance matrix illustrated in Fig. 2.

COMPUTER SIMULATION

In this section a comparison is made between the existing ISR method and the ISR-NC method. Due to the approximations made there could be a discrepancy between the approximated HCF function and the original even if the interference signals are free from noise. Thus, the performance of each method is compared as follows: the ISR method free of noise (ISR-NF) the ISR method with noise (ISR), and the ISR method with noise compensation (ISR-NC).

Simulation Setup

For the model, we assume that a nanometer-sized feature is buried under a thin film as shown in Fig. 4. The number of pixels used for the reference measurement M_{ref} is fixed as 32×32 1024 throughout the simulations. White Gaussian noise is added to the interferogram (time domain) and the light intensity is tuned to produce profiles similar to those obtained from the experimental data (Figs. 2 and 3) resulting in the noise variances shown in Fig. 5. Comparing the actual noise variances with those simulated as shown in Figs. 3 and 5, the Gaussian noise used in the simulations is considered to reasonably reproduce the actual noise in the frequency domain. Note that the wavelength range for the model is set from 400 to 730 nm, similar to that used previously [15].

Table 1 shows the results from all the models tested. To simulate a high-performance instrument having a 4M pixel camera, such as the CCI HD (Taylor Hobson), the signals will be averaged over every four pixels to create an interference signal, so that 1024 signals in each measurement become 256 averaged signals.

Results and Analysis

Comparisons between the noise-free ISR (ISR-NF), the ISR, and the noise robust ISR (ISR-NC) methods are made by examining the height of the buried substrate and the surface roughness (Sq) of the top surface and buried layer interface.

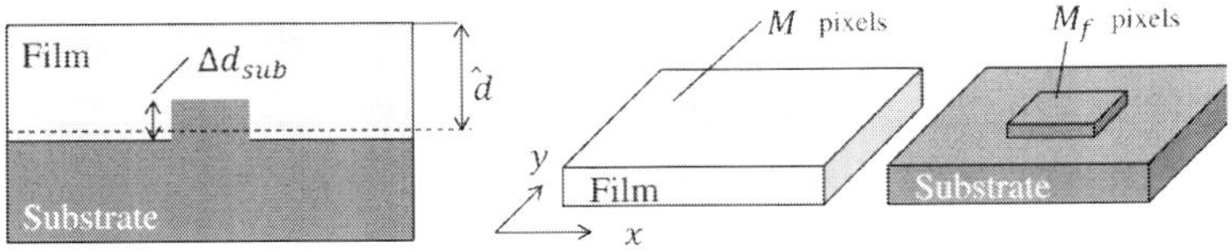

Figure 4: Schematic drawing of the model. The number of the pixels in the measurement area is M together with the pixels having the feature is M_f. Note that the global film thickness $\hat{d} \simeq d$.

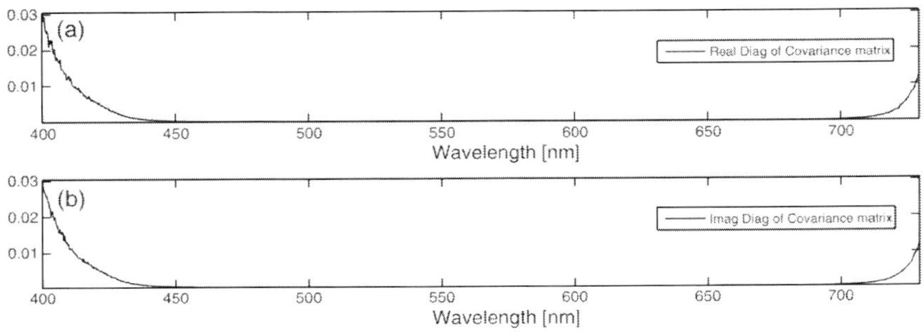

Figure 5: Noise variance in the frequency domain: The diagonal element of the (a) real and (b) imaginary parts of the noise variance-covariance matrix given to the simulations. Note that the signal-to-noise (S/N) ratio is 2000.

Table 1: Simulation Conditions Together with the Number of Pixels Allocated for the "Global" and "Featured" Areas

Sim #	Substrate Type	Film Type	Film Thickness (nm)	Feature (nm)	Noise S/N[a]	*MM*[b]	*MfMf*[b]
1-1	Si	SiO_2SiO_2	309, 514, 720, 925	5	103103	362362	8282
1-2	Si	ZrO_2ZrO_2	203, 339, 464, 610	5	103103	362362	8282
2-1	Si	SiO_2SiO_2	514	2.5, 5, 10, 20	103103	362362	8282
2-2	Si	ZrO_2ZrO_2	339	2.5, 5, 10, 20	103103	362362	8282
3-1	Si	SiO_2SiO_2	514	5	104–102104–102	362362	8282
3-2	Si	ZrO_2ZrO_2	339	5	104–102104–102	362362	8282
4-1	Si, SiC, BK7, Ge	SiO_2SiO_2	514	5	103103	362362	8282
4-2	Si, SiC, BK7, Ge	ZrO_2ZrO_2	339	5	103103	362362	8282

5-1	Si	SiO-$_2$SiO$_2$, ZrO-$_2$ZrO$_2$, Ta$_2$O-Ta$_2$O, AZO	350	5	103103	362362	8282
5-2	Si	SiO-$_2$SiO$_2$, ZrO-$_2$ZrO$_2$, Ta$_2$O-Ta$_2$O, AZO	700	5	103103	362362	8282

[a] The S/N level interval is set reasonably so that its effect on performance can be evaluated, particularly for 3-1 and 3-2. The noise added to the signals in the time domain follows a normal distribution.

[b] The number of pixels is reduced by a quarter after averaging the signal over four pixels.

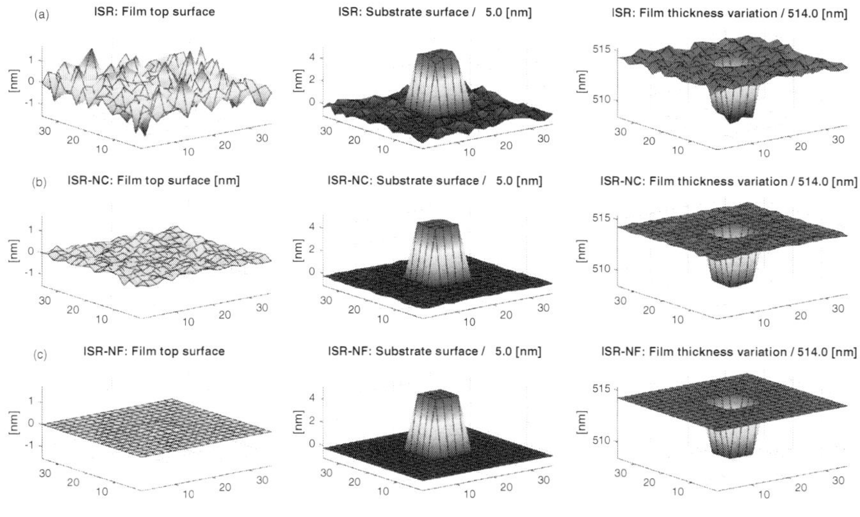

Figure 6: Comparisons between the three computational methods on the sample: SiO_2 (thicknes s = 514 nm) on a Si substrate; the feature height is 5 nm: (a) the ISR method (with noise), (b) the ISR-NC method (with noise), (c) the ISR-NF method (noise free). The S/N ratio is set at 102 to correspond to Sim 3-1 in Table 1 (color available online).

Figure 6 shows three-dimensional images of the resulting computations using the three methods. The ISR-NC method yields the smoothest surfaces, which is also the case for all the models tested since the method is free from noise.

Simulation 1: Performance Sensitivity to Variation in Thin Film Thickness

Thin films, of SiO_2 and ZrO_2, of varying thickness were investigated while other parameters remained unchanged, as shown in Table 1. The thicknesses in Table 1 correspond to odd integer multiples (3, 5, 7, 9) of the quarter-wavelength optical thickness (QWOT).

Although the feature heights determined by the ISR and ISR-NC methods were very similar, the height variance of the ISR-NC method was always less than that provided by the ISR method.

The surface roughness (Sq) of the top surface and the substrate determined by the ISR-NC method are smoother than those of the ISR method, as shown in Fig. 7. In addition, one simulation (Sim 1-2 with a 610 nm ZrO_2 film) did not work properly with the ISR method due to an inaccurate approximation. This is discussed later in Section 4.

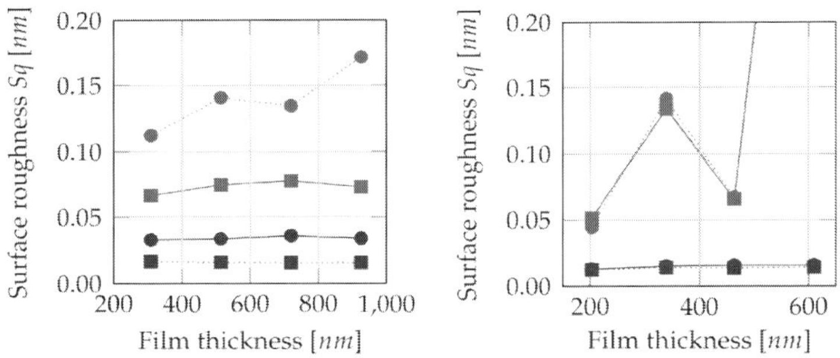

(a) Sim 1-1: SiO_2 film. (b) Sim 1-2: ZrO_2 film.

Figure 7: Surface roughness (Sq) as a function of film thickness: Red circles, top ISR; red squares, Sub ISR; blue circles, top ISR-NC; blue squares, sub ISR-NC.

Simulation 2: The Effect of Feature Height on Performance

The ISR methods, including ISR-NC, use a first-order Taylor expansion to the HCF function to make the problem linear [15,16,19]. This requires that the perturbation of the interfacial surface topography is "small." In this experiment, the noise is fixed at $S/N = 10^3$ and we evaluate the performance of the methods as a function of feature height with the fixed thin film thicknesses of 514 nm for SiO_2 and 339 nm for ZrO_2.

Figure 8 shows that the ISR methods with or without noise work well up to ~10 nm in feature height for the SiO_2 film and up to ~5 nm for the ZrO_2 film. The ISR-NC method gives a reasonable approximation up to ~10 nm for ZrO_2 film. These results, however, do not necessarily prove the superiority of the ISR-NC method. The root cause of the deterioration in performance, which is basically proportional to the feature height, is the quality of the first-order approximation to the HCF function of interest. The HCF function is poorly approximated in some wavelength regions which have a high noise variation, as shown in Fig. 5.

Simulation 3: Noise Compensation Performance

In this set of simulations, the performance of each method as a function of the signal-to-noise (S/N) ratio is investigated. The other parameters remain unchanged, as shown in Table 1.

The ISR-NF, ISR, and ISR-NC methods give similar mean feature height values regardless of the noise level for the SiO_2 film but this is not the case for ZrO_2 film. The ISR method determines the height as 4.3 nm compared to the actual value of 5 nm, as shown in Fig. 9(b). As in the previous Section 3.B.2, this is due to a poor first-order approximation of the amplitude reflection coefficient in the smaller variance wavelength regions.

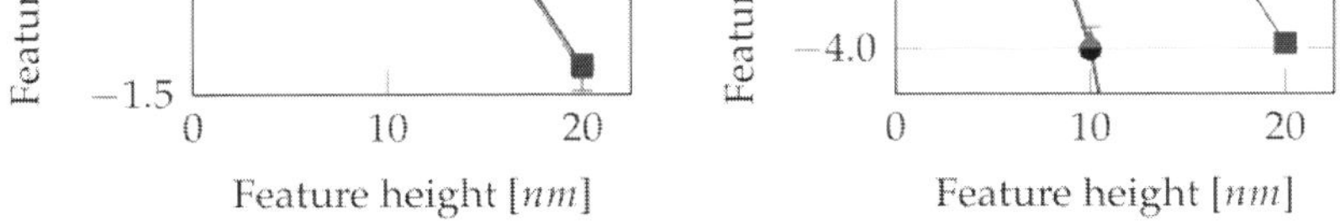

(a) Sim 2-1: SiO_2 film (514 *nm*). (b) Sim 2-2: ZrO_2 film (339 *nm*).

Figure 8: Feature height sensitivity as a function of feature height: black circles, ISR-NF; red triangles, ISR; blue squares, ISR-NC.

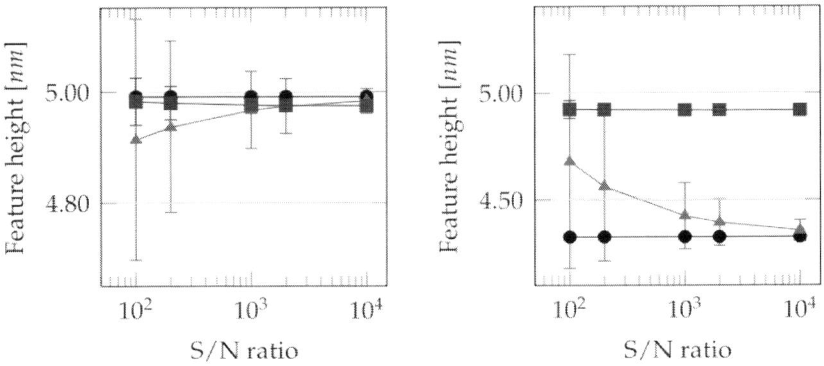

(a) Sim 3-1: SiO_2 film (514 nm). (b) Sim 3-2: ZrO_2 film (339 nm).

Figure 9: Signal-to-noise ratio sensitivity to the determined feature height: black circles, ISR-NF; red triangles, ISR; blue squares, ISR-NC.

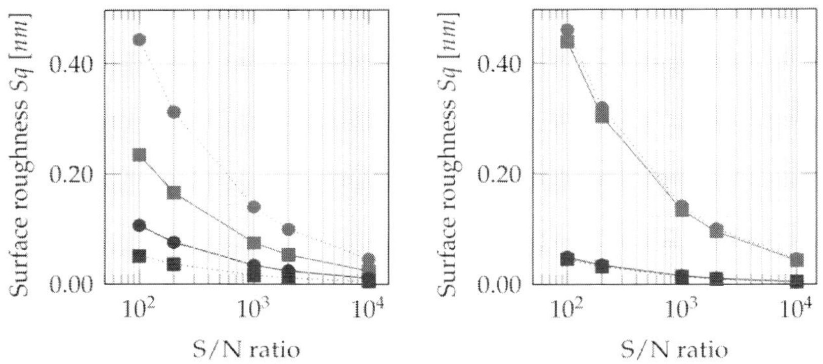

(a) Sim 3-1: SiO_2 film (514 nm). (b) Sim 3-2: ZrO_2 film (339 nm).

Figure 10: Surface roughness (Sq) as a function of the S/N ratio: red circles, top ISR; red squares, sub ISR; blue circles, top ISR-NC; blue squares, sub ISR-NC.

The thin film and substrate surface roughnesses determined by the ISR and ISR-NC methods are proportional to the increase in the S/N ratio, as shown in Fig. 10. However, the level of surface roughness determined by the ISR-NC method is lower for all noise levels.

Simulation 4: Effect of Substrate Materials

All the variables except for the substrate material are unchanged in the simulations 4-1 and 4-2, as shown in Table 1. The substrate materials used are Si, SiC, BK7 glass, and Ge.

Similar to previous results in Sections 3.B.1–3.B.3, the ISRNC method resulted in a smaller variance in feature height in both simulations 4-1 and 4-2. The film and substrate surfaces determined by the ISR-NC method are about an order of magnitude smoother than those from the ISR method.

Simulation 5: Effect of the Type of Deposited Film

Simulations 5-1 and 5-2 shown in Table 1 investigate the effect of different film materials for 350 nm and 700 nm thickness films. The ISR-NC method again provided more accurate buried surface topographies together with smaller variances.

The reconstructed surfaces were about an order of magnitude smoother irrespective of the thin film material using the ISRNC method.

Table 2: Corresponding Effective Quarter-Wavelength Optical Thickness Values at the Wavelength of 600 nm

Film Material	Index of Refraction at 600 nm	Film Thickness 350 nm	Film Thickness 700 nm
SiO_2	1.46	3.4	6.8
ZrO_2	2.21	5.2	10.3
Ta_2O_5	2.12	5.0	9.9
AZO	1.83	4.3	8.5

Table 2 shows the effective QWOT values of the films used in the simulations. Usually films with thickness greater than QWOT × 3 are considered to have enough features in the frequency domain for the HCF theory to work [15]. It follows that there should not be much difference in the simulated performance of the various films unless the first-order approximation to the HCF function is sufficiently accurate. However, the ISR method did not work for the Ta_2O_5 film in simulation 5-1 while the ISR-NC method did. This issue will be discussed in the following Section 4.

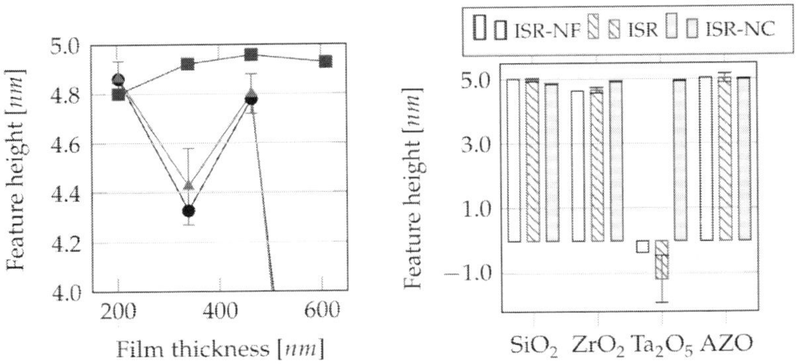

(a) Sim 1-2: ZrO_2 film. (b) Sim 5-1: 350 *nm* films.

Figure 11: Erroneous determined feature height (originally set as 5 nm): black circles, ISR-NF; red triangles, ISR; blue squares, ISR-NC.

DISCUSSION

As shown in Sections 3.B.1, 3.B.2 and 3.B.5, the ISR method does not work optimally resulting in erroneous buried feature heights such as those from simulation 2-2 with a feature height >10 nm, simulation 1-2 with a ZrO_2 600 nm film, and simulation 5-1 with a Ta_2O_5 350 nm film, as shown in Figs. 8(b) and 11, respectively. All these simulations show that the corresponding surfaces determined by the ISR-NC method are more accurately represented than those using the noise-free ISR method. The root cause of this problem lies in an inaccurate approximation to the HCF function. The lack of accuracy of the HCF function arises when the perturbation (feature height) of the interfacial topography is too large (~ > 10 nm) and when the approximated spectral amplitude reflection coefficient locally deviates from the true value resulting in a spike.

Consider first the simulation 2-2, which has a 20 nm feature height. If we compare the true HCF function with its firstorder approximation, then Fig. 12(a) shows that the first-order approximation does not hold, especially in the wavelength region between 400 and 475 nm. Fig. 12(b) shows the difference between the true HCF function HCF^{d}_{px} (without noise) and the approximated estimates by each method HCF^{s}_{px} (in the presence of noise). Prior knowledge of the noise variance-covariance matrix Σ allows the ISR-NC method to put less importance on the value of the HCF function in the specific wavelength domains where the noise is large, i.e., from ~400 to

~450 nm and from ~700 to ~730 nm, as shown in Fig. 5. This is not the case for the ISR method and is the reason why the ISRNC method provides more accurate determinations.

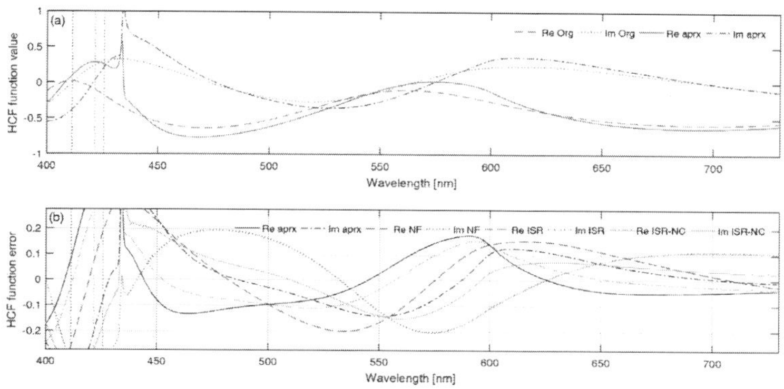

Figure 12: HCF functions generated at the feature pixel (simulation 2-2 with 20 nm feature height): (a) true HCF function (without noise) denoted by "Org" and its first-order approximation by "aprx"; (b) spectral difference between the true HCF function $\mathrm{HCF}^{d}_{\mathrm{px}}$ and the HCF functions produced by each method $\mathrm{HCF}^{s}_{\mathrm{px}}$ (noise occurs in both ISR and ISR-NC, and NF stands for ISR-NF) (color available online).

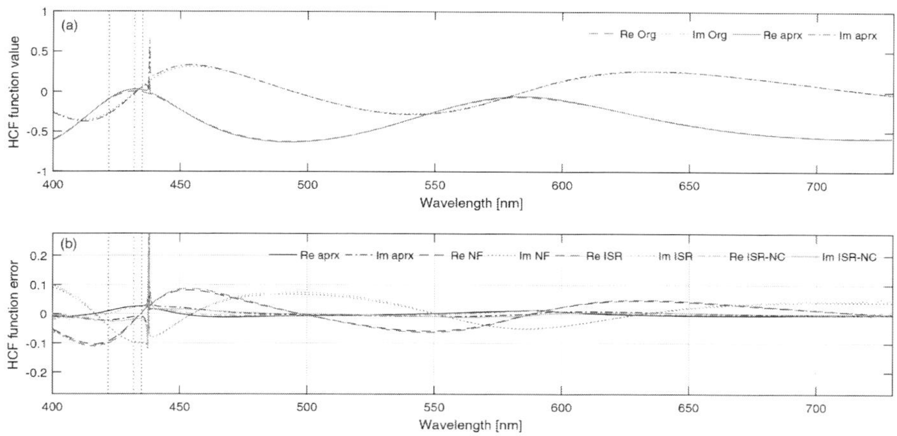

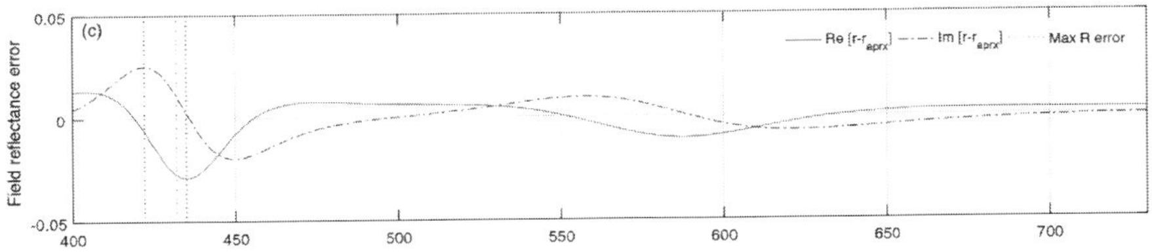

Figure 13: HCF functions generated at the feature pixel (simulation 5-1 with Ta_2O_5 film): (a) the true HCF function (without noise) denoted by "Org" and its first-order approximation by "aprx"; (b) the spectral difference between the true HCF function HCF^{d}_{px} and the HCF functions produced by each method HCF^{s}_{px} (noise exists for ISR and ISR-NC, and NF stands for ISR-NF); (c) the spectral difference between the real and imaginary parts of the true amplitude reflection coefficient and its first-order approximation. Note that the dotted lines (black and pink) represent the maximum deviations of the real $(\mathrm{Re}[r - r_{aprx}])$, imaginary $(\mathrm{Im}[r - r_{aprx}])$, and the reflectivity R, respectively (color available online).

The second cause of inaccurate surface reconstruction observed in simulations 1-2 and 5-1 is due to an inaccurate approximation to the amplitude reflection coefficient. Consider simulation 5-1 using the Ta_2O_5 thin film. The first-order approximation to the HCF function is successful, as shown in Fig. 13(a), except for a spike observed at ~440 nm wavelength denoted by "aprx". The solution provided by the ISR method defined in Eq. (5) is influenced by this spike, which reduces the fitting performance as shown in Fig. 13(b). The residual of $\|HCF^{s}_{px} - HCF^{d}_{px}\|^2$ at 435 nm wavelength is relatively small for the ISR method whereas that given by the ISR-NC method is large. It follows that the ISRNC method does not attempt to fit the spike feature due to the noise variance-covariance matrix Σ.

To confirm this further, an improvement in the performance of the ISR method was achieved by reducing the wavelength region used for numerical optimization to avoid the region in which the spikes occur To achieve a good fit between the determined and synthesized HCF functions in the frequency domain, there are two options: (1) using the ISR method with wavelength domains having less noise variance, such as from 430 to 700 nm in the examples above, or (2) using the ISR-NC method. The latter option enables the measurement of thinner films to be more stable owing to the wider wavelength domain for curve-fitting, irrespective of the noise characteristics.

CONCLUSIONS

Present methods for interfacial surface roughness measurement using CSI can be classified into two types: those that compute surface topographies in the time domain and those that determine surface topographies in the frequency domain. The methods belonging to the first group are used for films over ~1.5 μm in thickness whereas those in the second group are able to deal with thin films less than ~1.5 μm. The frequency domain methods usually use the least-squares optimization to fit the mathematical model to the measurement signal. However, the basic assumption for the method is that the noise is normally distributed and thus least-squares is not always suitable. In the examples above, the noise variance of the HCF function was not constant over the wavelength domain from 430 to 730 nm. The noise variance is always larger at the ends of the spectral region of interest where the light intensity is low. Therefore the noise variance-covariance matrix should be used in the numerical optimization of the ISR method. Such a matrix is obtained from the measurement of a known flat reference material and will vary depending on the environmental situation and the particular light source used.

Although the ISR method using the HCF function successfully determined the roughness of the thin film top surfaces and buried surfaces [15,16], spurious surface roughness in the determined substrate surfaces could be observed. This paper has presented an effective solution to that problem by introducing the noise variance-covariance matrix, which only involves a small computation when measuring the reference surface. Measurement of the reference surface is required anyway to counteract unknown changes in the phase and amplitude of the light provided by the optical system of the CSI instrument. Using these signals at the same time for the noise analysis is a beneficial side effect.

The reproducibility of the ISR-NC method was better than the existing ISR method for all the computer simulations in the presence of noise for determination of interfacial topography and surface roughness (Sq). The method was also effective over a wide wavelength range, thus allowing use of more features of the HCF function for the curve-fitting and hence better reproducibility. The noise used in the computer simulations is realistic since the noise variance-covariance matrix obtained from the flat silicon surface in Figs. 2 and 3 is similar to the noise in Fig. 5. Incorporation of noise compensation to the ISR method will improve the measurement accuracy.

ACKNOWLEDGMENT

Hirokazu Yoshino is grateful to Taylor Hobson Ltd. for funding a studentship at Loughborough University

REFERENCES

1. P. de Groot, "Coherence scanning interferometry," in *Optical Measurement of Surface Topography* (Springer, 2011), Chap. 9, pp. 187–208.
2. T. R. Thomas, "Other measurement topics," in *Rough Surfaces*, 2nd ed. (Imperial College, 1999), Chap. 3, pp. 35–61.
3. H. Yoshino, R. Smith, J. M. Walls, and D. Mansfield, "The development of thin film metrology by coherence scanning interferometry," Proc. SPIE 9749, 97490P(2016).
4. F. Blateyron, "The areal field parameters," in *Characterisation of Areal Surface Texture*(Springer, 2013), pp. 15–43.
5. B. S. Lee and T. C. Strand, "Profilometry with a coherence scanning microscope," Appl. Opt. 29, 3784–3788 (1990).
6. P. J. De Groot and X. C. de Lega, "Transparent film profiling and analysis by interference microscopy," Proc. SPIE 7064, 70640I (2008).
7. A. Bosseboeuf and S. Petitgrand, "Application of microscopic interferometry techniques in the MEMS field," Proc. SPIE 5145, 1–16 (2003).
8. S. Kim and G. Kim, "Method for measuring a thickness profile and a refractive index using white-light scanning interferometry and recording medium therefor," U.S. patent 6,545,763 (8 April 2003).
9. P. J. de Groot, "Interferometry method for ellipsometry, reflectometry, and scatterometry measurements, including characterization of thin film structures," U.S. patent 7,403,289 (22 July 2008).
10. Y. Ghim and S. Kim, "Spectrally resolved white-light interferometry for 3D inspection of a thin-film layer structure," Appl. Opt. 48, 799–803 (2009).
11. I. Abdulhalim, "Spectroscopic interference microscopy technique for measurement of layer parameters," Meas. Sci. Technol. 12, 1996–2001 (2001).
12. D. Mansfield, "Apparatus for and a method of determining characteristics of thin-layer structures using low-coherence interferometry," WO patent PCT/GB2005/002,783 (19 January 2006).
13. D. Mansfield, "The distorted helix: thin film extraction from scanning white light interferometry," Proc. SPIE 6186, 61860O (2006).
14. H. Yoshino, P. M. Kaminski, R. Smith, J. M. Walls, and D. Mansfield,

"Refractive index determination by coherence scanning interferometry," Appl. Opt. 55, 4253–4260 (2016).

15. H. Yoshino, A. Abbas, P. M. Kaminski, R. Smith, J. M. Walls, and D. Mansfield, "Measurement of thin film interfacial surface roughness by coherence scanning interferometry," J. Appl. Phys. 121, 105303 (2017).
16. D. Mansfield, "Extraction of film interface surfaces from scanning white light interferometry," Proc. SPIE 7101, 71010U (2008).
17. T. Lewis and P. Odell, "A generalization of the Gauss-Markov theorem," J. Am. Stat. Assoc. 61, 1063–1066 (1966).
18. K. Sekihara, *Introduction to Statistical Signal Processing* (Kyouritsu Shuppan, 2011).
19. D. I. Mansfield, "Apparatus for and a method of determining surface characteristics," U.S. patent 13/352,687 (12 July 2012).
20. A. Albert, "The Gauss-Markov theorem for regression models with possibly singular covariances," SIAM J. Appl. Math. 24, 182–187 (1973).

Chapter 13

Effect of Slurry Composition on the Chemical Mechanical Polishing of thin Diamond Films

Jessica M. Werrell [a], Soumen Mandal [a], Evan L. H. Thomas [a], Emmanuel B. Brousseau [b], Ryan Lewis[c], Paola Borri [c], Philip R. Davies [d] and Oliver A. Williams [a]

[a]School of Physics and Astronomy, Cardiff University, Cardiff, UK

[b]Cardiff School of Engineering, Cardiff University, Cardiff, UK

[c]School of Biosciences, Cardiff University, Cardiff, UK

[d]School of Chemistry, Cardiff University, Cardiff, UK

ABSTRACT

Nanocrystalline diamond (NCD) thin films grown by chemical vapour deposition have an intrinsic surface roughness, which hinders the

Citation: Jessica M. Werrell, Soumen Mandal, Evan L. H. Thomas, Emmanuel B. Brousseau, Ryan Lewis, Paola Borri, Philip R. Davies & Oliver A. Williams (2017) "Effect of slurry composition on the chemical mechanical polishing of thin diamond films", Science and Technology of Advanced Materials, 18:1, 654-663. https://dx.doi.org/10.1080%2F14686996.2017.1366815

development and performance of the films' various applications. Traditional methods of diamond polishing are not effective on NCD thin films. Films either shatter due to the combination of wafer bow and high mechanical pressures or produce uneven surfaces, which has led to the adaptation of the chemical mechanical polishing (CMP) technique for NCD films. This process is poorly understood and in need of optimisation. To compare the effect of slurry composition and pH upon polishing rates, a series of NCD thin films have been polished for three hours using a Logitech Ltd. Tribo CMP System in conjunction with a polyester/polyurethane polishing cloth and six different slurries. The reduction in surface roughness was measured hourly using an atomic force microscope. The final surface chemistry was examined using X-ray photoelectron spectroscopy and a scanning electron microscope. It was found that of all the various properties of the slurries, including pH and composition, the particle size was the determining factor for the polishing rate. The smaller particles polishing at a greater rate than the larger ones.

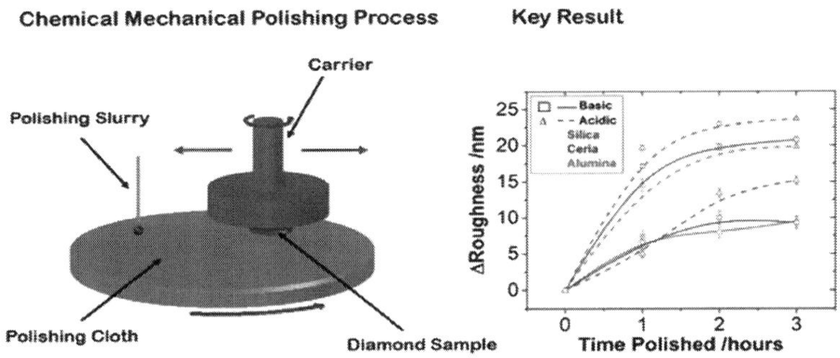

Keywords: Chemical mechanical polishing, nanocrystalline diamond, surface roughness, chemical vapour deposition

INTRODUCTION

Nanocrystalline diamond (NCD) thin films grown by chemical vapour deposition (CVD) are able to retain many of the properties of single crystal diamond and can be produced with high quality at relatively low cost [1Williams OA. Nanocrystalline diamond. Diamond Relat Mater. 2011;20:621–640. DOI:10.1016/j.diamond.2011.02.015]. These two properties have generated an interest for using NCD thin films in a number of areas such as tribology, optical coatings, electrochemistry, thermal management, surface acoustic wave (SAW) devices and micro-electro-mechanical systems (MEMS) [2Williams O, Kriele A, Hees J, et al. High young's modulus in ultra thin nanocrystalline diamond. Chem Phys Lett. 2010;495(1–3):84–89. DOI:10.1016/j.cplett.2010.06.054]. To align with current fabrication techniques, NCD thin films are grown on non-diamond substrates such as silicon. Successful and efficient growth on the majority of foreign substrates requires a nucleation enhancement step. One such method involves seeding the substrates with detonation nanodiamond (DND) particles [3Williams OA, Douhéret O, Daenen M, et al. Enhanced diamond nucleation on monodispersed nanocrystalline diamond. Chem Phys Lett. 2007;445(4–6):255–258. DOI:10.1016/j.cplett.2007.07.091]. During CVD growth these DND particles then grow via a Volmer–Weber model [4Jiang X, Schiffmann K, Klages C-P. Nucleation and initial growth phase of diamond thin films on (100) silicon. Phys Rev B. 1994;50:8402–8410. DOI:10.1103/PhysRevB.50.8402], i.e. the DND particles grow longitudinally and laterally in the plane on the surface of the substrate until they coalesce to form a film. This film will comprise diamond grains of varying sizes and facet orientations. CVD growth for each grain from this point onwards will be predominantly normal to the surface. The growth rate of diamond grains will depend upon their facet orientation in accordance with the Van der Drift growth model [5Smereka P, Li X, Russo G, et al. Simulation of faceted film growth in three dimensions: microstructure, morphology and texture. Acta Mater. 2005;53(4):1191–1204. DOI:10.1016/j.actamat.2004.11.013]. As has alreadybeen stated, there is a variation in facet orientation between each individual diamond grain. As a result there is a variation in growth rate between the different grains. This leads to competitive growth between the individual grains and results in an uneven final film layer, or surface roughness, which evolves with film thickness. This surface roughness acts as a significant barrier to the construction and performance of NCD thin films for various applications, such as SAW devices and MEMS to name but two [6Ergincan O, Palasantzas G, Kooi BJ. Influence of surface modification on the quality factor of microresonators. Phys Rev B. 2012;85:205420. DOI:10.1103/PhysRevB.85.205420].

One method to reduce this surface roughness involves interrupting the crystal growth and limiting the maximum size of the diamond grains. This is done by reducing the H2H2and/or increasing the CH4CH4 content of the plasma. This also leads to an increase in the sp2sp2 content of the material at the grain boundaries which in turn reduces the Young's modulus [1Williams OA. Nanocrystalline diamond. Diamond Relat

Mater. 2011;20:621–640. DOI:10.1016/j.diamond.2011.02.015,7Liu WL, Shamsa M, Calizo I, et al. Thermal conduction in nanocrystalline diamond films: effects of the grain boundary scattering and nitrogen doping. Appl Phys Lett. 2006;89(17):171915. DOI:10.1063/1.2364130,8Bhusari D, Yang J, Wang T, et al. Highly transparent nano-crystalline diamond films grown by microwave CVD. Solid State Commun. 1998;107(6):301–305. DOI:10.1016/S0038-1098(98)00226-9]. Also for films grown under these conditions it is difficult to achieve a root mean square (RMS) roughness lower than 5 nm [2Williams O, Kriele A, Hees J, et al. High young's modulus in ultra thin nanocrystalline diamond. Chem Phys Lett. 2010;495(1–3):84–89. DOI:10.1016/j.cplett.2010.06.054]. This figure still exceeds the accepted limit for the fabrication for their various applications [6Ergincan O, Palasantzas G, Kooi BJ. Influence of surface modification on the quality factor of microresonators. Phys Rev B. 2012;85:205420. DOI:10.1103/PhysRevB.85.205420,9Iriarte G, Rodríguez J, Calle F. Synthesis of c-axis oriented aln thin films on different substrates: a review. Mater Res Bull. 2010;45(9):1039–1045. DOI:10.1016/j.materresbull.2010.05.035] and so restricts the use of these films.

Another technique involves etching away the silicon substrate and using the nucleation diamond grains as the surface [10Rodriguez-Madrid JG, Iriarte GF, Pedros J, et al. Super-high-frequency saw resonators on aln/diamond. IEEE Electron Device Lett. 2012;33(4):495–497. DOI:10.1109/LED.2012.2183851]. The surface roughness will be significantly less this side of the diamond film than the other. Unfortunately the quality of this side of the diamond film is inferior to the upper coalesced film with reduced values of Young's modulus and thermal conductivity [10Rodriguez-Madrid JG, Iriarte GF, Pedros J, et al. Super-high-frequency saw resonators on aln/diamond. IEEE Electron Device Lett. 2012;33(4):495–497. DOI:10.1109/LED.2012.2183851]. This technique also requires thick free standing films and complex fabrication processes [10Rodriguez-Madrid JG, Iriarte GF, Pedros J, et al. Super-high-frequency saw resonators on aln/diamond. IEEE Electron Device Lett. 2012;33(4):495–497. DOI:10.1109/LED.2012.2183851].

Alternatively, surface roughness can be reduced through polishing the as-grown rough NCD thin films. Historically, diamond polishing has involved the use of diamond on diamond in a contact mechanical polishing mechanism [11Chen Y, Zhang L. Polishing of diamond materials. 1st ed. Portalnd: Springer Series; 2013.]. This process is impractical for NCD thin films because it produces uneven wear rates and the films are susceptible to shattering. The uneven wear rate is a result of the mechanical polishing of diamond being highly dependent upon both the crystallographic orientation of the diamond and the azimuthal angle of polishing [12Grillo SE, Field JE. The polishing of diamond. J Phys D Appl Phys. 1997;30(2):202. DOI:10.1088/0022-3727/30/2/007,13El-Dasher BS, Gray JJ, Tringea JW, et al. Crystallographic anisotropy of wear on a polycrystalline diamond surface. Appl Phys Lett. 2006;88:241915-1–241915-3. DOI:10.1063/1.2213180]. As stated, the surface roughness of NCD thin films is a result of crystallographic orientation variation; therefore, this type of polishing can lead to protruding crystals left behind on the film surface [14Bögli U, Blatter A, Pimenov S, et al. Tribological properties of smooth polycrystalline diamond films. Diamond Relat Mater. 1995;4(7):1009–1019. DOI:10.1016/0925-9635(95)00272-3]. Potential shattering is the result of the high pressures typical of mechanical polishing and the characteristic wafer bow of NCD thin films [11Chen Y, Zhang L. Polishing of diamond materials. 1st ed. Portalnd: Springer Series; 2013.]. There is an initial wafer bow always present in the substrate which is then compounded during cooling after CVD growth as a result of the difference in the coefficient of expansion between the diamond film and its non-diamond substrate [15Fan QH, Grácio J, Pereira E. Residual stresses in chemical vapour deposited diamond films. Diamond Relat Mater. 2000;9(9–10):1739–1743. DOI:10.1016/S0925-9635(00)00284-3,16Edwards MJ, Bowen CR, Allsopp DWE, et al. Modelling wafer bow in silicon-polycrystalline CVD diamond substrates for gan-based devices. J Phys D Appl Phys. 2010;43(38):385502. DOI:10.1088/0022-3727/43/38/385502]. The bow then places additional stress on the NCD thin film, especially when pushed against planar surfaces, making it more susceptible to damage whilst being subjected to heavy mechanical polishing.

To overcome these issues, the gentle chemical mechanical polishing (CMP) technique was adapted by Thomas et al. [17Thomas ELH, Nelson GW, Mandal S, et al. Chemical mechanical polishing of thin film diamond. Carbon. 2014;68:473–479. DOI:10.1016/j.carbon.2013.11.023]. This technique utilises a soft polyester based pad as the polishing base and a silica based colloid as the polishing slurry at room temperature. It is a commonly used technique in the integrated circuit (IC) fabrication industry for polishing silicon wafers [18Zantye PB, Kumar A, Sikder A. Chemical mechanical planarization for microelectronics applications. Mater Sci Eng R Rep. 2004;45(3–6):89–220. DOI:10.1016/j.mser.2004.06.002] and has been shown to polish both NCD [17Thomas ELH, Nelson GW, Mandal S, et al.

Chemical mechanical polishing of thin film diamond. Carbon. 2014;68:473–479. DOI:10.1016/j.carbon.2013.11.023] and bulk single crystal diamond (SCD) successfully [19Thomas ELH, Mandal S, Brousseau EB, et al. Silica based polishing of (100) and (111) single crystal diamond. Sci Technol Adv Mater. 2014;15(3):035013. DOI:10.1088/1468-6996/15/3/035013].

In both of the aforementioned studies, a basic polishing slurry containing silica (SiO2SiO2) particles was used. In the present study, three different polishing particles, ceria (CeO_2), alumina (Al_2O_3) and silica, all common to the IC fabrication polishing industry, were used to examine how composition affects the polishing rates. For each particle type there was an acidic (pH ≈6≈6) and a basic (pH ≈9≈9) slurry. The aim of this study was to compare and contrast the different slurries to understand better how silica, a material with a density of 2.4gcm−32.4gcm-3 and a hardness of 7 [20Oliver MR, Evans D, Stein D, et al. Chemical-mechanical, planarization of semiconductor materials. 1st ed. Portalnd: Springer Series; 2003. [Google Scholar]] on the Mohs scale, is able to polish diamond of density 3.52gcm−33.52gcm-3 [11Chen Y, Zhang L. Polishing of diamond materials. 1st ed. Portalnd: Springer Series; 2013.] and hardness 10 [20Oliver MR, Evans D, Stein D, et al. Chemical-mechanical, planarization of semiconductor materials. 1st ed. Portalnd: Springer Series; 2003. [Google Scholar]]. The significant difference in hardness and density does not make the polishing far-fetched, considering that harder materials can be polished by softer particles [21Neslen CL, Mitchel WC, Hengehold RL. Effects of process parameter variations on the removal rate in chemical mechanical polishing of 4h-sic. J Electron Mater. 2001;30:1271–1275. DOI:10.1007/s11664-001-0111-2–23Babu SV, Jindal A, Li Y. Chemical-mechanical planarization of cu and ta. JOM. 2001;53(6):50–52. DOI:10.1007/s11837-001-0104-x], e.g. tantalum by silica [23Babu SV, Jindal A, Li Y. Chemical-mechanical planarization of cu and ta. JOM. 2001;53(6):50–52. DOI:10.1007/s11837-001-0104-x].

EXPERIMENTAL PROCEDURE

For this study, a series of ≈≈360 nm thick NCD thin films were grown on a 500 nm thick buffer layer of silicon dioxide that coated a 500μm500μm thick p-type silicon (100) wafer 2-inches in diameter. Before growth, these wafers were cleaned using the standard SC-1 process [24Kern W. The evolution of silicon wafer cleaning technology. J. Electrochem. Soc.. 1990;137(6):1887–1892. DOI:10.1149/1.2086825] of 30% H_2O_2 : NH_4OH : deionised (DI) H_2O (1:1:5) at 75∘C75∘C for 10 minutes. The substrates were then rinsed in DI

H_2O in an ultrasonic bath for 10 minutes and spun dry. For the seeding step, the wafers were placed in a mono-dispersed nanodiamond (with a diameter of ~~5 nm) and DI H_2O colloid which was then agitated in an ultrasonic bath for 10 minutes. This processes encourages the nanodiamond particles to bond to the surface of the substrate via electrostatic attraction and is known to produce nucleation densities exceeding 1011cm−21011cm-2 [3Williams OA, Douhéret O, Daenen M, et al. Enhanced diamond nucleation on monodispersed nanocrystalline diamond. Chem Phys Lett. 2007;445(4–6):255–258. DOI:10.1016/j.cplett.2007.07.091]. After this the wafers were rinsed, spun dry at 3000 rpm, and then immediately placed inside the CVD chamber. CVD was carried out in a Seki 6500 series microwave plasma reactor under 3% CH_4/H_2 conditions at 47 Torr and 4.2 kW microwave power. Upon termination of growth, the films were cooled down in hydrogen plasma to ensure hydrogen termination and prevent deposition of non-sp33 material. Substrate temperatures were ≈830°C as determined by a dual-wavelength pyrometer, with substrate heating solely from the microwave induced plasma. Film thickness was determined *in situ* through the use of laserinterferometry, and *ex situ* with a Filmetrics F-20 Spectral Reflectance system. The system was modelled as roughness on diamond on silicon dioxide on silicon. Known values of the wave number k and refractive index n were used for the diamond, silicon dioxide and silicon from the Filmetrics database to determine the thickness. The RMS roughness of each film was measured using a Park Systems Park XE-100 atomic force microscope (AFM) with a Tespa-V2 tip in non-contact mode. The XY spatial resolution was 20 nm and the Z spatial resolution was 0.2 nm. The AFM images were analysed using the Park Systems software programme XEI Image Processing. An average of three 25μm^2-pagination areas of the as-grown films showed that they all had an approximate RMS average of 25.1±2.7nm.

Three different polishing particles were used: ceria (CeO_2), alumina (Al_2O_3) and silica (SiO_2). For each particle type there was an acidic (pH≈6) and a basic (pH≈9) slurry. The basic silica polishing slurry was SF1 Polishing Fluid from Logitech Ltd., which was made acidic in house through the addition of phosphoric acid (H_3PO_4). The alumina particle based polishing slurries were supplied by Saint Gobain Ltd., brand names Polycrystalline Alumina Polishing Slurries 9240 and 9245. Finally, the ceria based polishing slurry was supplied by Eminess, brand name Ultra-Sol®® Optiq, which arrived basic and was also acidified. Table 1 shows a summary of the specific properties of each individual slurry. The particle size and particle content was provided by the slurry manufacturers. The particle diameters were also measured using a dynamic light scattering (DLS) technique. To do this the slurries were diluted to a 1/1000 ratio and then measured using a Malvern Instruments Ltd. Zetasizer Nano Z device. pH values were measured using a Mettler Toledo™ FG2 FiveGo™ Portable pH Meter.

Table 1: Properties of slurries. Density and hardness values taken from [20Oliver MR, Evans D, Stein D, et al. Chemical-mechanical, planarization of semiconductor materials. 1st ed. Portalnd: Springer Series; 2003.]. The particle size and particle content were provided by the slurry manufacturers. DLS particle diameters and pH were measured in this work

Property	Silica	Alumina	Ceria
Density ($g\,cm^{-3}$)	2.2–2.6	4.0	7.1
Hardness (Mohs)	6–7	9	6
Particle diameter measured using DLS (μm)	0.1 (basic) 0.1 (acidic)	1.0 (basic) 0.2 (acidic)	0.5 (basic) 0.5 (acidic)
Particle size according to manufacturer (μm)	Unavailable	0.4 (basic) 0.3 (acidic)	0.4 (basic) 0.4 (acidic)
Particle content (%)	15–50	20 (basic) 20 (acidic)	20 (basic) 20 (acidic)
pH of solution	9.6 5.6	9.1 5.8	8.9 5.8

The NCD films were polished by CMP using a Logitech Ltd. Tribo CMP System in conjunction with a SUBA™ X polishing pad and the chosen slurry, at intervals of one hour for a total of three hours. Before use, the polyester polishing pad was conditioned for 30 minutes using an abrasive conditioning chuck consisting of a nickel plate embedded with diamond grit, and DI H_2O to ensure it had a high surface roughness for maximum polishing action and slurry distribution [25McGrath J, Davis C. Polishing pad surface characterisation in chemical mechanical planarisation. J Mater Process Technol. 2004;153–154:666–673. DOI:10.1016/j.jmatprotec.2004.04.094]. During film polishing, both the pad and carrier were kept at 60 rpm rotating in opposite directions, while the carrier swept across the pad. Down pressure was kept at 2 psi, while a backing pressure of 20 psi was applied in order to account crudely for the NCD film bow. After initial wetting of the plate, the feed slurry rate was kept at 40 ml/min. At hourly intervals the films underwent a clean using the standard SC−1 process and their roughness and thickness were measured.

For comparison of the surface chemistry of each film after the maximum duration of polishing, X-ray photoelectron spectroscopy (XPS) data was taken using a Thermo Scientific™K-Alpha++ spectrometer. Spectra were acquired using a monochromatic Al source operated at 72 W (6mA×12kV) and at pass energies of 40 and 150 eV for high resolution and survey scans, respectively, over an analysis area of 400μm400μm. Charge compensation was achieved using the K-Alpha charge neutralisation system, which employs a combination of both electrons and low energy argon ions. Spectra requiring charge neutralisation were subsequently calibrated to the C1s line at 285.0 eV. The results were analysed using the software CasaXPS. Scanning electron microscope (SEM) images were also taken of the final films using the in-lens detector of a Raith eLine system operated at 10 kV and a working distance of 10 mm.

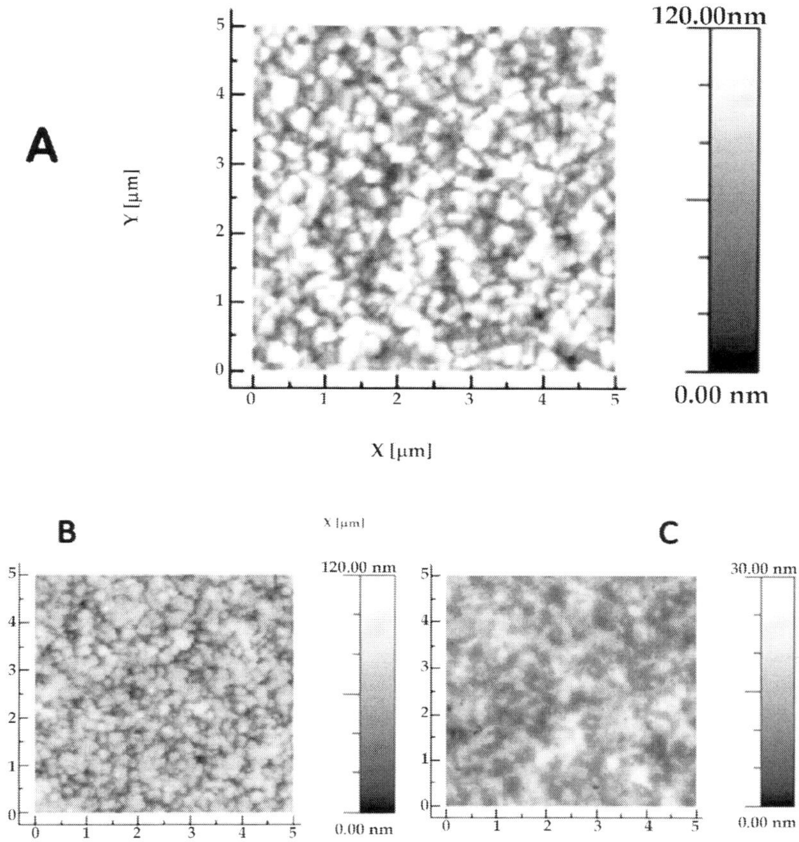

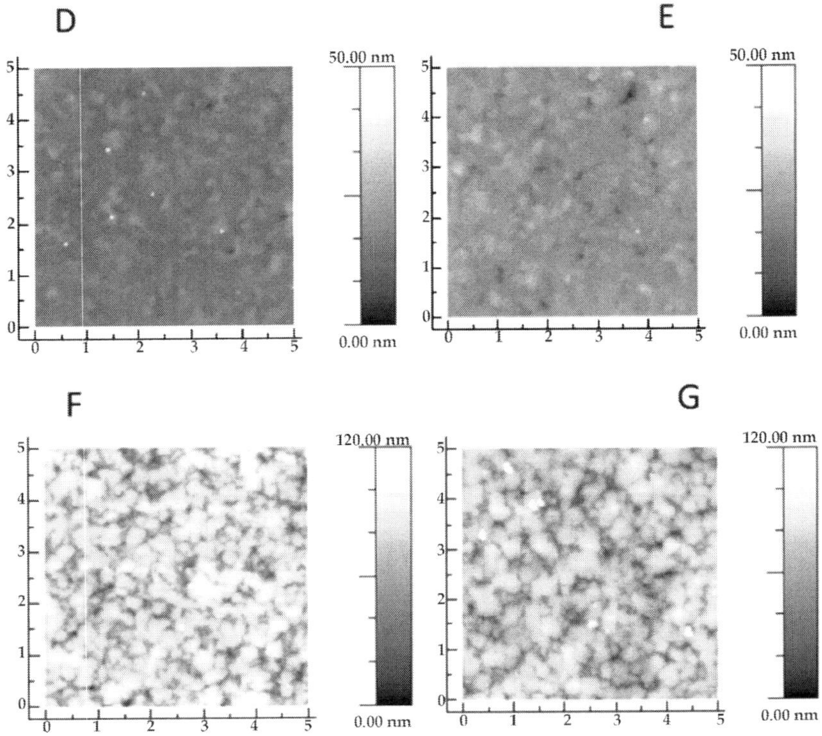

Figure 1: AFM images of the NCD films. (A) Shows the as-grown NCD film and is included here for comparison with the polished films. (B) NCD film after three hours of basic alumina polishing. (C) NCD film after three hours of acidic alumina polishing. (D) NCD film after three hours of basic silica polishing. The small white dots on the surface of Figure 1(D) are due to dust particles as a result of the samples exposure to air. They are only visible because the diamond surface is so smooth that minor particles can be observed as perturbations. They are few in number and sufficiently minor that they are unlikely to affect the AFM measurement significantly. (E) NCD film after three hours of acidic silica polishing. (F) NCD film after three hours of basic ceria polishing. (G) NCD film after three hours of acidic ceria polishing. There is a strong variation in the roughness reduction of each film for the different slurries. The films polished by basic and acidic silica – (D) and (E), respectively – are significantly smoother than the as-grown film (A). Whereas the surface polished by basic ceria (F) and the surface polished by acidic ceria (G) look similar to the as-grown film. Interestingly, the film polished by basic alumina looks similar to the surfaces polished by ceria – (G) and (F) – whilst the film polished by acidic alumina (C) looks smoother than these but not as smooth as the silica-polished surfaces – (D) and (E).

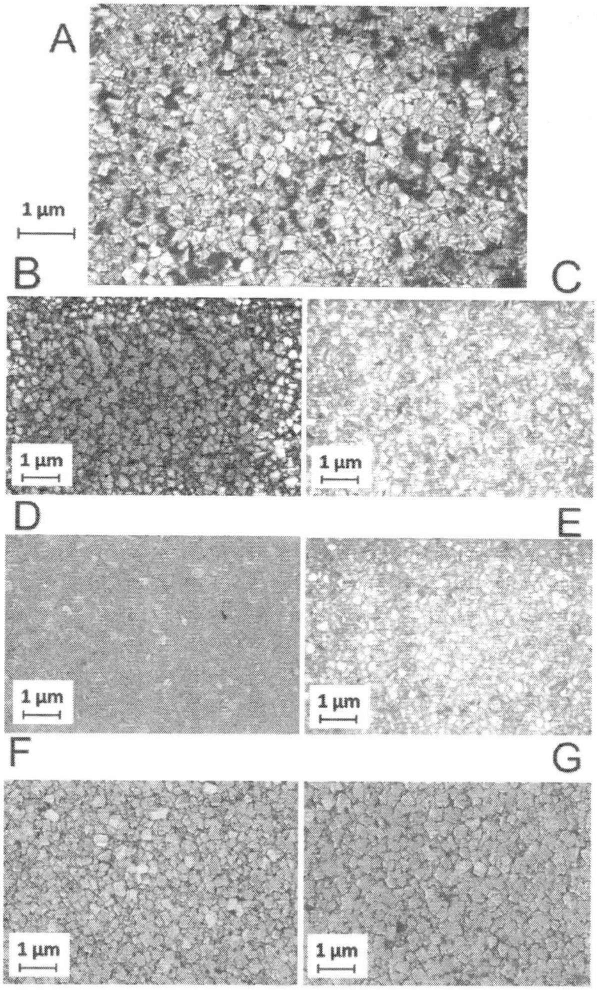

Figure 2: SEM images of the NCD films. (A) Shows the as-grown NCD film and is included here for comparison with the polished films. (B) NCD film after three hours of basic alumina polishing. (C) NCD film after three hours of acidic alumina polishing. (D) NCD film after three hours of basic silica polishing. (E) NCD film after three hours of acidic silica polishing. (F) NCD film after three hours of basic ceria polishing. (G) NCD film after three hours of acidic ceria polishing. It is clear just from these images that there is a variation in the roughness reduction of the different slurries. The films polished by basic silica (D), acidic silica (E) and acidic alumina (C) are significantly smoother than the as-grown film (A). Whereas the surfaces polished by basic ceria (F), acidic ceria (G) and basic alumina (B) look similar to the as-grown film. Interestingly,

there is a variation between the acidic and basic slurries containing alumina particles but not for the other two particles. These results agree with the AFM measurements.

RESULTS AND DISCUSSION

Morphology

Figure 1 shows AFM images of seven NCD thin films and Figure 2 shows SEM images of the same set of films. Image (A) for each case is the as-grown film and is included in the set for comparison against six polished films. As stated in the caption to Figure 1, the small white dots on the surface of Figure 1(D) are dust particles; they are only visible because the surface is so smooth that minor particles can be observed as perturbations. Rougher AFMs do not show these dots due to tip convolution and the dust particles' being very small compared to surface roughness. The dust particles are few in number and sufficiently minor that they are unlikely to affect the AFM measurement significantly. If there were any other type of defect on the surface they would have appeared on the SEM image of the same sample but, as can be seen from Figure 2(D), they do not. It is clear just from these images that there is a significant variation in the roughness reduction of the different slurries. The surface of the NCD thin film polished by basic silica (D) and the surface of the NCD thin film polished by acidic silica (E) are significantly smoother than the as-grown film (A). Whereas the surface polished by basic ceria (F) and the surface polished by acidic ceria (G) look similar to the as-grown film. Interestingly the surface polished by basic alumina looks similar to the surfaces polished by ceria – (G) and (F) – whilst the surface polished by acidic alumina (C) looks smoother than these but not as smooth as the silica-polished surfaces (D) and (E). This can be seen from both the AFM and the SEM images.

Figure 3 graphically shows the reduction in the RMS roughness of each film after each hour of polishing up to a total of three hours. The average RMS of three25μm225μm2 areas of each polished film is plotted with their respective standard deviations. Figure 4 shows the corresponding thickness reduction. For those slurries that polished with the greater rate there is also a corresponding greater reduction in the thickness of the diamond. This is to be expected as the polishing is removing the roughness layer in order to produce a more uniform film.

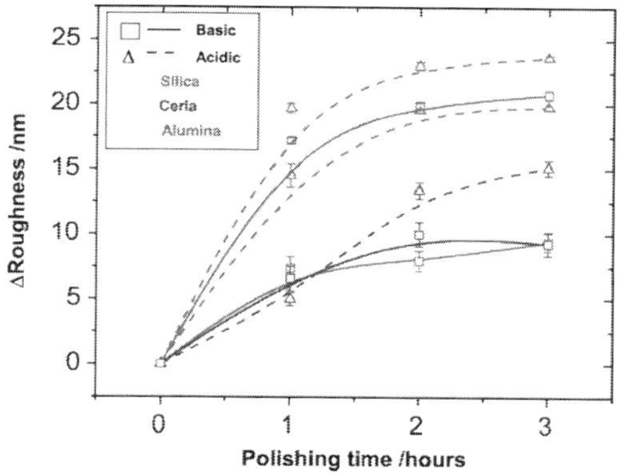

Figure 3: The reduction in the RMS roughness of each film after each hour of polishing up to a total of three hours. The average RMS of three 25μm^2 areas of each polished film is plotted with their respective standard deviations, the zero mark being the roughness of the as-grown diamond film. It can be seen clearly that, regardless of particle type, the acidic slurry always leads to a greater RMS roughness reduction compared to its basic counterpart. However, the variation between basic and acidic is inconsistent between particles, suggesting that pH alone cannot be a deciding polishing factor. The slurries containing silica particles polish at greater rates than the ceria particles but there is an inconsistency here with the alumina slurries. This large variation between the alumina slurries means particle composition is not the only deciding polishing factor like pH.

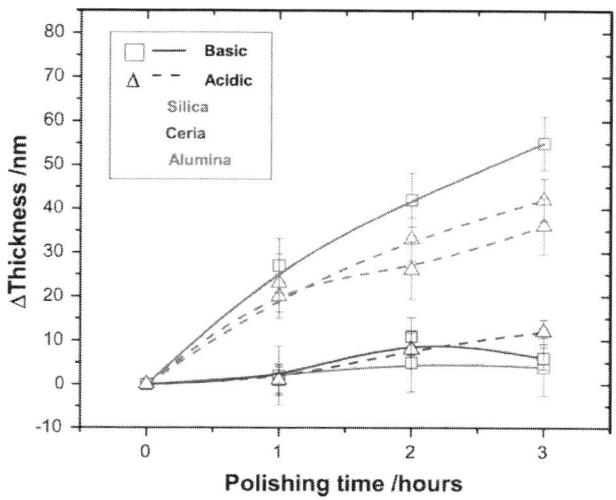

Figure 4: The change in the thickness of the NCD films during the three hours of polishing, the zero mark being the thickness of the as-grown diamond film.

For those slurries that polished with the greater rate (see Figure 3) there is also a corresponding greater reduction in the thickness of the diamond. This is to be expected as the polishing removes the diamond peaks.

The zero mark for each graph being the roughness and thickness respectively of the as-grown diamond film. Only three hours of polishing were performed for this study because wear rate was seen to drop significantly after the first hour and an observable difference in polishing was recorded within this time period.

From Figure 3, it is clear that the acidic slurries polish at a greater rate than their basic counterparts; however, the degree to which they polish varies widely. In the three hours, the acidic silica polishing slurry reduced the surface roughness by 23.7±0.4nm, the basic version of this slurry reduced surface roughness in the same time by 20.8±0.6nm20.8±0.6nm. Whilst not within errors this is at most a difference of 3.8 nm. In contrast, the alumina particles show a difference of 10.4±1.0 nm. The ceria particles show 5.8±1.1 nm. The variation is not consistent between the particles, showing that the pH alone cannot be a deciding polishing factor.

A comparison of the results considering only the particle composition shows that the silica particles polish at greater rates than the ceria particles but there is an inconsistency here with the alumina slurries. The acidic alumina polishes at a rate similar to the silica but the basic alumina at a rate comparable to ceria. This large variation between the alumina slurries means particle composition, and therefore their particles properties such as Mohs hardness and density, is not the only deciding polishing factor.

Comparing Figure 3 with the other known slurry properties in Table 1, a relationship between polishing rate and particle size is revealed. Figure 5 plots the diameter of the respective particles and their standard deviation, as measured by DLS, against the change in roughness after two hours of polishing. There is a clear correlation between the diameter and rate, with the smaller particles polishing faster than the larger particles. The faster rate is most likely due to an increase in contact area as a result of the use of smaller sized particles. The percent particle contents of the slurries are within an order of magnitude of each other, which means that those slurries with smaller particles will also contain a greater number of the said particles.

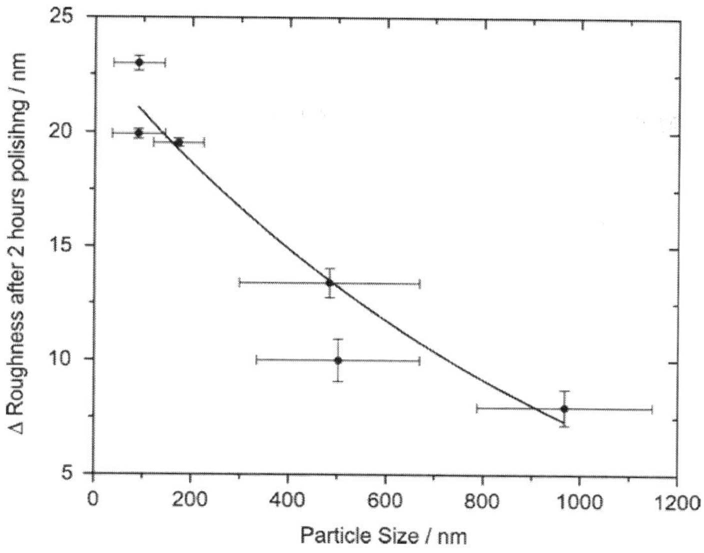

Figure 5: Change in the RMS roughness after two hours polishing compared against the size of the polishing particles and their respective standard deviation, measured by DLS. There is a clear correlation between particle size and rate. Smaller particles, with a diameter less than 200 nm, polish NCD thin films at a greater rate than the larger ones with a diameter greater than 500 nm.

X-ray Photoelectron Spectroscopy

XPS was taken of the six polished films and an as-grown film in order to compare the surface chemistry before and after polishing, the aim being to observe if the components from the slurry were bonding to the surface to facilitate CMP. The dust particles previously noticed by the AFM are highly unlikely to be the components of the slurries (ceria, silica or alumina) so there was no concern regarding contamination of the XPS data, the results of which are displayed in Figure 6.

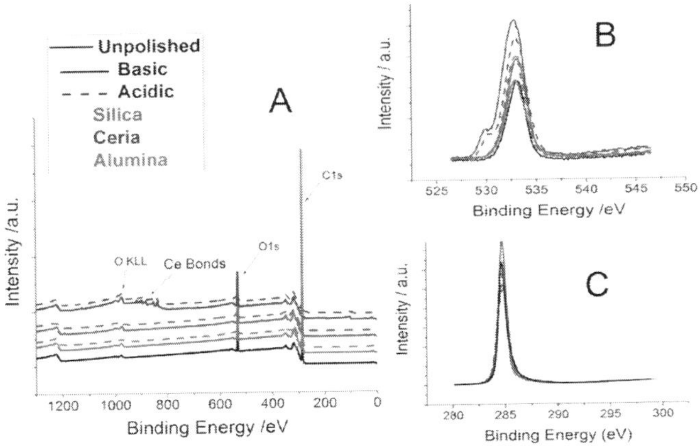

Figure 6: XPS data of the six polished films and an as-grown film. Panel (A) shows a sweep of the binding energy from 0 to 1200, revealing, on a broad spectrum, what is present on the surface of the diamond films. Panel (B) shows the O1s peaks and Panel (C) the C1s peaks.

Panel (A) shows that the only NCD thin films whose surface chemistry significantly changed were those polished by the ceria particle slurries. In the case of these two films, ceria particles remained bonded to the surface, detected around 900 eV, even after cleaning using the standard SC-1 process. There was a very weak Al2Al2 signal on the film that was polished using the basic alumina slurry but not for the acidic alumina. It is possible that the cleaning removed any trace of the Al and Si that was previously bonded to the surface.

Panel (B) depicts only the O1s signal of the surface; a sweep of binding energy from 520 up to 550 eV. The major peak at 532 eV corresponds to the C–O bond. A second peak can be observed at 530 eV corresponding to the Ce–O bond. Panel (B) shows that aside from ceria there is no other change to the type of oxygen bonds present on the surface. Panel (C) shows only the C1s signal with a sweep from 280 to 300 eV. Only the C–C bond is present with no other change observed before and after polishing.

Table 2 shows the O1s/C1s ratio for the unpolished film and the six polished films. Comparing these values shows how the oxygen environment relative to the carbon environment has changed after polishing. For all the films polished, there was an increase in the oxygen environment on the surface – implying the chemical nature of the technique. There was some variation between the oxidation levels. Those films polished by ceria show the highest increase to its oxygen environment. This is not surprising considering the ceria particles are still connected to the surface as well. The film polished with basic alumina shows very little oxygen increase.

Table 2: The O1s/C1s ratios for the films before and after polishing. The oxygen environment is seen to increase after polishing for all the films. For all the films polished, there was an increase in the oxygen environment on the surface – implying the chemical nature of the technique. There was some variation between the oxidation levels

Polishing slurry	O1s/C1s ratio
None	0.1613
Basic alumina	0.1680
Acidic alumina	0.2022
Basic silica	0.2033
Acidic silica	0.2247
Acidic ceria	0.3123
Basic ceria	0.3567

Discussion

Thomas et al. proposed that the CMP of NCD thin films followed a mechanism similar to the CMP of silicon dioxide [17Thomas ELH, Nelson GW, Mandal S, et al. Chemical mechanical polishing of thin film diamond. Carbon. 2014;68:473–479. DOI:10.1016/j.carbon.2013.11.023]. In this more traditional polishing process, the hydroxide ions within the polishing fluid react with the surface siloxane (Si–O–Si) bonds, creating a silanol based passivation layer ($Si(OH)_4$) [26Krishnan M, Nalaskowski JW, Cook LM. Chemical mechanical planarization: slurry chemistry, materials, and mechanisms. Chem Rev. 2010;110(1):178–204. DOI:10.1021/cr900170z,27Hochenga H, Tsaia HY, Sub YT. Modeling and experimental analysis of the material removal rate in the chemical mechanical planarization of dielectric films and bare silicon waferss. Electrochem Soc. 2001;148(10):G581–G586. DOI:10.1149/1.1401087]. Silica particles within the polishing fluid will then attach themselves to the hydrated groups of the passivation layer. The polishing pad then introduces a shearing force on the said silica particles. If the energy from this shearing action is larger than the binding energy, the molecule will be removed, resulting in polishing of the surface. Thus, relating this process to diamond, a passivation layer would be formed by the increase of carbonyl and hydroxyl groups on the surface allowing the polishing slurry particles to bond to its surface.

Table 2 of this study shows that there is an increase in the oxygen content on the surface of all the diamond films, this is regardless of whether an acidic or basic slurry was used. In the case of ceria which remains on the surface, there is the greatest increase in oxygen content. The increase in oxygen suggests there is a reaction occurring on the surface of the NCD films which could be facilitating the polishing particles bonding to the surface, although there is no evidence for a passivation layer forming due to carbonyl and hydroxyl groups, otherwise there would have been a greater oxygen content on the surface of the films that were polished with the basic slurries. The basic slurries had a pH of only ≈9≈9, so it is possible a stronger basic solution could produce a passivation layer.

The polishing mechanism then requires, according to previous literature [17Thomas ELH, Nelson GW, Mandal S, et al. Chemical mechanical polishing of thin film diamond. Carbon. 2014;68:473–479. DOI:10.1016/j.carbon.2013.11.023,26Krishnan M, Nalaskowski JW, Cook LM. Chemical mechanical planarization: slurry chemistry, materials, and mechanisms. Chem Rev. 2010;110(1):178–204. DOI:10.1021/cr900170z–28Penta NK, Veera PRD, Babu SV. Role of poly(diallyldimethylammonium chloride) in selective polishing of polysilicon over silicon dioxide and silicon nitride films. Langmuir. 2011;27:3502–3510. DOI:10.1021/la104257k], the following relative differences in binding energies: C–C << C–O ≤≤ Ce–O/Si–O/Al–O. The current accepted values for the various bond strengths are shown in Table 3. Comparing the bond strength of the Al–O bond and the C–C bond (see Table 3) the C–C bond is clearly stronger and therefore Al–O should not be capable of polishing the diamond films if this were the polishing mechanism.

Table 3: The bond strength of the different bond types for each polishing particle [17Thomas ELH, Nelson GW, Mandal S, et al. Chemical mechanical polishing of thin film diamond. Carbon. 2014;68:473–479. DOI:10.1016/j.carbon.2013.11.023,29Lide DR. 'Section 9' CRC handbook of chemistry and physics. 84th ed. Boca Raton (FL): CRC Press; 2003--2004.]

Bond	Strength ($kJ\ mol^{-1}$)
Ce–O	795
Al–O	511
Si–O	800
O–C	1077
C–C	610

This anomaly can be explained by recent results from Peguiron et al. [30Peguiron A, Moras G, Walter M, et al. Activation and mechanochemical breaking of c-c bonds initiate wear of diamond (110) surfaces in contact with silica. Carbon. 2016;98:474–483. DOI:10.1016/j.carbon.2015.10.098]. The authors performed density functional theory calculations on the degradation of the diamond surface when in contact with silica and silicon, duplicating the model described by Thomas et al. [17Thomas ELH, Nelson GW, Mandal S, et al. Chemical mechanical polishing of thin film diamond. Carbon. 2014;68:473–479. DOI:10.1016/j.carbon.2013.11.023]. They discovered that in a silicon particle bonded to diamond system, once a shearing force was applied this would always result in deformation of the silicon. However in the silica bonded to diamond system, once a shearing force was applied C–C bond breakage was shown, although this was not with great regularity. They determined that this is the result of activation (weakening) of C–C bonds between the terminating zigzag carbon chains and the underlying diamond bulk atoms. This weakening is two-fold requiring first that a pilot atom such as Si, O or H bond to the surface. The results of this study show oxygen bonding to the surface. It is possible that if the silicon is bonding to the surface it is being removed after being cleaned by the standard SC-1 process and hydrogen is not detectable by XPS. Then the weakening process requires that the pilot atom be attached to, or replaced by, a covalent and highly polar material, such as silica not silicon. This connection further weakens the C–C bond resulting in a breakage when a shearing force is applied. They concluded that polishing by this mechanism could also be achieved with alumina. This study has shown that it is possible to polish NCD films using alumina as predicted. This mechanism could provide an explanation for why ceria does not polish to the same rate as silica. Ceria contains both ionic and covalent bonding, with the ionic bond dominating [31Nakamatsu H, Mukoyama T, Adachi H. Ionic and covalent bonds in ceo2 crystal. Chem Phys Lett. 1995;247(1):168–172. DOI:10.1016/0009-2614(95)01185-7]. This variation in bonding could also be the reason that the ceria remains bonded to the diamond film surface where the alternatives do not. It is possible that the covalent bonding, whilst not dominant, might be enough to provide the C -C bond weakening - since ceria still polishes at a similar (and in the case of the acidic ceria a greater) rate than the basic alumina slurry. The model proposed by Peguiron et al. [30Peguiron A, Moras G, Walter M, et al. Activation and mechanochemical breaking of c-c bonds initiate wear of diamond (110) surfaces in contact with silica. Carbon. 2016;98:474–483. DOI:10.1016/j.carbon.2015.10.098] explains why alumina polishes, and possibly explains why ceria polishes, but it does not explain the variation between the alumina basic and acidic slurries. This variation can be explained if their particle size is considered in relation to the polishing rate, as Figure 5 clearly shows. This implies that the CMP of NCD thin films follows a contact-area mechanism where the roughness reduction is determined by the contact area of the particles with the sample, hence being inversely proportional to the particle size. This relationship has been observed in the CMP of other materials [32Choi W, Abiade J, Lee S-M, et al. Effects of slurry particles on silicon dioxide CMP. J Electrochem Soc. 2004;151:G512–G522. DOI:10.1149/1.1768133], although in cases apart from diamond the density and hardness of the polishing particles play a more significant role [32Choi W, Abiade J, Lee S-M, et al. Effects of slurry particles on silicon dioxide CMP. J Electrochem Soc. 2004;151:G512–G522. DOI:10.1149/1.1768133].

CONCLUSIONS

This study has shown that the CMP of NCD thin films is a chemical process that increases the O1s oxygen content on the surface of the films. Further changes have been proposed to the mechanism of the CMP of NCD thin films. Firstly it has been shown that CMP is not limited to slurries containing silica polishing particles. Three oxide polishing particles, ceria (CeO_2), alumina (Al_2O_3) and silica (SiO_2), were tested under acidic and basic conditions. It was found that the acidic alumina slurry polished at a similar rate as the basic silica and acidic silica slurries, which agrees with separate theoretical predictions. However, the basic alumina polished at a significantly lower rate. This study shows that this is a result of the order of magnitude difference between the diameters of the polishing particles. It was found that of all the various properties of the slurries, including pH and composition, the particle size was a determining factor for the polishing rate, particles with a smaller diameter being capable of a greater RMS roughness reduction than those with a large diameter. This implies that the CMP of NCD thin films follows a contact-area mechanism where the rate of roughness reduction is determined by the contact area of the particles with the sample. Smaller particles are able to bond in greater numbers to the NCD thin film surface leading to more instances of C–C bond weakening and therefore successful polishing.

ACKNOWLEDGEMENTS

The authors wish to acknowledge the financial support of the European Research Council (ERC) Consolidator Grant for the development of 'Superconducting Diamond Quantum Nano-Electro-Mechanical Systems', Project ID: 647471.

REFERENCES

1. Williams OA. Nanocrystalline diamond. Diamond Relat Mater. 2011;20:621–640. DOI:10.1016/j.diamond.2011.02.015
2. Williams O, Kriele A, Hees J, et al. High young's modulus in ultra thin nanocrystalline diamond. Chem Phys Lett. 2010;495(1–3):84–89. DOI:10.1016/j.cplett.2010.06.054
3. Williams OA, Douhéret O, Daenen M, et al. Enhanced diamond nucleation on monodispersed nanocrystalline diamond. Chem Phys Lett. 2007;445(4–6):255–258. DOI:10.1016/j.cplett.2007.07.091
4. Jiang X, Schiffmann K, Klages C-P. Nucleation and initial growth phase of diamond thin films on (100) silicon. Phys Rev B. 1994;50:8402–8410. DOI:10.1103/PhysRevB.50.8402
5. Smereka P, Li X, Russo G, et al. Simulation of faceted film growth in three dimensions: microstructure, morphology and texture. Acta Mater. 2005;53(4):1191–1204. DOI:10.1016/j.actamat.2004.11.013
6. Ergincan O, Palasantzas G, Kooi BJ. Influence of surface modification on the quality factor of microresonators. Phys Rev B. 2012;85:205420. DOI:10.1103/PhysRevB.85.205420
7. Liu WL, Shamsa M, Calizo I, et al. Thermal conduction in nanocrystalline diamond films: effects of the grain boundary scattering and nitrogen doping. Appl Phys Lett. 2006;89(17):171915. DOI:10.1063/1.2364130
8. Bhusari D, Yang J, Wang T, et al. Highly transparent nano-crystalline diamond films grown by microwave CVD. Solid State Commun. 1998;107(6):301–305. DOI:10.1016/S0038-1098(98)00226-9
9. Iriarte G, Rodríguez J, Calle F. Synthesis of c-axis oriented aln thin films on different substrates: a review. Mater Res Bull. 2010;45(9):1039–1045. DOI:10.1016/j.materresbull.2010.05.035
10. Rodriguez-Madrid JG, Iriarte GF, Pedros J, et al. Super-high-frequency saw resonators on aln/diamond. IEEE Electron Device Lett. 2012;33(4):495–497. DOI:10.1109/LED.2012.2183851
11. Chen Y, Zhang L. Polishing of diamond materials. 1st ed. Portalnd: Springer Series; 2013.
12. Grillo SE, Field JE. The polishing of diamond. J Phys D Appl Phys. 1997;30(2):202. DOI:10.1088/0022-3727/30/2/007
13. El-Dasher BS, Gray JJ, Tringea JW, et al. Crystallographic anisotropy of wear on a polycrystalline diamond surface. Appl Phys Lett.

2006;88:241915-1–241915-3. DOI:10.1063/1.2213180
14. Bögli U, Blatter A, Pimenov S, et al. Tribological properties of smooth polycrystalline diamond films. Diamond Relat Mater. 1995;4(7):1009–1019. DOI:10.1016/0925-9635(95)00272-3
15. Fan QH, Grácio J, Pereira E. Residual stresses in chemical vapour deposited diamond films. Diamond Relat Mater. 2000;9(9–10):1739–1743. DOI:10.1016/S0925-9635(00)00284-3
16. Edwards MJ, Bowen CR, Allsopp DWE, et al. Modelling wafer bow in silicon-polycrystalline CVD diamond substrates for gan-based devices. J Phys D Appl Phys. 2010;43(38):385502. DOI:10.1088/0022-3727/43/38/385502
17. Thomas ELH, Nelson GW, Mandal S, et al. Chemical mechanical polishing of thin film diamond. Carbon. 2014;68:473–479. DOI:10.1016/j.carbon.2013.11.023
18. Zantye PB, Kumar A, Sikder A. Chemical mechanical planarization for microelectronics applications. Mater Sci Eng R Rep. 2004;45(3–6):89–220. DOI:10.1016/j.mser.2004.06.002
19. Thomas ELH, Mandal S, Brousseau EB, et al. Silica based polishing of (100) and (111) single crystal diamond. Sci Technol Adv Mater. 2014;15(3):035013. DOI:10.1088/1468-6996/15/3/035013
20. Oliver MR, Evans D, Stein D, et al. Chemical-mechanical, planarization of semiconductor materials. 1st ed. Portalnd: Springer Series; 2003.
21. Neslen CL, Mitchel WC, Hengehold RL. Effects of process parameter variations on the removal rate in chemical mechanical polishing of 4h-sic. J Electron Mater. 2001;30:1271–1275. DOI:10.1007/s11664-001-0111-2
22. Ahn Y, Yoon J, Baek C, et al. Chemical mechanical polishing by colloidal silica-based slurry for micro-scratch reduction. Wear. 2004;257:785–789. DOI:10.1016/j.wear.2004.03.020
23. Babu SV, Jindal A, Li Y. Chemical-mechanical planarization of cu and ta. JOM. 2001;53(6):50–52. DOI:10.1007/s11837-001-0104-x
24. Kern W. The evolution of silicon wafer cleaning technology. J. Electrochem. Soc.. 1990;137(6):1887–1892. DOI:10.1149/1.2086825
25. McGrath J, Davis C. Polishing pad surface characterisation in chemical mechanical planarisation. J Mater Process Technol. 2004;153–154:666–673. DOI:10.1016/j.jmatprotec.2004.04.094
26. Krishnan M, Nalaskowski JW, Cook LM. Chemical mechanical

planarization: slurry chemistry, materials, and mechanisms. Chem Rev. 2010;110(1):178–204. DOI:10.1021/cr900170z

27. Hochenga H, Tsaia HY, Sub YT. Modeling and experimental analysis of the material removal rate in the chemical mechanical planarization of dielectric films and bare silicon waferss. Electrochem Soc. 2001;148(10):G581–G586. DOI:10.1149/1.1401087
28. Penta NK, Veera PRD, Babu SV. Role of poly(diallyldimethylammonium chloride) in selective polishing of polysilicon over silicon dioxide and silicon nitride films. Langmuir. 2011;27:3502–3510. DOI:10.1021/la104257k
29. Lide DR. 'Section 9' CRC handbook of chemistry and physics. 84th ed. Boca Raton (FL): CRC Press; 2003--2004.
30. Peguiron A, Moras G, Walter M, et al. Activation and mechanochemical breaking of c-c bonds initiate wear of diamond (110) surfaces in contact with silica. Carbon. 2016;98:474–483. DOI:10.1016/j.carbon.2015.10.098
31. Nakamatsu H, Mukoyama T, Adachi H. Ionic and covalent bonds in ceo2 crystal. Chem Phys Lett. 1995;247(1):168–172. DOI:10.1016/0009-2614(95)01185-7
32. Choi W, Abiade J, Lee S-M, et al. Effects of slurry particles on silicon dioxide CMP. J Electrochem Soc. 2004;151:G512–G522. DOI:10.1149/1.1768133

SECTION V:
LASER DAMAGE RESISTANCE

Chapter 14

Laser Damage Resistance of Polystyrene Opal Photonic Crystals

Lei Pan[1], Hongbo Xu[1], Ruizhen Lv[2], Jun Qiu[3], Jiupeng Zhao[1] & Yao Li[2]

[1] MIIT Key Laboratory of Critical Materials Technology for New Energy Conversion and Storage, School of Chemistry and Chemical Engineering, Harbin Institute of Technology, Harbin, 150001, China

[2] Center for Composite Materials and Structure, Harbin Institute of Technology, Harbin, 150001, China

[3] School of Energy Science and Engineering, Harbin Institute of Technology, Harbin, 150001, China

ABSTRACT

The laser damage behavior of three-dimensional photonic crystals (3D-PCs) with an opal structure is investigated using both experimental

Citation: Lei Pan, Hongbo Xu, Ruizhen Lv, Jun Qiu, Jiupeng Zhao & Yao Li "Laser damage resistance of polystyrene opal photonic crystals" Scientific Reports volume 8, Article number: 4523 (2018). https://doi.org/10.1038/s41598-018-22831-7

and simulation methods. Polystyrene (PS) colloidal crystals films with a reflection peak at 1064 nm are used as the model material. Disordered films (DF) are fabricated with PS microspheres as contrast samples. The laser-induced damage threshold (LIDT) of 3D-PCs is tested, which is as 2.35 times high as the LIDT of DF. All laser damages are derived from defects in 3D-PCs, implying that the LIDT of ideal 3D-PCs will be significantly increased. The simulation results show that the electric field is contained in the pores of 3D-PCs while it is reduced in the PS microspheres, which may decrease the absorption of laser energy by 3D-PCs. In contrast, the electric field distribution is irregular in DF. Enhanced electric field areas are located in both the pores and microspheres of DF. Considering higher electric field intensity causes more energy absorption and higher temperature, the DF have a lower LIDT. The 3D-PCs structure uses ordered vacancy to contain and strike back laser energy and can increase the LIDT without changing the chemical composition of materials.

INTRODUCTION

Photonic crystals (PCs) are a class of artificial materials with photonic band gaps (PBGs) that originate from their periodic dielectric structures[1]. Many laser-related applications of PCs are being proposed or demonstrated, such as all-optical circuits[2], omnidirectional reflectors[3], high power laser applications[4], high-Q nanocavity lasers[5,6], and low-threshold lasers[7,8,9], owing to its ability to manipulate light. For a versatile tool to generate and manipulate laser, laser-induced damage threshold (LIDT) of PCs is a key property that decides the whole performance[10].

Traditional optical coatings such as high reflection coatings, anti-reflection coatings, and beam splitters are composed of periodic dielectric films and can be regarded as one-dimensional PCs (1D-PCs). The origin, nature, and characteristics of the LIDT of 1D-PCs have been extensively studied[10,11,12,13,14,15,16,17]. Contamination and structural defects give rise to most failures in 1D-PCs[10,16] and lead to a much lower LIDT than the LIDT of bulk materials.

PBG fiber, a kind of two dimensional PCs (2D-PCs), attracted attention as candidates for the transport of very high power laser pulses recently[18]. It is reported that a PBG fiber carries most of the transmitted power in its hollow core and as little as 0.2% of the energy in the silica[19], which results in a LIDT representing an order of 4 improvement compared to solid-core photonic crystal fibers20. Similar results are reported for two dimensional periodic arrays of nano-horns as anti-reflection layers21·22·23·24. The chalcogenide glass with 2D-PCs shows almost a factor of 2 improvement in the LIDT compared to untreated bulk glass22 mainly because no foreign material is introduced21. The strong electric field region for 1064 nm laser is distinctly distributed in the hollow part of the periodic array, which may also promote the LIDT21.

Three dimensional ordered25·26 and disordered25·27·28 nano-structures were also manufactured for high power laser applications. Li *et al.* observed a slightly higher LIDT in ordered porous silica coating than in the disordered one25. Despite the difference in absorption, the authors believe that the ordered structure is the key to interpret the high LIDT. However, *Du et al.* argued that for a small period-to-wavelength ratio, electromagnetic fields do not vary spatially and the periodicity of nanostructures would be less important to their performance27. In this sense, the ordered nanoporous coating was not a typical photonic crystal for the laser used in the testing. High LIDT was reported for three dimensional disordered nanoporous antireflective coatings as well and the influence of orderliness on LIDT was not discussed27·28.

An interesting result can be found if one compares the LIDT investigation of 1D and 2D photonic crystals, despite different testing methods, materials, and substrates, the LIDT of 1D-PCs is generally much lower than that of the substrates10·11·12·13·14·15·16·17 while the LIDT of 2D-PCs is close to or higher than the LIDT of substrates or bulk materials18·19·20·21·22·23·24. Many reasons may underlie this result, the ways by which the two kinds of ordered structures were constructed being one of them. 1D-PCs must consist of at least two dielectric materials (both solid usually) layer by layer and laser needs to propagate through them in sequence. The layer with the lowest LIDT, which may be defined by its chemical nature, contamination, or defects in the interface, will be the shortest stave in the barrel. However, in the 2D-PCs, the second dielectric material is usually a vacancy (air or vacuum) whose LIDT is much higher than all solid materials. Besides, the manner in which laser passes through the 2D-PCs may be different from 1D-PCs. By elaborately designing the arrangement of solid material, the bulk of laser energy can be confined within the vacancy to avoid the damage of the solid part such that the LIDT of the whole 2D-PCs is promoted. This mechanism accords with the old Chinese philosophy that vacancy makes tools more useful (Tao-te-ching, chapter 11, about 500BC).

Vacancies exist in 3D-PCs as well; however, solid materials in three dimensions baffle them periodically. Until now, the influence three-dimensional ordered structures with periods falling in the range of laser wavelength exert on the LIDT was unknown. Hereby we carried out experimental and simulative research on this problem.

RESULTS AND DISCUSSION

Experiments Results

Among the various preparation methods for 3D-PCs, self-assembly of PS microspheres is most widely used, as it is convenient and low cost. Moreover, the low melting point of PS is in favor of the LIDT test. Hence, we choose PS 3D-PCs to conduct the laser-induced damage investigation. To format a pseudo band gap centered at 1064 nm, the average diameter of PS microspheres was carefully adjusted to 465 nm. The PDI of the diameter was 1.04. PS 3D-PCs were grown on glass slides by vertical deposition[29,30]. The SEM images of PS 3D-PCs are shown in Fig. 1. It is observed in the Fig. 1a that the deposited PS 3D-PCs have a well-ordered nano-architecture consisting of close-packed spheres. However, as shown in Fig. 1b, point defects and planar defects also exist in a larger range.

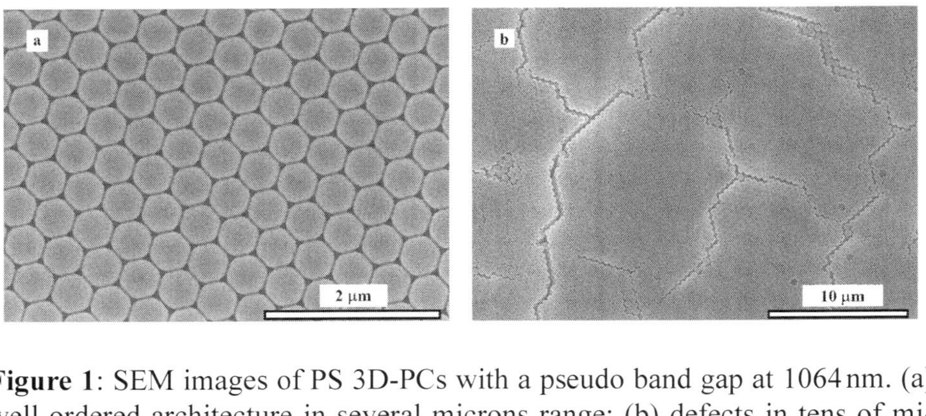

Figure 1: SEM images of PS 3D-PCs with a pseudo band gap at 1064 nm. (a) well-ordered architecture in several microns range; (b) defects in tens of microns range.

The experimental reflectance spectrum of the PS 3D-PCs is given in Fig. 2. A reflective peak near 90% can be found for PS 3D-PCs in Fig. 2, whose location is very close to the wavelength of Nd:YAG laser, indicating the highly ordered structure of the PS 3D-PCs.

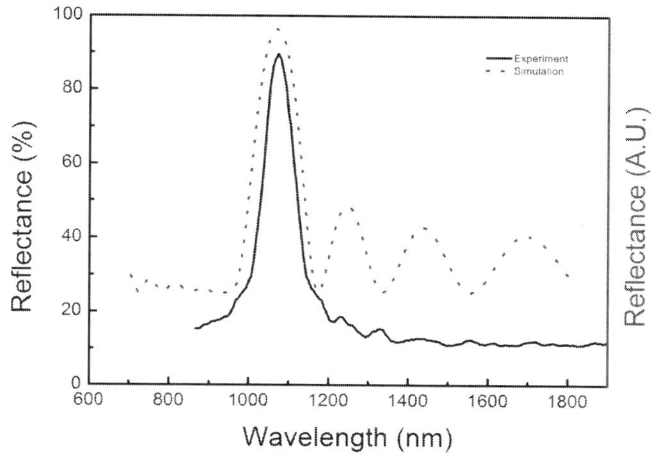

Figure 2: Experiment and simulation reflectance spectra of PS 3D-PCs.

LIDT Test

An Nd:YAG laser with a wavelength of 1064 nm was used as laser source for the LIDT test in continuous wave mode. The $1/e^2$ spot diameters measured via the knife-edge method were 3.0 mm and the effective spot sizes were 7.1 mm^2. The beam of laser was collimated instead of focused because the high power density is not necessary for the damage of PS.

The damage testing was performed as per the R-on-1 regime (i.e. multi tests per sample with increasing laser power until the occurring of damage). Ten 3D-PCs samples were irradiated and each sample was irradiated atone spot only. The damage was estimated via the visual inspection of the melting of PS membrane firstly and then confirmed by SEM examination. The LIDT of each sample was defined as the minimum laser power density for which the damage occurred within 30 s irradiation. In fact, all damage took place between 10–12 s under the irradiation of LIDT power. If damage did not occur in 30 s, it would not ensue for at least 2 min. Once the melting was detected, the laser was switched off immediately so that the initial stage of laser-induced damage can be observed subsequently. Eventually, the LIDT of 3D-PCs was obtained by averaging each sample. Moreover, to identify the influence of ordered structure on LIDT, two kinds of PS microspheres with different diameters were mixed together on purpose and disordered films (DF) were fabricated with these mixed microspheres using the vertical deposition method under the same fabrication conditions as of 3D-PCs. Ten

DF were tested via the same procedures. The results of LIDT test for3D-PCs and DF are given in Fig. 3.

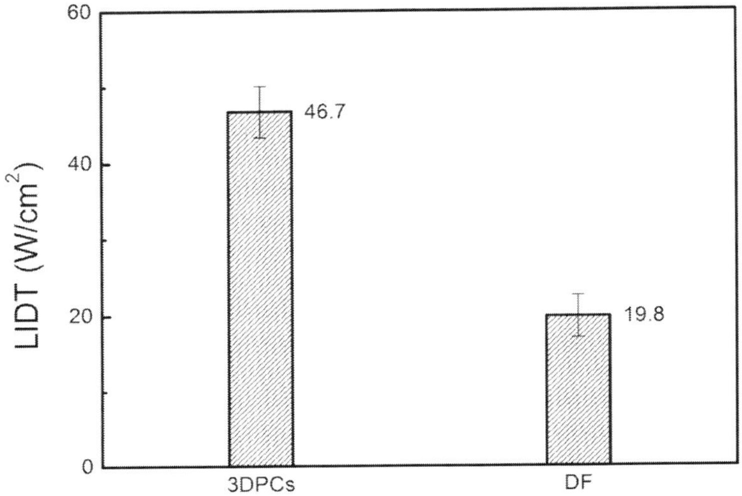

Figure 3: LIDT for PS 3D-PCs and DF.

The test results show that the DF starts to melt under the laser illumination at 19.8 W/cm^2 while 3D-PCs films start melting at a considerably high value, i.e., ~46.7 W/cm^2. The LIDT of materials comprised of same PS microspheres increased for more than 135% with ordered structure. The 3D-PCs structure can realize excellent resistance to laser damage without changing the chemical component of materials.

The SEM images of damaged samples are given in Fig. 4. Figure 4a is the whole irradiation spot on 3D-PCs. It is observed from Fig. 4e that the melting of PS occurs first on the top layer of the photonic crystals, while the lower-layers of the photonic crystal structure are relatively well preserved. Defects such as sphere missing, surface contamination, dislocations, and cracks are inevitable in 3D-PCs prepared by vertical deposition. The illuminated defective areas are exhibited in Fig. 4b–e. As seen from these figures, more serious melting occurs at defective regions than surrounding ordered regions, which is consistent with our expectations. Defects are generally regarded as a source of electromagnetic field disturbance, causing micro-focus effect and leading to the damage of materials. The fact that laser damages of 3D-PCsare derived from defects further implies that if the defects could be effectively removed, the LIDT of perfect 3D-PCs would be significantly increased. Moreover, it can be speculated reasonably that the redistribution of electromagnetic field caused by the ordered structure of 3D-PCs neither gives rise to the harmful micro-focus in the PS microspheres lattice nor promotes the melting of PS like disordered structures. Near-field simulation will be employed to prove this speculation.The morphology of DF before irradiation is exhibited in Fig. 4g. Microspheres are piled together in a jumble as expected. Homogeneous melting occurs in the DF under laser irradiation and aggravates uniformly with the time, as noted from Fig. 4h and i.

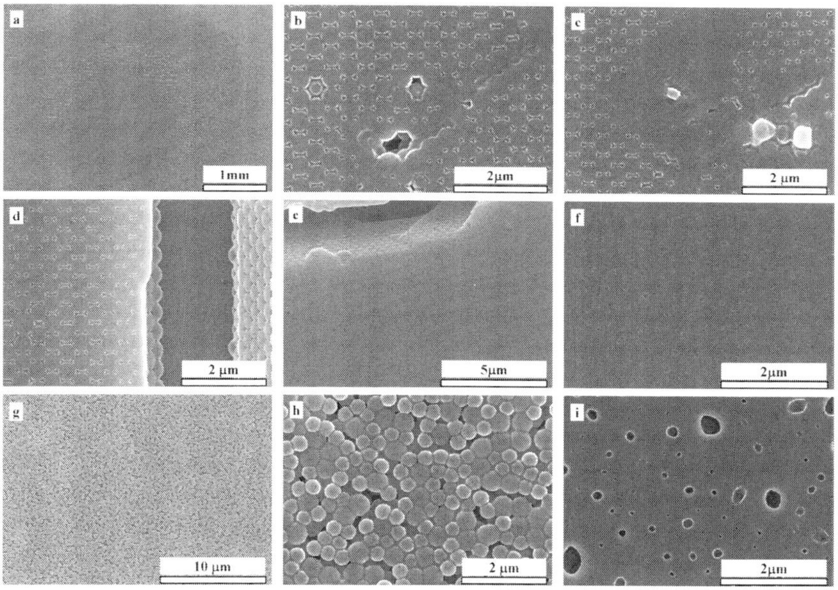

Figure 4: SEM images of 3D-PCs with a pseudo band gap at 1064 nm (a–f) and DF (g–i) before or after 1064 nm laser irradiation. (a) The whole irradiation spot on 3D-PCs, (b–e) different defects after irradiation, (f) severely melted 3D-PCs, (g) DF before irradiation, (i) lightly melted DF (h), severely melted DF.

Simulation

Various investigations showed that the LIDT of optical materials is closely related to the electric field intensity(EFI) in the materials and the distribution of the electric field (DEF) throughout the materials is taken as an indication of susceptibility to damage31. This is because although the materials used in these researches are almost transparent, very small but non-negligible absorptance for laser still exists owing to impurities, which gives rise to thermal effect and is responsible for most laser induced damages. Absorption of laser is positively related to the EFI. Therefore, it is necessary to investigate the DEF in 3D-PCs for further interpreting how ordered structures affect the LIDT of materials. The DEF intensity of 3D-PCs and DF was simulated numerically using finite-difference time-domain method.

The unit cell of both 3D-PCs and DF were designed in rectangle lattices, as illustrated in Fig. 5a and b, respectively. Periodic boundary conditions were set around a unit cell, while perfectly matched layer absorbing boundary conditions were used at the top and bottom boundaries of the cell. The incidence optical source was set as unpolarized plane waves at normal incidence.

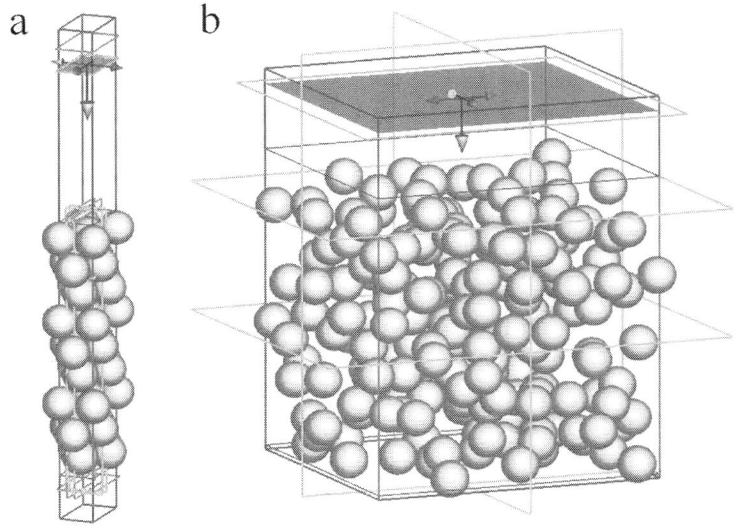

Figure 5: Simulation unit of (a) 3D-PCs and (b) DF. 3D-PCs consist of 9 layers of PS microspheres in opal structure. DF consist of random distributed PS microspheres. Periodic boundary conditions were set around a unit cell, while perfectly matched layer absorbing boundary conditions were used at the top and bottom boundaries of the cell.

The reflective spectrum of 3D-PCs was simulated using a source with a wavelength range from 700 nm to 1800nm and was plotted in Fig. 2. The position of the reflective peak agrees well with the experimental result. The DEF was simulated using a source with a wavelength at 1064 nm. Figure 6a and b depict the normalized DEF images of the x-z plane, where y=0 and y-z plane, where x=0, respectively. Black lines highlight the boundaries of PS microspheres. It is immediately noticed that the DEF in 3D-PCsis highly ordered. The EFI is mainly distributed in the vacancies among PS microspheres, and the EFI inside the PS microspheres is significantly lower than that in the vacancies, which is obviously beneficial to elevate the LIDT of the entire film. The ordered vacancies firstly contain the energy of laser without damaging the material and then reflect it backwards. This principle is similar to the traditional Shadow Boxing conceived from Tao-te-ching, which fights back the strongest strike with nothing. Besides, it can be seen that the EFI is rapidly attenuated due to the pseudo forbidden band in the z-direction; the EFI in the upper most level, both inside and outside the microspheres, is the highest, which is consistent with the results that the first-level microspheres melt the most, as shown in Fig. 4e.

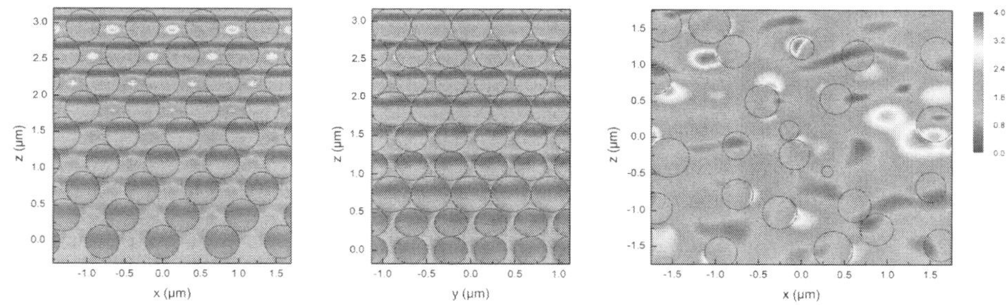

Figure 6: Distribution of electric field in (a) x-z plane of 3D-PCs, where y=0, (b) y-z plane of 3D-PCs, where x=0 and (c) x-z plane of DF, where y=0.

On the contrary, the DEF of disordered PS microsphere films is very disordered and nonuniform, as illustrated in Fig. 6c. Owing to the "micro-focus" effect of disordered structures, the highest EFI is almost 4 times that of the source. In addition, the "micro-focuses" locate randomly both in vacancies and PS microspheres. The average EFI inside the microspheres of disordered PS films is significantly higher than that of 3D-PCs, which naturally leads to higher absorption and lower LIDT. Besides, laser propagates through the disordered PS film without evident attenuation.

METHODS

All reagents except water used were commercial Sigma-Aldrich products. High-purity water with a resistivity of 18.2 MΩ·cm at 25 °C was prepared by a water purification system (CSR-1-30T).

Monodisperse polystyrene (PS) microspheres were synthesized by an emulsifier-free emulsion polymerization technique32. Scanning electron microscopy (SEM) images were obtained using a Hitachi SU 8000 microscope operated at 15 kV. The average diameter of the PS microspheres was calculated from more than 100 microspheres. Polydispersity index (PDI) = D_m/D_{n-1}, where D_m and D_n were the weight-average and number-average diameters, respectively33. The normal-incident reflectance spectra of 3D-PCswere obtained by using an Ocean Optics NIR Quest instrument equipped with a reflection probe. A specular reflectance standard(PN A338-MS-1) with a reflectance of 85–90% was used as measured reference.

An Nd:YAG laser was used as a laser source for LIDT test on continuous wave mode. The $1/e^2$ spot diameters measured via the knife-edge method were 3 mm and the effective spot sizes were 7 mm^2.

CONCLUSIONS

PS 3D-PCs with a pseudo forbidden band centered at 1064 nm were prepared by vertical deposition. The LIDT of 3D-PCs is 2.35 times high as the LIDT of DF fabricated with the same material. The laser damages of 3D-PCsare derived from defects. The simulation results show that the ordered structure of the 3D-PCs plays a very interesting role of modulating the DEF of laser. The EFI is mainly contained in the vacancies among microspheres during its short propagation in 3D-PCs and the EFI inside the microspheres is significantly lower than that outside. Besides, the EFI rapidly decreases with the increase of penetration depth in 3D-PCs. Such a DEF is very conducive to preventing laser damage. In contrast, "micro-focuses" with high EFI locate randomly both in vacancies and PS microspheres in DF. This explains the higher LIDT of 3D-PCs than DF. The laser damages observed in the experiment are derived from a variety of defects. Therefore, it can be reasonably speculated that ideal flawless opal photonic crystals will have a much higher LIDT.

In short, 3D-PCs structure uses ordered vacancy to contain and strike back laser energy and can increase the LIDT without changing the chemical composition of materials.

REFERENCES

1. Yablonovitch, E. Inhibited spontaneous emission in solid-state physics and electronics. *Phys. Rev. Lett.* 58, 2059–2062 (1987).
2. Joannopoulos, J. D., Villeneuve, P. R. & Fan, S. Photonic crystals: putting a new twist on light. *Nature* 386, 143–149 (1997).
3. Zhang, H., Liu, S., Kong, X., Bian, B. & Dai, Y. Omnidirectional photonic band gap enlarged by one-dimensional ternary unmagnetized plasma photonic crystals based on a new Fibonacci quasiperiodic structure. *Phys. Plasmas* 19, 112102, https://doi.org/10.1063/1.4765063 (2012).
4. Benoit, F. *et al.* In *Nanoengineering: Fabrication, Properties, Optics, and Devices XII.* (eds Campo, E. M., Dobisz, E. A. & Eldada, L. A.) (Spie-Int Soc Optical Engineering).
5. Akahane, Y., Asano, T., Song, B. & Noda, S. High-Q photonic nanocavity in a two-dimensional photonic crystal. *Nature* 425, 944–947 (2003).
6. Iwamoto, S. *et al.* In *Ieee Photonics Conference*. 866–867 (2012).
7. Painter, O. *et al.* Two-dimensional photonic band-gap defect mode laser. *Science* 284, 1819–1821 (1999).
8. Furumi, S., Fudouzi, H., Miyazaki, H. T. & Sakka, Y. Flexible polymer colloidal-crystal lasers with a light-emitting planar defect. *Adv. Mater.* 19, 2067–2072, https://doi.org/10.1002/adma.200602855 (2007).
9. Zhong, K. *et al.* Defect Mode Passband Lasing in Self-Assembled Photonic Crystal. *Acs Photonics* 3, 2330–2337, https://doi.org/10.1021/acsphotonics.6b00511 (2016).
10. Shah, R. S., Rey, J. J. & Stewart, A. F. In *Laser-Induced Damage in Optical Materials*: *2006*. 640305 (2007).
11. Cheng, X. *et al.* Optimal coating solution for a compact resonating cavity working at Brewster angle. *Opt. Express* 24, 24313–24320, https://doi.org/10.1364/oe.24.024313 (2016).
12. Cheng, X. B., Jiao, H. F., Lu, J. T., Ma, B. & Wang, Z. S. Nanosecond pulsed laser damage characteristics of HfO2/SiO2 high reflection coatings irradiated from crystal-film interface. *Opt. Express* 21, 14867–14875, https://doi.org/10.1364/oe.21.014867 (2013).
13. Cheng, X. B. *et al.* Physical insight toward electric field enhancement at nodular defects in optical coatings. *Opt. Express*23, 8609–8619, https://

doi.org/10.1364/oe.23.008609 (2015).

14. Neauport, J. *et al.* Effect of electric field on laser induced damage threshold of multilayer dielectric gratings. *Opt. Express* 15, 12508–12522 (2007).
15. Cheng, X., Jiao, H., Lu, J., Ma, B. & Wang, Z. Nanosecond pulsed laser damage characteristics of HfO 2/SiO 2 high reflection coatings irradiated from crystal-film interface. *Opt. Express* 21, 14867–14875 (2013).
16. Cheng, X. *et al.* In *SPIE Optical Systems Design*. 816816 (International Society for Optics and Photonics).
17. Yan, L. *et al.* Coupling effect of multi-wavelength lasers in damage performance of beam splitters at 355 nm and 1064 nm. *Appl. Opt.*51, 3243–3249 (2012).
18. Dumitrache, C., Rath, J. & Yalin, A. P. High Power Spark Delivery System Using Hollow Core Kagome Lattice Fibers. *Materials* 7, 5700–5710, https://doi.org/10.3390/ma7085700 (2014).
19. Humbert, G. *et al.* Hollow core photonic crystal fibers for beam delivery. *Opt. Express* 12, 1477–1484 (2004).
20. Michaille, L. F. *et al.* In *Integrated Optical Devices, Nanostructures, and Displays*. 30–38 (2004).
21. Ye, X. *et al.* High power laser antireflection subwavelength grating on fused silica by colloidal lithography. *J. Phys. D: Appl. Phys.* 49, 265104 (2016).
22. Florea, C., Sanghera, J., Busse, L., Shaw, B. & Aggarwal, I. In *Photonic and Phononic Properties of Engineered Nanostructures*. 794610–794618 (2011).
23. Jing, X., Shao, J., Jin, Y. & Fan, Z. Near-field distribution of broadband antireflective nanostructure arrays. *Optik* 123, 527–533 (2012).
24. Hobbs, D. S. & MacLeod, B. D. In *Laser-Induced Damage in Optical Materials*. 67200L (2007).
25. Li, X., Zou, L., Wu, G. & Shen, J. Laser-induced damage on ordered and amorphous sol-gel silica coatings. *Opt. Mater. Express* 4, 2478–2483, https://doi.org/10.1364/OME.4.002478 (2014).
26. Jonušauskas, L. *et al.* Optically Clear and Resilient Free-Form µ-Optics 3D-Printed via Ultrafast Laser Lithography. *Materials* 10, 12 (2017).
27. Du, Y. *et al.* Laser-induced damage properties of antireflective porous

glasses. *Opt. Commun.* 285, 5512–5518 (2012).

28. Ye, X. *et al.* Subwavelength porous silica antireflection coating. *J. Optoelectron. Adv. M.* 13, 532–536 (2011).
29. Zhao, J., Li, Y., Xin, W. & Li, X. Preparation and characterization of three-dimensionally ordered macroporous yttria-stabilized zirconia by aqueous organic gel route. *J. Solid State Chem.* 181, 239–244 (2008).
30. Li, Z. *et al.* Magnetron sputtering SiC films on nickel photonic crystals with high emissivity for high temperature applications. *Appl. Surf. Sci.* 259, 811–815 (2012).
31. Ling, X., Wang, G., Zhao, Y., Liu, X. & Shao, J. Laser-induced damage of the optical coatings due to organic contamination in vacuum. *Appl. Surf. Sci.* 270, 346–351, https://doi.org/10.1016/j.apsusc.2013.01.028 (2013).
32. Wang, H., Zhang, S., Wang, M. & Ge, X. Preparation of monodisperse polystyrene particles from emulsifier-free miniemulsion polymerization. *Chem. Lett.* 37, 1158–1159 (2008).
33. Pan, L. *et al.* Synthesis of silica particles with precisely tailored diameter. *Chin. J. Chem. Phys.* 27, 563–567 (2014).

Chapter 15

Surface Molecular Structure Defects And Laser-Induced Damage Threshold of Fused Silica During A Manufacturing Process

Yuan Li[1,2], Hongwei Yan[2] , Ke Yang[2] , Caizhen Yao[2] , Zhiqiang Wang[2] , Xinshu Zou[2] , Chunyan Yan[1] , Xiaodong Yuan[2] , Xin Ju[1] & Liming Yang[3]

[1]Department of Physics, University of Science and Technology Beijing, Beijing, 100083, China

[2]Laser Fusion Research Center, China Academy of Engineering Physics, Mianyang, 621900, China

[3]Fine Optical Engineering Research Center, Chengdu, 610041, China

ABSTRACT

Laser induced damage of fused silica is a serious problem for high power laser systems, and damage precursors are mainly induced by

Citation: Yuan Li, Hongwei Yan, Ke Yang, Caizhen Yao, Zhiqiang Wang, Xinshu Zou, Chunyan Yan, Xiaodong Yuan, Xin Ju & Liming Yang "Surface molecular structure defects and laser-induced damage threshold of fused silica during a manufacturing process" Scientific Reportsvolume 7, Article number: 17870 (2017). https://doi.org/10.1038/s41598-017-18249-2

manufacturing processes. In this work, fused silica samples were prepared from a manufacturing process including grinding, polishing and etching procedures. The chemical disorder of the prepared samples was inspected by using fluorescence microscopy and ultra-violet fluorescence spectrometer. The physical disorder was characterized by using Infrared and Raman spectrometer. Laser induced damage thresholds (LIDTs) were measured in R-on-1 mode by 355 nm 6.4 ns laser pulse. Results showed that with the manufacturing processes transforming from grinding to etching, the magnitude of fluorescence point defects reduced while their types did not change, the Si-O-Si bonds of prepared samples were strained and the strained bonds were mitigated. The LIDTs increased with the reducing of fluorescence defects and strained Si-O-Si bonds. However, these structural defects can not be eliminated by the current manufacturing process. Improvements may be needed to eliminate the structural defects for a higher LIDT of fused silica.

INTRODUCTION

Fused silica optics are widely used in high power laser systems such as National Ignition Facility (NIF) and Laser Megajoule (LMJ) [1,2]. When subjected to 355 nm laser pulse irradiation at high fluence, fused silica optics may suffer from laser induced damage[3,4,5]. Inpurities and fracture defects were found to be the main damage precursors which were responsible for the initial of laser damage[6,7]. These damage precursors were usually induced in manufacturing process, and can be mitigated by the improvement of the manufacturing technologies. Recently, laser damage initiation density has been reduced significantly by using improved fabrication processes and the advent of whole optics mitigation strategies[8,9,10,11]. Based on the elastic-plastic deformation theory, Shi feng et al.[12] analyzed status of abrasives and workpiece in magnetorheological finishing (MRF) process and the feasibility of elastic polishing. Results showed that laser induced damage threshold (LIDT) was improved from 9.77 to 19.2 J/cm^2 after MRF elastic polishing. J. Bude et al.[13] reduced damage density in fused silica at high fluence (>40 J/cm^2) by more than 100 times by minimizing the presence of precipitates during chemical etching process. P. E. Miller et al.[6]combined MRF polishing and chemical leaching to explore the defects influence on the damage of fused silica and found that a thin defect layer on fracture surfaces is the dominant source of laser damage initiation. Though the LIDTs of fused silica optics have been successfully improved, it is still far below the intrinsic threshold (~150 J/cm^2) of fused silica14·15. While, the damage precursors are still quite vague at the fluence above ~40 J/cm^2. It is difficult to locate impurities in nanoscale on a sample surface with a size of several centimeters. While, a tiny variation of molecular structures can be detected by a fluorescence, infrared or Raman spectroscopies. Fused silica may contain two types of structural defetcs: chemical disorder (point defetcs) and physical disorder (Si-O-Si bond angle and $(Si\text{-}O)_n$ ring structure)16·17. In this work, efforts were made to inspect surface molecular structures of fused silica samples during a manufacturing process. The influence of molecular structural defects (chemical and physical disorders) on LIDTs of fused silica optics were explored.

RESULTS AND DISSCUSION

Chemical Disorder

Figure 1 shows bright field images (a), fluoresence images (b), and inversed images (c) of prepared samples. Samples A and B were ground by different size of abrasive. Samples C and D were polished by ceria and MRF polishing, respectively. Sample E was prepared by MRF polishing and HF etching. In the bright field images, the bright points indicated the scattering lights due to the surface fractures induced by the processing. While the bright points in the fluoresence images were corresponding to the fluorescence by the surface fractures. It is obvious that surface fractures of ground samples are more than those of polished samples, and the etched sample possesses less fractures than the polished samples. To quantify the magnitude of fluoresence points in these images, the figures of Fig. 1(b) were transformed to gray value images (range of the value, −255~255 (a.u.)) and then filtered by the background (value of 100 (a.u.), in our situation). The contrast was inversed to give a better visual effect, as shown in Fig. 1(c), and the black points represented the positions where the fluoresence occured. As shown in Fig. 2, the surface fluorescence defects magnitude of ground samples A and B is about one order more than those of other samples.With the precedures transforming from grinding to etching, the surface fluorescence defects magnitude reduced.

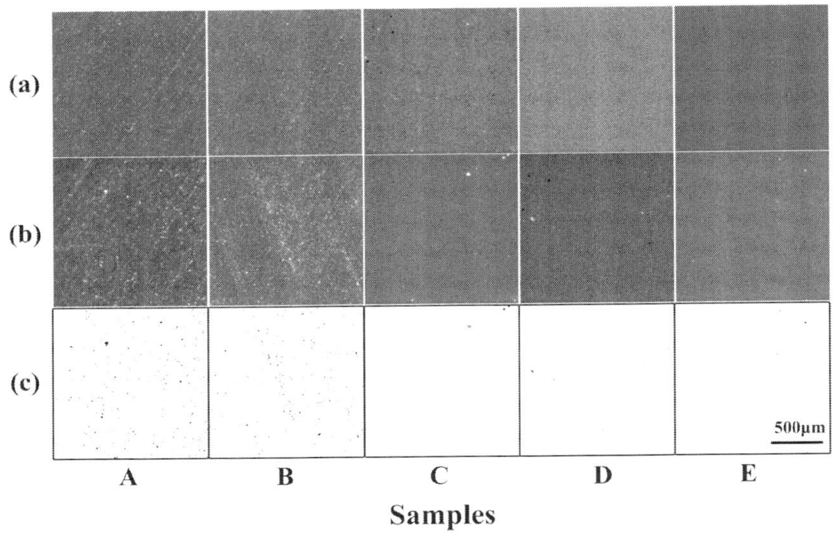

Figure 1: Images of samples: (a) bright field images, (b) fluorescence images and (c) inversed images. All images share the same scale bar. (Please refer to the methods section for processing parameters of samples A, B, C, D, and E).

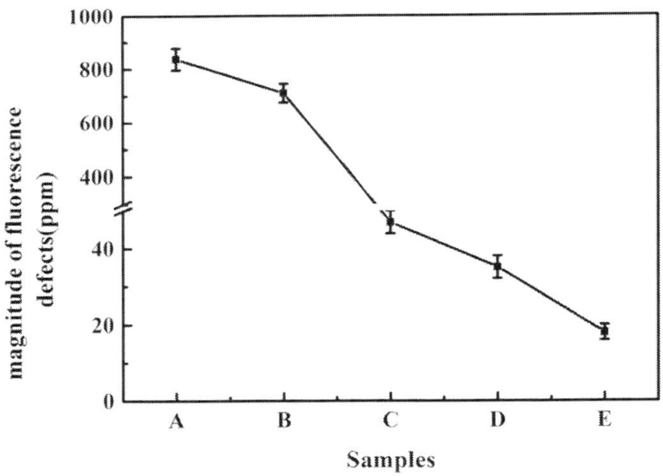

Figure 2: fluorescence defects magnitude of treated samples. The unit of ppm refers to part per million.

Figure 3 shows the comparison of normalized fluorescence emission spectra of treated samples. There are three main peaks in the spectra at around 310 nm, 390 nm and 590 nm bands, corresponding to oxygen-deficiency centers (II) (ODC(II), around 290 nm for α band and 390 nm for β band), and E_{δ}' center defects (590 nm band), respectively[18,19,20]. The characterized peak at around 310 nm band is an incomplete peak of ODCs (around 290 nm band) due to the application of filter (290 nm).

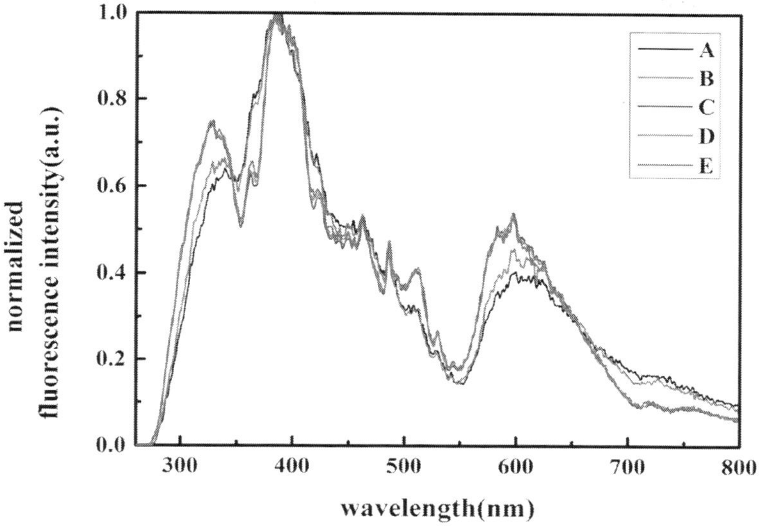

Figure 3: The normalized fluorescence emission spectra of samples.

In the gringding procedures, the decrease of abrasive particles size hardly changed the species and the relative ratios of the fluorescence defects. The polished and etched samples possessed the same defects species with the ground samples. However, the relative ratios of 310 nm and 590 nm defects were higher than those of ground samples. In other words, though the improvement of the manufacturing procedures, the types of fluorescence defets were remained the same while the relative ratio of 390 nm defect was reduced. The fluorescence defects with lifetime shorter than the laser pulse correlate to damage propensity by the single-photon absorption of sub-band gap light[21]. According to L. Skuja et al.[18,20,22], the lifetimes of these fluorescence defects discussed above are 4 ns (α-ODC(II)), 110 μs (β-ODC(II)) and 300 ns (E_{δ}'), respectively. In our case (laser pulse is 6.4 ns), α-ODC(II) might be the main laser damage precursor. Unfortunately, it was not removed by the current manufacturing processes. However, a detailed understand about this 310 nm defect is still open to the further work.

Physical Disorder

The infrared spectra are carried out to find the strained Si-O-Si bonds of the prepared samples surfaces as shown in Fig. 4. Under the irradiation of ns laser, the strained Si-O-Si bond is an absorptive precursor, which induces damages and leads to the reduction of laser damage threshold of fused silica. These spectra typically include four peaks corresponding to four vibrational modes of Si-O-Si bond: ν_R(rocking mode, ~470 cm^{-1}), ν_B (bending vibration, ~800 cm^{-1}), ν_S(TO)(transverse optical mode of asymmetric stretching vibration, ~1020 cm^{-1}), and ν_S(LO) (longitudinal optical mode of asymmetric stretching vibration, ~1130 cm^{-1})23. The wavenumbers of these peaks can be theoretically derived by using a single nearest neighbor central force assumption24 with a short-range Born potential25, and can be calculated by the following equations26·27:

$$\nu_R = \frac{1}{2\pi c}\sqrt{\frac{2\beta}{m_O}} \tag{1}$$

$$\nu_B = \frac{1}{2\pi c}\sqrt{\frac{4(\alpha + 2\beta)}{3m_{Si}}} \tag{2}$$

$$\nu_S = \frac{1}{2\pi c}\sqrt{\frac{2}{m_O}\left(\alpha\, sin^2\frac{\theta}{2} + \beta\, cos^2\frac{\theta}{2}\right)} \tag{3}$$

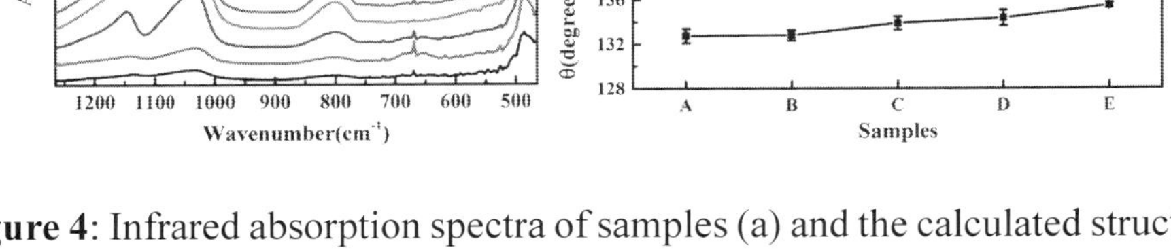

Figure 4: Infrared absorption spectra of samples (a) and the calculated structure parameters of the Si-O-Si bonds (b).

In which, α is the central force constant, β is the non-central force constant, and θ is the Si-O-Si bond angle. m_{Si} and m_O are the atomic mass of silicon and oxygen, respectively. c is the velocity of light. The force constants (α and β) are related to elastic constants: α determines Young's modulus, and β determines the shear modulus26. The angle of Si-O-Si bond (θ) is related to the density of surface materials28. These structure parameters can be derived from the wavenumbers of the infrared spectra by using equations mentioned above.

The calculated values of structure parameters are shown in Fig. 4(b). The central force constant values of the polished samples are higher than those of grinded samples and lower than those of etched samples ($\alpha_{grinded} < \alpha_{polished} < \alpha_{etched}$). The tend of the non-central force constant is opposite ($\beta_{grinded} > \beta_{polished} > \beta_{etched}$). Young's modulus increases with decreasing α, while shear modulus decreases with decreasing β. Thus, Young's modulus and shear modulus of grinded samples are higher than those of polished samples, and the etching process make the elasticity moludus smaller.

The estimated Si-O-Si bond angles (θ) are also shown in Fig. 4(b). The angles of all prepared samples are lower than the average value estimated by X-ray diffraction (~145°)29·30. This discrepancy indicates that Si-O-Si bonds in the surface layer are strained and the densities of the surface materials after grinding, polishing and etching are slightly higher than that of the body material of fused silica. The densification of surface materials may due to the normal pressure employed during the grinding and polishing processes.

Raman spectra of treated samples are shown in Fig. 5. The two sharp contributions at 490 cm^{-1} (D_1) and 606 cm^{-1} (D_2) correspond to the symmetric stretching vibrations ("breathing mode") in four- and three- member rings of SiO_4 tetrahedra31.

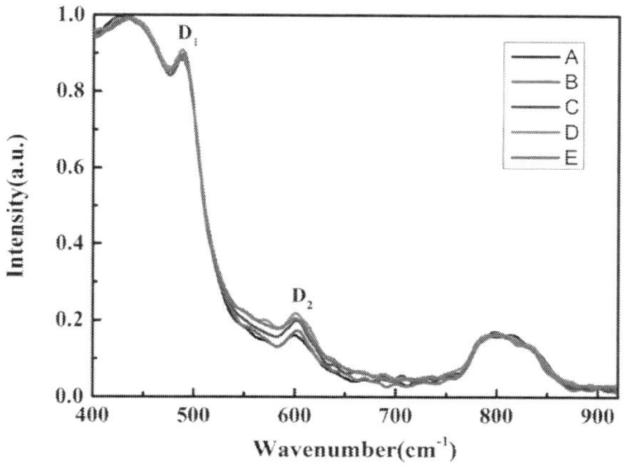

Figure 5: The Raman spectra of samples. D_1 and D_2 indicate the four- and three-member rings of SiO_4tetrahedra, respectively.

Si-O-Si bonds in these smalll ring structures are strained and may result in defect pairs of E' and NBOHC under the laser irradiation[16]. Results of both Infrared and Raman spectra indicate that the current manufacturing processes can reduce the strained Si-O-Si bonds but still can not eliminate them.

LIDTS

The R-on-1 LIDTs of samples are shown in Fig. 6. The LIDTs of grinded samples are 5.0 J/cm^2 and 5.6 J/cm^2, which are less than those of polished samples (13.5 J/cm^2 for ceria polishing and 16.1 J/cm^2 for MRF finishing, respectively). The etched Sample E has the highest LIDT of 19.8 J/cm^2. Combining with the discussions above of fluorescence defects and strained Si-O-Si bonds, the sample with more fluorescence defects and strained Si-O-Si bonds has a lower LIDT, as seen in Fig. 6.

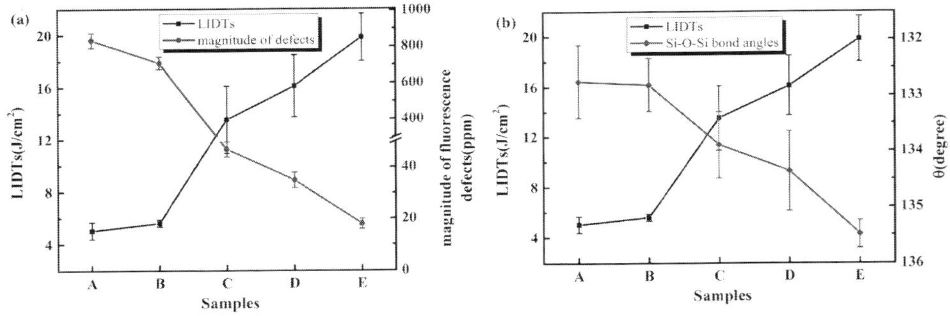

Figure 6: Comparison of R-on-1 LIDTs of the samples, and the influence of fluorescence defects and Si-O-Si bonds on LIDTs of fused silica optics.

The effect of abrasive particle size may due to the difference of average load on per particle[8]. The average load grows with the increase of abrasive particle size, and the fractured cracks occur under high average load during grinding process. Consequently, for samples A and B, the usage of smaller abrasive particles brings less fluorescence defects and strained Si-O-Si bonds. It has been reported that the MRF process can mitigate surface and subsurface defects of conventional polished samples while leave a thin surface Beilby layer[32,33], which can be mitigated after etching process. However, the fluorescence defects and strained Si-O-Si bonds still exist and limit the LIDT of fused silica optics.

CONCLUSIONS

This paper presents fluorescence defects, surface molecular structures and laser damage performances of fused silica samples prepared by grinding, polishing and etching processes. Results showed that the fluorescence defects and strained Si-O-Si bonds reduced with the improvement of the manufacturing process from gringing to etching and LIDTs of fused silica increased with the reducing of the fluorescence defects and strained Si-O-Si bonds. However, these structural defects can not be eliminated by the manufacturing process disscused here, and limited the LIDT of fused silica under 6.4 ns 355 nm laser irradiation. Both the grinding and polishing processes will induce fluorescence defects and strained Si-O-Si bonds to the surfaces of fused silica samples. Though the HF acid etching process can mitigate thoses precursors and improve the LIDTs of fused silica samples, but improvements are still needed to eliminate these structural defects for a higher laser damage resistance of fused silica.

METHODS

Sample Preparation

Fused silica samples (50 mm × 50 mm × 10 mm; Corning 7980, Corning, NY) were firstly polished down to 1 nm (RMS) on both sides. One face of each sample was then treated by one of the manufacturing procedures with 1.5 μm removed:

- Sample A was ground on a Strasbaugh Model 6Y2 grinder using 15 μm Al_2O_3 abrasive in water on a Pyrex glass lap (load = 25 N, lap rotation rate = 16 rpm).
- Sample B was ground on a Strasbaugh Model 6Y2 grinder using 9 μm Al_2O_3 abrasive in water on a Pyrex glass lap (load = 25 N, lap rotation rate = 16 rpm).
- Sample C was conventionally polished by continuous pitch polishing techniques using 10 μm ceria (CeO_x) abrasive.
- Sample D was subjected to MRF polishing using standard ceria (CeO_x) slurry contained in an iron based MR fluid media.
- Sample E was firstly polished by the same MRF polishing as the Sample D, and then etched in 5%wt HF etchant with ultrasonic

agitation of 40–270 kHz for 5 min to remove the surface Beilby layer.

Before LIDT tests, all samples were clearned in ultrapure water with ultrasonic agitation of 40–270 kHz for 30 min, and allowed to air dry in a clean room.

R-on-1 LIDTs

R-on-1 LIDTs tests were carried out at ambient conditions by using a 6.4 ns 355 nm beam from a Q-switched Nd-YAG laser. The beam profile was Gaussian with a $1/e^2$ area of 0.6 mm^2 at the sample plane. The test laser fluence was started at 1 J/cm^2 and then increased by 0.1 J/cm^2each shot until the damage occurred (determined by in situ optical microscopy). Schematic of the experimental setup for laser damage tests is shown in Fig. 7. The surfaces prepared by different procedures were rear surfaces of samples.

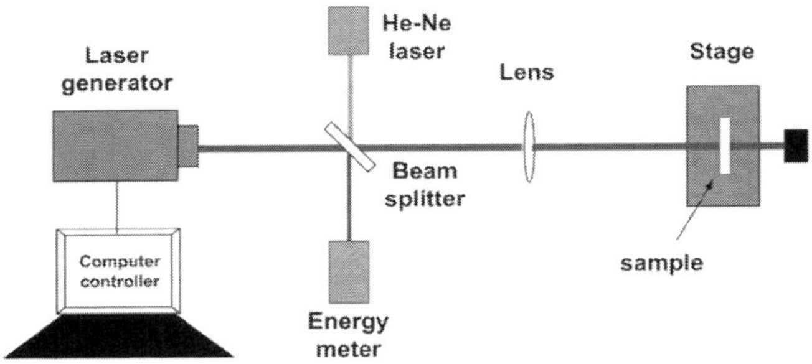

Figure 7: Schematic of the experimental setup for laser damage tests.

UV Confocal Fluorescence Image

An integrated confocal fluorescence microscope system consists of a fluorescence microscope, laser light sources, and a scan head. The laser was directed on the sample and the emission was collected, a computer with software was used to control the scan head and display the acquisition. The excitation beam was 355 nm laser and the images were 20 × objective. Both fluorescence and bright field images can be obtained.

PL Spectra

PL spectra were carried out on a fluorescence spectrometer equipped with a photomultiplier detector. Excitation was realized at 248 nm by a Xe lamp at ambient temperature. The focal spot size on the sample was about 50 mm^2, and the slit width for exitation and emission was 7 nm and 4 nm, respectively. A 290 nm high-pass filter was laid in front of the detector.

Infrared Spectra

Fourier transform infrared absorption spectra were obtained by a Nicolet 5700 spectrometer. A reflection mode was employed to measure the spectra in a frequency region of 400–1300 cm^{-1}, associating with the Si-O-Si stretching, bending, and rocking vibrations. The depth of penetration into the sample is in the order of a few micrometers. All spectra were taken at ambient temperature with more than 200 scans at a resolution of 0.96 cm^{-1}.

Raman Spectra

Raman spectra were obtained with more than 300 scans at ambient conditions by using the Nicolet spectrometer. The spectral resolution was ~0.96 cm^{-1}, and the spot diameter was 50 μm.

REFERENCES

1. Lowdermilk, W. H. Status of the National Ignition Facility project. Proc. SPIE 3047, 16–37 (1996).
2. Andre, M. L. Status of the LMJ project. Proc. SPIE 3047, 38–42 (1996).
3. Salleo, A. et al. Laser-driven formation of a high-pressure phase in amorphous silica. Nat. Mater. 2, 796–800 (2003).
4. Wong, J. et al. Morphology and microstructure in fused silica induced by high fluence ultraviolet 3x (355 nm) laser pulses. J. Non-Cryst. Solids 352, 255–272 (2006).
5. Papernov, S. & Schmid, A. W. Laser-induced surface damage of optical materials: Absorption sources, initiation, growth, and mitigation. Proc. SPIE 7132, 71321J (2008).
6. Miller, P. E. et al. Fracture-induced subbandgap absorption as a precursor to optical damage on fused silica surfaces. Opt. Lett. 35, 2702–2704 (2010).
7. Hongjie, L. et al. Subsurface defects of fused silica optics and laser induced damage at 351 nm. Opt. Express 21, 12204–12217 (2013).
8. Suratwala, T. et al. Sub-surface mechanical damage distributions during grinding of fused silica. J. Non-Cryst. Solids 352, 5601–5617 (2006).
9. Bude, J. D. et al. Silica laser damage mechanisms, precursors and their mitigation. Proc. SPIE 9237, 92370S (2014).
10. Suratwala, T. I. et al. HF-Based Etching Processes for Improving Laser Damage Resistance of Fused Silica Optical Surfaces. J. Am. Ceram. Soc. 94, 416–428 (2011).
11. Ye, X. et al. Advanced Mitigation Process (AMP) for Improving Laser Damage Threshold of Fused Silica Optics. Scientific reports6, 31111 (2016).
12. Shi, F., Shu, Y., Dai, Y., Peng, X. & Li, S. Magnetorheological elastic super-smooth finishing for high-efficiency manufacturing of ultraviolet laser resistant optics. Opt. Eng. 52, 075104 (2013).
13. Bude, J. et al. High fluence laser damage precursors and their mitigation in fused silica. Opt. Express 22, 5839–5851 (2014).
14. Wang, J. et al. Producing Fused Silica Optics with High UV-Damage Resistance to Nanosecond Pulsed Lasers. Proc. SPIE9532, 95320H (2015).

15. Stuart, B. C. et al. Nanosecond-to-femtosecond laser-induced breakdown in dielectrics. Phys. Rev. B 53, 1749–1761 (1996).
16. Hosono, H., Ikuta, Y., Kinoshita, T., Kajihara, K. & Hirano, M. Physical Disorder and Optical Properties in the Vacuum Ultraviolet Region of Amorphous SiO2. Phys. Rev. Lett. 87, 175501 (2001).
17. Wright, A. C. Neutron scattering from vitreous silica. V. The structure of vitreous silica: What have we learned from 60 years of diffraction studies? J. Non-Cryst. Solids 179, 84–115 (1994).
18. Skuja, L. Optically active oxygen-deficiency-related centers in amorphous silicon dioxide. J. Non-Cryst. Solids 239, 16–48 (1998).
19. Meinardi, F. & Paleari, A. Native and radiation-induced photoluminescent defects in SiO2: Role of impurities. Phys. Rev. B58, 3511–3514 (1998).
20. Fournier, J. et al. Green luminescence in silica glass: A possible indicator of subsurface fracture. Appl. Phys. Lett. 100, 114103 (2012).
21. Laurence, T. A. et al. Metallic-like photoluminescence and absorption in fused silica surface flaws. Appl. Phys. Lett. 94, 151114 (2009).
22. Nishikawa, H. et al. Visible photoluminescence from Si clusters in g-irradiated amorphous SiO_2. J. Appl. Phys. 80, 3513–3517 (1996).
23. Li, C. H. et al. Structural modification in amorphous silica after exposure to low fluence 355 nm laser irradiation. Nucl. Instrum. Meth. B 269, 544–549 (2011).
24. Galeener, F. L. Band limits and the vibrational spectra of tetrahedral glasses. Phys. Rev. B 19, 4292–4297 (1979).
25. Lehmann, A., Schumann, L. & Hubner, K. Asymmetric peak line shape of infrared dielectric function spectra for thermally grown silicon dioxide films. Phys. Status Solidi B 117, 689 (1983).
26. Tomozawa, M., Hong, J.-W. & Ryu, S.-R. Infrared (IR) investigation of the structural changes of silica glasses with fictive temperature. J. Non-Cryst. Solids 351, 1054–1060 (2005).
27. Ishikawa, K., Suzuki, K. & Okamura, S. Asymmetric peak line shape of infrared dielectric function spectra for thermally grown silicon dioxide films. J. Appl. Phys. 88, 7150–7156 (2000).
28. Agarwal, A. & Tomozawa, M. Correlation of silica glass properties with the infrared spectra. J. Non-Cryst. Solids 209, 166–174 (1997).
29. Mozzi, R. L. & Warren, B. E. Correlation of silica glass properties with

the infrared spectra. J. Appl. Crystallogr. 2, 164 (1969).

30. Uchino, T., Takahashi, M. & Yoko, T. E' Centers in Amorphous SiO2 Revisited: A New Look at an Old Problem. Phys. Rev. Lett. 86, 5522 (2001).
31. Varkentina, N. et al. High repetition rate femtosecond laser irradiation of fused silica studied by Raman spectroscopy. Opt. Mater. Express 6, 79–90 (2016).
32. Catrin, R. et al. Magnetorheological finishing for removing surface and subsurface defects of fused silica optics. Opt. Eng. 53, 092010 (2014).
33. Miao, C. Shear stress in magnetorheological finishing for glasses. Appl. Opt. 48, 2585–2594 (2009).

Chapter 16

Impact of the Polishing Suspension Concentration on Laser Damage of Classically Manufactured and Plasma Post-Processed Zinc Crown Glass Surfaces

Christoph Gerhard [1] and Marco Stappenbeck [2]

[1]Faculty of Engineering and Natural Sciences, Technical University of Applied Sciences Wildau, Hochschulring 1, 15745 Wildau, Germany

[2]Faculty of Natural Sciences and Technology, University of Applied Sciences and Arts, Von-Ossietzky-Straße 99, 37085 Göttingen, Germany

ABSTRACT

The laser-induced damage threshold of optics is an issue of essential importance in high-power laser applications. However, the complex and partially interacting mechanisms as well as the underlying reasons for laser damage of glass surfaces are not yet fully understood. The aim of

Citation: Gerhard, Christoph; Stappenbeck, Marco. 2018. "Impact of the Polishing Suspension Concentration on Laser Damage of Classically Manufactured and Plasma Post-Processed Zinc Crown Glass Surfaces." Appl. Sci. 8, no. 9: 1556. https://doi.org/10.3390/app8091556

the present work is to contribute to a better understanding of such damage mechanisms by providing original results on the impact of classical glass surface machining on the laser-induced damage threshold. For this purpose, glass samples were prepared with well-defined process conditions in terms of the used lapping and polishing agents and suspensions. Further, the samples were post-processed by atmospheric pressure plasma for precision cleaning. The laser-induced damage threshold and surface contamination by residues from the manufacturing process were determined before and after plasma post-processing. It is shown that the polishing suspension concentration has a certain impact on the laser-induced damage threshold and surface contamination by residues from used working materials. The highest damage threshold of 15.2 J/cm^2 is found for the lowest surface contamination by carbon which occurs in the case of the highest polishing suspension concentration. After plasma treatment for merely 60 s, this value was increased to 20.3 J/cm^2 due to the removal of surface-adherent carbon. The results thus imply that the laser-induced damage threshold can notably be increased by first choosing appropriate process parameters during classical manufacturing and second plasma post-processing for surface finishing.

Keywords: Polishing, glass, optics manufacturing, surface contamination, laser-induced damage threshold, plasma cleaning

INTRODUCTION

Optical components with high laser-induced damage threshold (LIDT) are of essential importance for the realisation of optical systems used for high-power laser setups and facilities. However, classically manufactured optics exhibit comparatively poor laser resistance which can be directly attributed to effects arising in the course of different optics manufacturing process steps. Conventional optical manufacturing involves rough grinding, fine grinding or lapping, and polishing. By these steps, material removal is achieved by the application of different abrasives where the mechanical impact of such abrasives on the glass surface comes along with the formation of either open or closed subsurface micro cracks as shown in Figure 1. The crack depth is approximately 0.3 times the mean diameter of the used abrasive grains and ranges from 1 to 20 μm for polished surfaces [1]. Such cracks were identified to play a key role for laser damage due to the formation of interference effects, an increased absorption and a reduction in mechanical strength of a glass surface [2,3]. Moreover, residues from operation materials used in optics manufacturing can accumulate within existing micro cracks [4]. This leads to a complex possible contamination consisting of (i) mineral oil-containing cooling lubricants used during rough grinding, (ii) wear debris from abrasive grains, e.g., silicon carbide (SiC) or diamond (C) from the rough and fine grinding or lapping process, and (iii) polishing agents as for example cerium oxide, zirconium oxide, or aluminium oxide as well as wear debris from the used polishing pad.

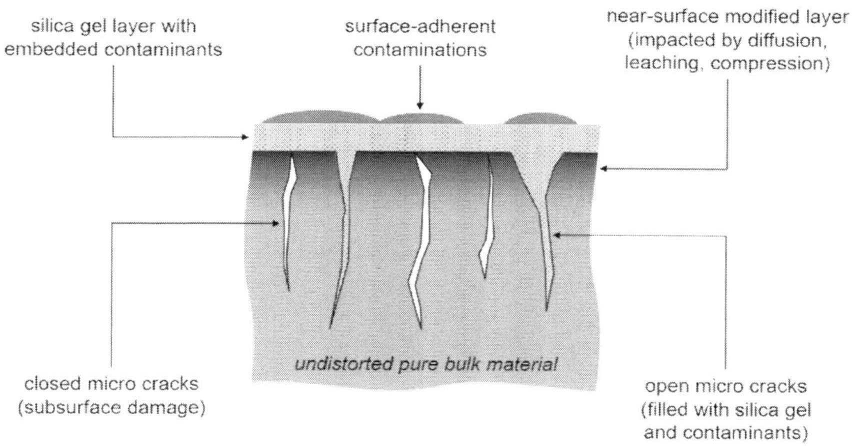

Figure 1: Overview on possible defects of classically manufactured silicon dioxide-based glass surfaces [5].

Since cooling lubricants and lapping and polishing suspensions are water-based operating materials, the involved hydrogen further impacts the glass surface, for example by hydrolytic scission. This leads to the formation of silica gel or hydrated silica during mechanical chemical polishing (MCP) in the case of silicon dioxide-based glasses [6]. As shown in Figure 1, this medium further accumulates within cracks, roughness valleys and digs and can act as matrix for the above-mentioned contaminants and preferentially residues from polishing agents as shown by *Liao* and co-workers [7]. Moreover, it was recently reported that an accumulation of such residues within surface and subsurface damage, e.g., micro cracks, gives rise to surface inhomogeneity in terms of chemical composition [4,8]. These residues or contaminants act as precursors for laser damage [9,10,11] and consequently reduce the LIDT of polished glass surfaces [12]. Against this background, a considerable number of investigations on surface and subsurface contamination by polishing agents were performed in the past, e.g., [7,9,13,14]. *Krol* and co-workers have further shown that the LIDT of fused silica depends on the polishing method and the particularly corresponding density of such precursors, respectively [15]. It is further known that severe reduction in LIDT also results from surface contamination by carbonaceous compounds [16]. For instance, *Bien-Aimé* and co-workers have shown that organic contaminants, here the hydrocarbon dioctylphthalate (DOP), can lower the LIDT dramatically [17]. In practice, optics surfaces may exhibit severe carbon contamination [18] which might result from the use of carbon-based cleaning agents such as ethanol (C_2H_6O) or acetone (C_3H_6O) or originate from adsorption of hydrocarbons from ambient air in the course of storing or usage [19].

In order to increase the LIDT of optics surfaces, different approaches and techniques for monitoring, controlling, and removing surface contaminants were investigated and developed in the past [20,21]. This includes different post-processing steps after classical polishing such as carbon dioxide laser treatment [22], hydrofluoric acid (HF) etching [1], ion beam etching (IBE) [23] or the combination of HF etching, subsequent precision polishing and final ion beam etching [24,25]. In the first case, an increase in LIDT is mainly due to surface relaxation via laser melting whereas the application of etching techniques allows the removal of polishing compounds, metal impurities and subsurface damages [26].

The latter issue can efficiently be addressed with the aid of plasmas. As shown by *Juškevi ius* and co-workers, the removal of near-surface layers—and subsurface damage, respectively—with a thickness in the range from 50 to 200 nm via etching by argon radio frequency plasma in vacuum allows a significant increase in LIDT of fused silica surfaces by a factor of 8.4 [27]. Moreover, *Zhang* and co-workers have successfully applied oxygen plasmas for reducing the number of micro surface defects on zirconium dioxide films. Thanks to such surface healing, the LIDT was increased by a factor of 1.45 (from 15.9 J/cm^2 to 23.1 J/cm^2) [28] due to the reduction in absorption of incoming laser irradiation. However, further work by the authors on hafnium dioxide films showed that oxygen plasma treatment may also not impact the LIDT if subsurface nanoscale absorber defects are not affected or removed by the plasma [29].

A special type of absorber defects contributing to a decrease in LIDT are surface-adherent carbonaceous compounds as mentioned above. In this context, atmospheric pressure plasmas have turned out to be suitable and powerful tools for surface cleaning [18,30,31,32]. Most recently, it was shown that the use of direct dielectric barrier discharge (DBD) plasmas at atmospheric pressure using argon as process gas allows an increase in LIDT of super-polished sapphire window surfaces by a factor of 1.5 (from 7.84 J/cm^2 to 11.89 J/cm^2) [16]. This effect is not due to surface etching or material removal but selective removal of surface-adherent contaminants by plasma species. Against this background, the impact of such plasma treatment on zinc crown glass samples was investigated in the present work. Moreover, special attention was also paid to the influence of process parameters applied during sample preparation, namely the particular polishing suspension concentration.

MATERIALS AND METHODS

Sample Preparation

In this work, plane samples made of zinc crown glass (glass type N-ZK7 from Schott AG, Mainz, Germany) were investigated. The samples were produced by classical optical manufacturing via rough grinding, fine grinding and polishing as described in more detail in [33]. For fine grinding, the method of loose abrasive grinding, also referred to as lapping, was applied. Here, a lapping suspension consisting of water and silicon carbide abrasives with a grain size in the range from 2 to 14 µm (F800 FEPA standard) was used. As usual in optics manufacturing, the samples were lapped slightly concave in order to ensure edge support of the tool employed in the subsequent and final manufacturing step, polishing. For this step, a plane polishing tool with a polishing pad made of polyurethane was used since this type of polishing pad allows reducing the number of micro cracks formed during polishing [34]. The polishing agent was a premixed water-based polishing suspension (type Hastilite PO from Universal Photonics, Inc., NY, USA), where the actual polishing agent is cerium oxide (CeO_2) with a mean grain size of 0.8 µm. The initial concentration of this premixed polishing suspension was $c_s = 60\%$. In order to obtain further polishing suspensions with defined concentrations, the premixed suspension was thinned down using distilled water. According to

$$c_S = \frac{m_{pa}}{m_S} \cdot 100\%, \qquad (1)$$

where m_{pa} is the mass of the polishing agent within the suspension and m_s is its total mass [33], two further suspensions with a concentration of $c_s = 44\%$ and $c_s = 19\%$, respectively, were realised in this way.

Measurement Techniques and Approaches

In order to determine the LIDT, a diameter regression fit method as suggest by *Liu* [35] was applied. This 1-on-1 laser damage testing approach is an easy and distinct method for the determination of the LIDT in the picosecond pulse duration range on the basis of the diameter of ablated spots [36]. Here, the sample surface is sectioned into a number of test sites and each site is irradiated with a single laser pulse with a defined and known fluence. The squared diameters of the damaged sites are then plotted vs. the laser fluence. The resulting straight line is extrapolated and the intercept of this regression line with the logarithmically scaled *x*-axis (i.e., the fluence values) finally gives the LIDT as shown schematically in Figure 2.

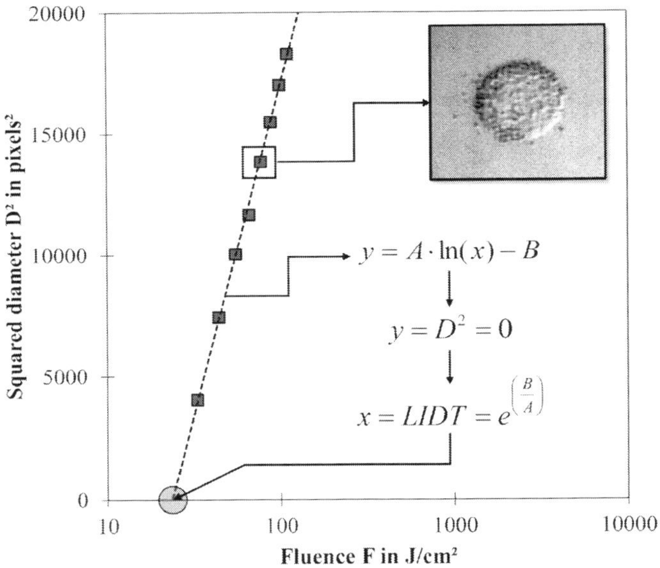

Figure 2: Principle of the diameter regression technique applied for the determination of the laser-induced damage threshold (LIDT) including an example for the observed damage morphology (inset).

In the present case, 25 test sites per sample were investigated. The used laser source was a Yb:YAG solid state picosecond laser (model TruMicro 5025 from Trumpf GmbH & Co. KG, Ditzingen, Germany) with a pulse duration of = 8 ps, an emission wavelength of = 1030 nm, and a maximum pulse energy of E = 125 μJ. The raw beam with a diameter of $2w_r$ = 5 mm was focused onto the sample surface by a convex lens with a focal length of f = 50 mm, resulting in a laser beam waist diameter of $2w_0$ = 17 μm. As a consequence, the initial laser fluence at the maximum pulse energy was 110 J/cm^2. This value was successively reduced in eight steps down to a minimum fluence of 22 J/cm^2 where the decrement per step was 11 J/cm^2. For the determination of the actual LIDT, the diameters of the laser-induced damaged sites on the sample surfaces (given in pixels) were measured with the aid of a light microscope (model Axioskop 2 MAT from Carl Zeiss Microscopy GmbH, Jena, Germany).

In addition to the LIDT, the chemical composition of the sample surfaces was measured using a scanning X-ray photoelectron spectroscopy (XPS) apparatus (model PHI VersaProbe II from ULVAC-PHI, Inc., Hagisono, Japan) in order to determine and identify impurities induced by the manufacturing process. Here, the elements of interest were carbon and cerium, representing possible contaminants occurring in the course of sample preparation by rough grinding, lapping and polishing as well as silicon and oxygen since the investigated zinc crown glass is a high-silica glass (60–70 wt %) [37]. The particular elemental fractions were determined via evaluation of the C1s-peak of carbon, the Ce3d5-peak of cerium, the Si2p-peak of silicon, and the O1s-peak of oxygen in the measured spectrum.

All measurements were performed on samples prepared at three different concentrations of the used polishing agent (19, 44, and 60%, see above). In addition, the impact of a plasma post-processing as described in the following section on the measured parameters was investigated.

Plasma Post-Processing

For plasma post-processing, a plasma source as described in more detail in previous work [38,39,40] was used. This plasma source basically consists of a conical rotation-symmetric high-voltage copper electrode embedded in a plastic housing made of polyoxymethylene (POM). The gas channel for feeding the plasma process gas is found between the high-voltage electrode and the housing. In the present work, argon (Ar 5.0 from Linde Gas AG, Munich, Germany) was used as process gas where the gas flow rate was 4 standard litres per minute. The plasma was ignited between the high-voltage electrode and an external plane ground electrode, both made of copper. The glass samples were placed between the high-voltage electrode and the ground electrode; the working distance between the outlet nozzle of the plasma source and the work piece surface was approx. 10 mm. This setup leads to the formation of a direct dielectric barrier discharge (DBD) plasma where the sample itself is an essential part of the plasma ignition which consequently occurs directly on the sample surface as visualised in Figure 3.

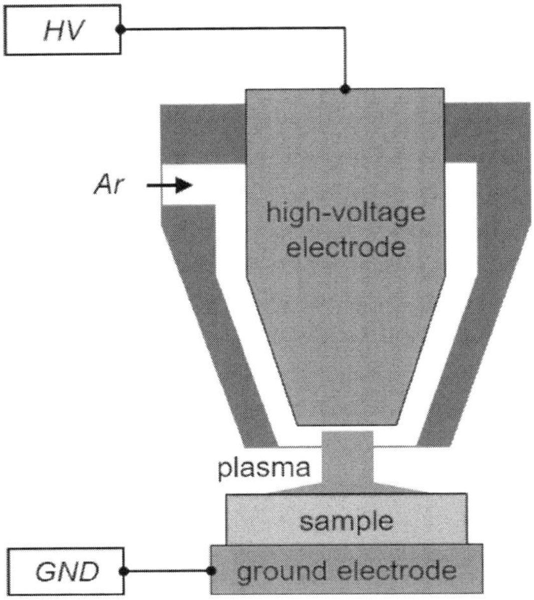

Figure 3: Experimental setup for plasma post-processing of polished glass samples applying a direct dielectric barrier discharge (DBD) plasma.

The plasma source was driven by a pulsed power supply where the voltage of each pulse was $U = 11$ kV, the nominal pulse repetition rate was $f = 7$ kHz and the pulse duration of each pulse was = 80 μs. Due to the functional principle—the DBD—the plasma gas temperature and sample surface heating, respectively, are quite moderate and amount to approximately 90 °C as determined via infrared camera measurements in previous work [41]. This is due to the fact that extremely low currents in the range of some microamperes to milliamperes occur during DBD plasma operation. In contrast to classical arc discharges, burn-off of the electrode material is moreover reduced or even inhibited, avoiding the deposition of electrode material on the sample surface. In the present work, the plasma treatment duration was 60 s. The impact of such short-term direct DBD plasma post-processing was investigated since it was observed in previous work that applying this type of plasma for merely one minute allows considerable smoothing of optical media [39,40] as well as surface cleaning and an accompanying increase in LIDT of sapphire substrates [16].

RESULTS

Polished Samples

The investigation of the chemical composition of the samples after polishing revealed that the cerium content was approximately 0.25 ± 0.01 at% for all samples, so no dependency of surface contamination by the polishing agent on the polishing suspension concentration was detected. In contrast, an obvious impact of the polishing suspension concentration on first the surface contamination by carbon, and second the LIDT, was observed as shown in Figure 4.

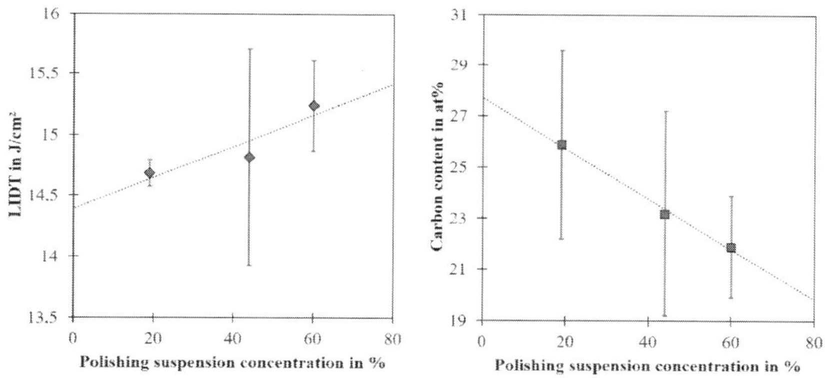

Figure 4: Laser-induced damage threshold (LIDT, left) and carbon content (right) vs. polishing suspension concentration including trend (dotted line).

It turns out that the LIDT increases with increasing polishing suspension concentration, whereas the carbon content exhibits the reverse behaviour. As a consequence, these two considered parameters indicate a certain correlation; it can be stated that the lower the carbon content, the higher the LIDT. At an average, an LIDT of the polished zinc crown glass of approximately 15 J/cm^2 was determined. This value cannot be compared to literature since the LIDT of this specific glass is not reported. However, it is in a typical range for transparent optical materials exposed to focussed laser irradiation at a wavelength around 1 micron and a pulse duration of some picosconds. For fused silica, quite different values ranging from approximately 5 J/cm^2 [42] to even 44 J/cm^2 [43] are reported. For sapphire, the LIDT amounts to approximately 8 J/cm^2 [16,44]. Finally, dielectric coatings feature LIDT in the range from about 2 to 5 J/cm^2 [45,46].

Polished and Plasma Post-Processed Samples

As ascertained via XPS measurements, plasma post-processing did not notably impact the cerium content at the sample surfaces. With respect to the initial average value of 0.25 ± 0.01 at%, a marginal increase to 0.26 ± 0.09 at% was detected. In contrast, a significant reduction in carbon content was detected as shown by the comparison in Figure 5. At an average, the carbon content was 23.67 ± 2.04 at% before and 7.82 ± 0.86 at% after plasma post-processing, corresponding to a decrease by a factor of approximately 3.

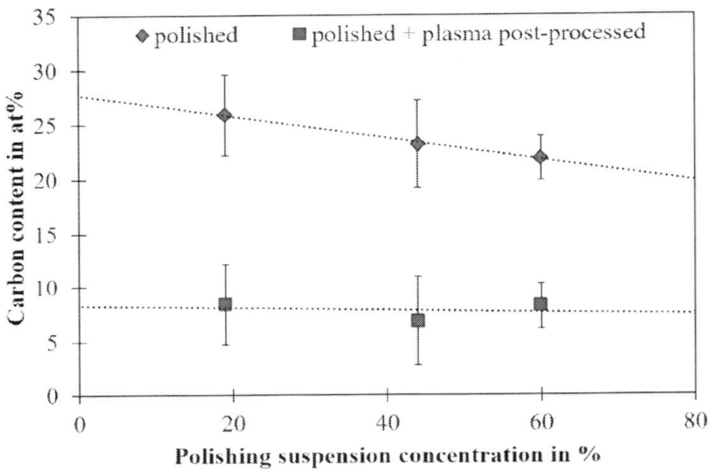

Figure 5: Carbon content of zinc crown glass samples before and after plasma post-processing vs. polishing suspension concentration including trend (dotted line).

It further turns out that, in contrast to the initial state before plasma post-processing, no dependency of the particular carbon content on the polishing suspension concentration is observed after plasma treatment as visualised by the trend lines in Figure 5. The residual carbon content is rather constant. Another observed effect is a general increase in LIDT due to plasma post-processing as shown in Figure 6.

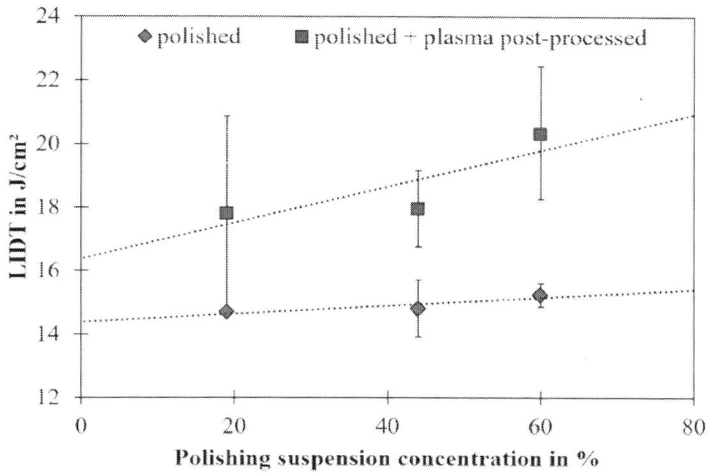

Figure 6: Laser-induced damage threshold (LIDT) of zinc crown glass samples before and after plasma post-processing vs. polishing suspension concentration including trend (dotted line).

In contrast to the carbon content presented in Figure 5, the LIDT shows a slightly stronger dependency by trend on the polishing suspension concentration after plasma post-processing.

DISCUSSION

Impact of the Polishing Suspension Concentration

As a general point, it should be noted that the measurements presented in Section 3 feature relatively large error bars. Regarding the LIDT measurements, statistically distributed absorbing surface defects may in some cases not be affected by the incident laser irradiation due to the small spot size of the laser focus. This fact might contribute to the relatively large error bars of the LIDT values which result from the particular values measured at 25 test sites per sample. The observed dependencies as discussed in the present subsection have thus to be treated as a general trend.

As presented in Section 3.1, the cerium content at the surfaces of the investigated samples was approximately 0.25 at% after polishing. This value was detected for all polishing suspension concentrations, a correlation between the cerium content within the suspension and surface contamination by this element or polishing agent, respectively, was thus not observed. However, the use of such polishing suspension led to a constant attachment or implantation of cerium at/into the surface as discussed in more detail in Section 4.2. This effect, a saturation of the interaction between the polishing pad or suspension and the glass surface during polishing once a certain concentration of the polishing suspension is reached, was already reported in 2008 by *Rogov* and co-workers [47].

An interesting point is the trend of a decreasing carbon content with increasing polishing suspension concentration. For the origin of the detected carbon, several sources from the manufacturing process should be considered since the glass itself does not contain any carbon according to the manufacturer's datasheet [37]:

- First, the carbon could originate from the rough grinding process which is usually performed with the aid of cylindrical cup wheels where the bound abrasives are usually made of diamond, i.e., pure carbon.
- Second, the application of cooling lubricants during rough grinding is another possible source for carbon since such lubricants are a suspension of water and carbonaceous mineral oil.
- Third, the lapping agent used for sample preparation, silicon carbide (SiC), could accumulate within digs and cracks of the glass surface and finally be embedded within hydrated silica which is formed during polishing as introduced in Section 1. In the course of this terminal production step, carbonaceous contaminants could finally arise in the form of wear debris from the used polishing pad made of polyurethane foil. This last aspect could explain the observed behaviour, i.e., an increased carbon content in the case of thin suspensions. Here, wear preferentially occurs between the polishing pad and the glass surface due to direct contact. The higher the concentration of the suspension, the more polishing agent grains are found in between the pad and the surface, consequently reducing direct wear.

The presence of carbon turns out to be the main reason for a decrease in LIDT. Since the cerium content is quite constant for all polishing suspension concentrations as mentioned in Section 3.1, the impact of this parameter on the LIDT can be assumed to be constant as well. In contrast, the LIDT and the carbon content tends to depend on the polishing suspension concentration, see Figure 4. The increase in LIDT with decreasing carbon content can be explained by the fact that a reduction in surface-adherent carbon and carbonaceous compounds corresponds to a reduction in absorbing contaminants which may act as precursor for laser damage. This is visualised by the comparison of the absorption coefficients of the abovementioned involved media (or a particular reference material) in Table 1. For example, the absorption coefficient of the used lapping agent, silicon carbide, is more than four million times higher than the absorption coefficient of the investigated zinc crown glass.

Table 1: Absorption coefficients of the investigated zinc crown glass, silicon carbide, glycol, and immersion oil at a wavelength of = 1030 nm

Material	Absorption Coefficient α in cm^{-1}	Reference
Zinc crown glass N-ZK7	0.0024	[37]
Silicon carbide [1]	10,070	[48]
Glycol [2]	0.22	[49]
Immersion oil [3]	1.013	[50]

[1] Lapping agent used in this work, [2] reference material for polyurethane (i.e., the polishing pad material used in this work), [3] reference material for mineral oil-based cooling lubricants.

It should be noted that this comparison of the linear absorption coefficients of the investigated glass and possible surface-adherent and subsurface contaminants is only one aspect of laser-induced damage. Especially in the considered pulse duration range of some picoseconds, a number of further effects contribute to the LIDT. This includes nonlinear optical phenomena such as self-focussing, a dynamic increase in absorption due to multiphoton absorption and the formation of charge carriers by laser-induced avalanche ionisation. However, it can finally be stated that the polishing suspension concentration and the polishing process in general may have a noticeable impact on the LIDT. This effect was also qualitatively reported by *Zawilski* and co-workers where the LIDT of optical crystals was increased via optimising the applied polishing process [51].

Impact of Plasma Post-Processing

As mentioned in Section 3.2, cerium was not removed by plasma post-processing; its content was even slightly increased. This indicates that the detected cerium was not surface-adherent, but implanted within a subsurface zone during polishing. Considering that the information depth of XPS is approximately 10 nm and that the mean thickness of hydrated silica layers is in this order of magnitude (1 to 20 nm [6]), the cerium content measured via XPS was most likely embedded in such a silica layer which was formed in the course of the polishing process. Apparently, this layer was not removed by the short-term plasma treatment due to its chemical stability. The slight increase in cerium content rather indicates a plasma-induced uncovering of this element by the removal of surface-adherent carbonaceous contaminants as discussed in more detail below. It can finally be stated that even though cerium is known to be a precursor for laser damage [12], its impact on the observed increase in LIDT can be neglected in the present case. In contrast, the correlation of the LIDT and the near-surface carbon content is obvious; as shown in Figure 7, the LIDT increases with decreasing carbon content.

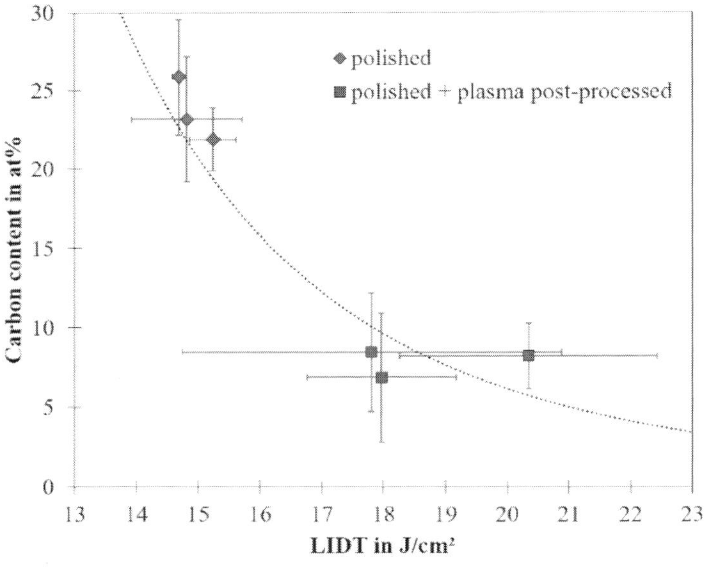

Figure 7: Correlation of the carbon content and the laser-induced damage threshold (LIDT) of polished zinc crown glass samples with and without plasma post-processing including general trend (dotted line).

The impact of the presence of carbon on surface absorption and the resulting LIDT was already discussed in Section 4.1. The removal of carbon by the plasma thus allows reducing the effects and mechanisms mentioned there. Plasmas are well known to be suitable for the removal of surface-adherent hydrocarbons (C_xH_y, where x<y) or other carbonaceous compounds. For instance, decomposition of hydrocarbons can be initiated by atomic oxygen (O) which is provided via dissociation within the plasma volume by collisions of oxygen molecules (O_2) with free electrons. According to

$$2CH_2(s) + 6O(p) \rightarrow 2CO_2(g) + 2H_2O(g) \tag{2}$$

the resulting volatile by-products are then gaseous carbon dioxide (CO_2) and water (H_2O). This process is very efficient when working with "cold" dielectric barrier discharges [52] and when adding marginal fractions of oxygen (i.e., approximately 1 to 2%) to argon as carrier gas [30,31]. In the present case, the plasma is driven at atmospheric pressure without any housing. Oxygen from the surrounding air is thus taken in by the plasma, subsequently mixed with the used argon and finally subject to the electric discharge within the discharge gap between the high voltage electrode and the sample surface. Apart from such pure chemical plasma-induced reaction of oxygen species with contaminants, further effects could contribute to the removal of carbon and the cleaning process, respectively. *Swain* and co-workers have reported on the increase in LIDT of crown glass due to laser cleaning at low fluence below the ablation threshold. The authors attributed this effect to a removal of absorbers from the glass surface by irradiation-induced desorption [53]. This mechanism could also play a role in the present case. Even though the plasma fluence is quite low, i.e., $F_{plasma} = 0.54$ J/cm^2 [39,40], the plasma used in this work is ignited directly on the glass surface and thus provides a considerable amount of ultraviolet irradiation which might induce desorption. Another possible underlying mechanism, the emission of electrons or ions from the surface and the resulting alteration of electrical forces within the surface which bind the contaminants, was reported by *Logothetis* and *Hartman* [54]. This effect finally leads to a removal of contaminants. As a result of direct plasma ignition on the sample surface, the impact of charged particles on the surface is very high and might support the removal of carbon as observed in this work.

Plasma-induced surface cleaning by the removal of carbon can also be expressed in the form of surface coverage as shown in Figure 8. Here, the surface coverage as given by the ratio of carbon and the glass component silicon (C/Si) is shown. Silicon was chosen for calculating this ratio since silicon dioxide is the main compound of the investigated glass, representing 60–70 wt % of the entire glass weight [37].

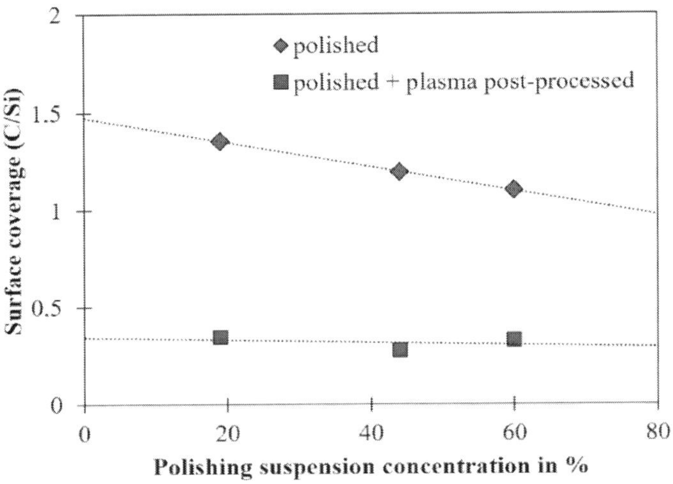

Figure 8: Comparison of the surface coverage by carbon (i.e., the ratio of carbon and silicon) of zinc crown glass samples before and after plasma post-processing vs. polishing suspension concentration including trend (dotted line).

It turns out that a high surface coverage is found for untreated samples where the amount of carbon is even higher than the amount of silicon since the surface coverage value exceeds one in any case. It can further be seen that the degree of surface coverage decreases with increasing polishing suspension concentration (compare also Figure 4). After plasma post-processing, the surface coverage approaches a saturation value of approximately 0.3, corresponding to a total content of carbon of about 7.82% on average. This value does not depend on the initial carbon content before plasma treatment and is thus independent on the polishing suspension concentration. It can consequently be stated that surface-adherent carbon was efficiently removed in all cases in the course of plasma post-processing. However, there is a residual constant carbon content which was not removed by the plasma. It can be assumed that this remaining carbon is given by contaminants from the used tools and working materials which have accumulated within digs and micro cracks underneath and/or within the hydrated silica gel layer. The plasma is thus capable of removing surface-adherent contamination by carbon such as wear debris from the polyurethane polishing pad or residues from cleaning agents but does not affect deeper regions of the polished glass surface.

Even though the final amount of remaining carbon after plasma post-processing does not depend on the initial carbon content, the LIDT shows a certain dependency on this value and the polishing suspension concentration, respectively, where the LIDT increases with increasing polishing suspension concentration. This indicates that the observed increase in LIDT does not exclusively follow from plasma-induced removal of surface-adherent carbon, but further subsurface effects and mechanisms are involved additionally. Such possible mechanisms are constructive interference of incident laser light, an increased absorption and a reduced mechanical strength of the glass due to the presence of micro cracks [3] where absorbing and scattering residues from working materials can moreover accumulate [4]. It is known that the crack density increases with increasing polishing suspension concentration [55]. It can thus be assumed that the crack depth features the same behaviour since, in the case of increasing polishing suspension concentration, the high number of polishing grains between the polishing pad and the glass surface leads to a reduction in local pressure and load. Consequently, surface stress as expressed by the stress intensity factor K is reduced as shown by *Swain* and co-workers [56]. Since this factor is inversely proportional to the crack depth [57], a higher crack depth and crack volume arise in this case, so the available volume for the accumulation of laser light-absorbing contaminants is increased. Based on this theoretical approach, the LIDT should finally feature a systematic dependency on the polishing suspension concentration. A trend towards this behaviour was observed in the present work, see Figure 4. One has to consider that this dependency was not only observed for polished and plasma post-processed samples, but also for polished ones, see Figure 6. This fact supports the assumption that the polishing suspension concentration and the resulting crack density may play an additional and essential role for the observed behaviour of the LIDT.

CONCLUSIONS

The results presented in this work clearly show that the laser-induced damage threshold of the investigated glass is directly related to surface contamination by carbon. The highest LIDT was found for the lowest contamination which occurred in the case of the highest polishing suspension concentration. In contrast, surface contamination by the actual polishing agent, cerium, does not depend on the polishing suspension concentration. The results thus suggest that classical polishing of laser optics surfaces with high polishing suspension concentration allows increasing the LIDT. However, this assumption has to be verified in the frame of further ongoing investigations due to the extraordinarily high number of possible configurations during manufacturing in terms of the glass type, the used abrasives (especially carbon-based ones such as silicon carbide or diamond), the pH-value of the polishing suspension, the used tools and polishing pads, etc.

The increase in carbon contamination with decreasing polishing suspension concentration is most likely due to a higher wear between the polishing pad made of carbonaceous polyurethane and the glass surface, due to a reduced number of polishing grains within the suspension. Thus, the effective direct contact area of the pad and the glass is bigger, consequently leading to increased formation of carbonaceous wear debris from the polishing pad according to the Preston equation [6]. Such debris can adhere on the surface and also accumulate in the formed silica layer or even penetrate into the glass bulk material by diffusion. In this context, the pH-value of the polishing suspension gains importance [58]. This value moreover influences the magnitude of near-surface forces during polishing [59] and the accompanying wear and formation of wear debris. As mentioned above, this point is thus to be investigated in future work.

Apart from an increased carbon contamination, another possible mechanism for the decrease in LIDT with decreasing polishing suspension concentration could be material densification due to local load induced by the polishing tool. Such densification leads to an increase in index of refraction [60] and absorption, respectively. Actually, such an increase in index of refraction with decreasing polishing suspension concentration was qualitatively observed and verified by ellipsometric measurements.

The significant increase in LIDT after plasma post-processing is obviously directly related to the removal of carbon from the glass surfaces. The effect of plasma-induced cleaning of glass by the removal of carbonic contaminations surfaces is well-known and used in quite different fields of application [30]. However, the use of plasmas for increasing the LIDT of optics is a comparatively new area of research (compare Section 1). Plasma post-processing as presented in this work stands out due to the operation at atmospheric pressure and comparatively low temperature, the use of the standard process gas argon and the short treatment duration of 60 s. This approach can be integrated into existing optics production chains quite easily and thus features high potential for industrial processes for surface finishing of high-power laser optics.

AUTHOR CONTRIBUTIONS

C.G. conceived and designed the experiments, analysed the data and wrote the paper; M.S. prepared the samples, performed the experiments and analysed the data.

ACKNOWLEDGMENTS

The authors thank the German Research Foundation (Deutsche Forschungsgemeinschaft, DFG) and the Ministry for Science and Culture of Lower Saxony (Ministerium für Wissenschaft und Kultur, MWK) for funding the laser source which was used for LIDT measurements. Further, the authors thank Lutz Müller, Robert Köhler, and Daniel Tasche from the University of Applied Sciences and Arts in Göttingen for their help during the preparation and analysis of the investigated samples. In addition, the support by the German Research Foundation (Deutsche Forschungsgemeinschaft, DFG) and the Technical University of Applied Sciences Wildau for covering open access publication costs is gratefully acknowledged.

REFERENCES

1. Suratwala, T.I.; Miller, P.E.; Bude, J.D.; Steele, W.A.; Shen, N.; Monticelli, M.V.; Feit, M.D.; Laurence, T.A.; Norton, M.A.; Carr, C.W.; et al. HF-based etching processes for improving laser damage resistance of fused silica optical surfaces. *J. Am. Ceram. Soc.* 2011, *94*, 416–428.
2. Bloembergen, N. Role of cracks, pores, and absorbing inclusions on laser induced damage threshold at surfaces of transparent dielectrics. *Appl. Opt.* 1973, *12*, 661–664.
3. Feit, M.D.; Rubenchik, A.M. Influence of subsurface cracks on laser-induced surface damage. *Proc. SPIE*2004, *5273*, 264–272.
4. Pfiffer, M.; Longuet, J.-L.; Labrugère, C.; Fargin, E.; Bousquet, B.; Dussauze, M.; Lambert, S.; Cormont, P.; Néauport, J. Characterization of the polishing-induced contamination of fused silica optics. *J. Am. Ceram. Soc.* 2017, *100*, 96–107.
5. Ye, X.; Huang, J.; Liu, H.; Geng, F.; Sun, L.; Jiang, X.; Wu, W.; Qiao, L.; Zu, X.; Zheng, W. Advanced mitigation process (AMP) for improving laser damage threshold of fused silica optics. *Sci. Rep.* 2016, *6*, 31111.
6. Cook, L.M. Chemical processes in glass polishing. *J. Non-Cryst. Solids* 1990, *120*, 152–171.
7. Liao, D.; Chen, X.; Tang, C.; Xie, R.; Zhang, Z. Characteristics of hydrolyzed layer and contamination on fused silica induced during polishing. *Ceram. Int.* 2014, *40*, 4479–4483.
8. Gerhard, C.; Tasche, D.; Uteza, O.; Hermann, J. Investigation of nonuniform surface properties of classically-manufactured fused silica windows. *Appl. Opt.* 2017, *56*, 7427–7434.
9. Camp, D.W.; Kozlowski, M.R.; Sheehan, L.M.; Nichols, M.A.; Dovik, M.; Raether, R.G.; Thomas, I.M. Subsurface damage and polishing compound affect the 355-nm laser damage threshold of fused silica surfaces. *Proc. SPIE* 1998, *3244*, 356–364.
10. Bertussi, B.; Natoli, J.Y.; Commandre, M. Effect of polishing process on silica surface laser-induced damage threshold at 355 nm. *Opt. Commun.* 2004, *242*, 227–231.
11. Krol, H.; Gallais, L.; Grèzes-Besset, C.; Natoli, J.-Y.; Commandré, M. Investigation of nanoprecursors threshold distribution in laser-damage testing. *Opt. Commun.* 2005, *256*, 184–189.

12. Neauport, J.; Lamaignere, L.; Bercegol, H.; Pilon, F.; Birolleau, J.-C. Polishing-induced contamination of fused silica optics and laser-induced damage density at 351 nm. *Opt. Express* 2005, *13*, 10163–10171.

13. Kozlowski, M.R.; Carr, J.; Hutcheon, I.D.; Torres, R.A.; Sheehan, L.M.; Camp, D.W.; Yan, M. Depth profiling of polishing-induced contamination on fused silica surfaces. *Proc. SPIE* 1998, *3244*, 365–375.

14. Bude, J.; Miller, P.; Baxamusa, S.; Shen, N.; Laurence, T.; Steele, W.; Suratwala, T.; Wong, L.; Carr, W.; Cross, D.; et al. High fluence laser damage precursors and their mitigation in fused silica. *Opt. Express* 2014, *22*, 5839–5851.

15. Krol, H.; Gallais, L.; Commandré, M.; Grèzes-Besset, C.; Torricini, D.; Lagier, G. LIDT improvement of multilayer coatings by accurate analysis of fabrication steps. *Proc. SPIE* 2005, *5963*, 596311.

16. Gerhard, C.; Tasche, D.; Munser, N.; Dyck, H. Increase in nanosecond laser-induced damage threshold of sapphire windows by means of direct dielectric barrier discharge plasma treatment. *Opt. Lett.* 2017, *42*, 49–52.

17. Bien-Aimé, K.; Néauport, J.; Tovena-Pecault, I.; Fargin, E.; Labrugère, C.; Belin, C.; Couzi, M. Laser induced damage of fused silica polished optics due to a droplet forming organic contaminant. *Appl. Opt.* 2009, *48*, 2228–2235.

18. Gerhard, C.; Weihs, T.; Tasche, D.; Brückner, S.; Wieneke, S.; Viöl, W. Atmospheric pressure plasma treatment of fused silica, related surface and near-surface effects and applications. *Plasma Chem. Plasma Process.* 2013, *33*, 895–905.

19. Langmuir, I. The adsorption of gases on plane surfaces of glass, mica and platinum. *J. Am. Chem. Soc.*1918, *40*, 1361–1403.

20. Giuliano, C.R. Laser-induced damage in transparent dielectrics: Ion beam polishing as a means of increasing surface damage thresholds. *Appl. Phys. Lett.* 1972, *21*, 39–41.

21. Cheng, X.; Miao, X.; Wang, H.; Qin, L.; Ye, Y.; He, Q.; Ma, Z.; Zhao, L.; He, S. Surface contaminant control technologies to improve laser damage resistance of optics. *Adv. Cond. Matter Phys.* 2014, *2014*, 974245.

22. Temple, P.A.; Lowdermilk, W.H.; Milam, D. Carbon dioxide laser

polishing of fused silica surfaces for increased laser-damage resistance at 1064 nm. *Appl. Opt.* 1982, *21*, 3249–3255.

23. Xu, S.; Zheng, W.; Yuan, X.; Lv, H.; Zu, X. Recovery of fused silica surface damage resistance by ion beam etching. *Nucl. Instr. Meth. Phys. Res. B* 2008, *266*, 3370–3374.
24. Kamimura, T.; Akamatsu, S.; Horibe, H.; Shiba, H.; Motokoshi, S.; Sakamoto, T.; Jitsuno, T.; Okamoto, T.; Yoshida, K. Enhancement of subsurface-damage resistance by removing subsurface damage in fused silica and its dependence on wavelength. *Jpn. J. Appl. Phys.* 2004, *43*, L1229–L1231.
25. Kamimura, T.; Akamatsu, S.; Yamamoto, M.; Yamato, I.; Shiba, H.; Motokoshi, S.; Sakamoto, T.; Okamoto, T.; Yoshida, K. Enhancement of surface-damage resistance by removing a subsurface damage in fused silica. In Proceedings of the SPIE Conference 35th Annual Symposium on Optical Materials For High Power Lasers: Boulder Damage Symposium, Boulder, CO, USA, 22–24 September 2003; Volume 5237, pp. 224–250.
26. Liu, H.; Ye, X.; Zhou, X.; Huang, J.; Wang, F.; Zhou, X.; Wu, W.; Jiang, X.; Sui, Z.; Zheng, W. Subsurface defects characterization and laser damage performance of fused silica optics during HF-etched process. *Opt. Mater.* 2014, *36*, 855–860.
27. Juškevičius, K.; Buzelis, R.; Abromavičius, G.; Samuilovas, R.; Abbas, S.; Belosludtsev, A.; Drazdys, R.; Kičas, S. Argon plasma etching of fused silica substrates for manufacturing high laser damage resistance optical interference coatings. *Opt. Mater. Express* 2017, *7*, 3598–3607.
28. Zhang, D.; Shao, J.; Zhang, D.; Fan, S.; Tan, T.; Fan, Z. Employing oxygen-plasma posttreatment to improve the laser-induced damage threshold of ZrO_2 films prepared by the electron-beam evaporation method. *Opt. Lett.* 2004, *29*, 2870–2872.
29. Zhang, D.; Wang, C.; Fan, P.; Cai, X.; Zheng, Z.; Shao, J.; Fan, Z. Influence of plasma treatment on laser-induced damage characters of HfO_2 thin films at 355 nm. *Opt. Express* 2009, *17*, 8246–8252.
30. Shun'ko, E.V.; Belkin, V.S. Cleaning properties of atomic oxygen excited to metastable state $2s^2 2p^4 (^1S_0)$. *J. Appl. Phys.* 2007, *102*, 083304.
31. Iwasaki, M.; Inui, H.; Matsudaira, Y.; Kano, H.; Yoshida, N.; Ito, M.; Hori, M. Nonequilibrium atmospheric pressure plasma with ultrahigh

electron density and high performance for glass surface cleaning. *Appl. Phys. Lett.* 2008, *92*, 081503.

32. Buček, A.; Homola, T.; Aranyosiová, M.; Velič, D.; Plecenik, T.; Havel, J.; Stahel, P.; Zahoranová, A. Atmospheric pressure nonequilibrium plasma treatment of glass surface. *Chem. Listy* 2008, *102*, S1459–S1462.
33. Gerhard, C. *Optics Manufacturing: Components and Systems*, 1st ed.; CRC Taylor & Francis: Boca Raton, FL, USA, 2017; ISBN 978-1-4987-6459-9.
34. Li, Y.; Hou, J.; Xu, Q.; Wang, J.; Yang, W.; Guo, Y. The characteristics of optics polished with a polyurethane pad. *Opt. Express* 2008, *16*, 10285–10293.
35. Liu, J.M. Simple technique for measurements of pulsed Gaussian-beam spot sizes. *Opt. Lett.* 1982, *7*, 196–198.
36. Sanner, N.; Utéza, O.; Bussiere, B.; Coustillier, G.; Leray, A.; Itina, T.; Sentis, M. Measurement of femtosecond laser-induced damage and ablation thresholds in dielectrics. *Appl. Phys. A* 2009, *94*, 889–897.
37. Schott, A.G. *Technical Safety Information N-ZK7*, version 5; Schott AG: Mainz, Germany, 2015; pp. 1–7.
38. Brückner, S.; Rösner, S.; Gerhard, C.; Wieneke, S.; Viöl, W. Plasma-based ionisation spectroscopy for material analysis. *Mater. Test.* 2011, *53*, 639–642.
39. Gerhard, C.; Roux, S.; Brückner, S.; Wieneke, S.; Viöl, W. Low-temperature atmospheric pressure argon plasma treatment and hybrid laser-plasma ablation of barite crown and heavy flint glass. *Appl. Opt.* 2012, *51*, 3847–3852.
40. Gerhard, C.; Weihs, T.; Luca, A.; Wieneke, S.; Viöl, W. Polishing of optical media by dielectric barrier discharge inert gas plasma at atmospheric pressure. *J. Eur. Opt. Soc. Rapid Publ.* 2013, *8*, 13081.
41. Gredner, A.; Gerhard, C.; Wieneke, S.; Schmidt, K.; Viöl, W. Increase in generation of poly-crystalline silicon by atmospheric pressure plasma-assisted excimer laser annealing. *J. Mater. Sci. Eng. B* 2013, *3*, 346–351.
42. Perry, M.D.; Stuart, B.C.; Banks, P.S.; Feit, M.D.; Yanovsky, V.; Rubenchik, A.M. Ultrashort-pulse laser machining of dielectric materials. *J. Appl. Phys.* 1999, *85*, 6803–6810.
43. Yang, L.; Yuan, X.D.; Deng, H.X.; Xiang, X.; Zheng, W.G.; He, S.B.;

Jiang, Y.; Lv, H.B.; Ye, L.; Wang, H.J.; et al. Influence of ambient temperature on nanosecond and picosecond laser-induced bulk damage of fused silica. *Adv. Cond. Matter Phys.* 2014, *2014*, 308918.

44. Uteza, O.; Bussière, B.; Canova, F.; Chambaret, J.-P.; Delaporte, P.; Itina, T.; Sentis, M. Laser-induced damage threshold of sapphire in nanosecond, picosecond and femtosecond regimes. *Appl. Surf. Sci.* 2007, *254*, 799–803.
45. Gallais, L.; Mangote, B.; Zerrad, M.; Commandré, M.; Melninkaitis, A.; Mirauskas, J.; Jeskevic, M.; Sirutkaitis, V. Laser-induced damage of hafnia coatings as a function of pulse duration in the femtosecond to nanosecond range. *Appl. Opt.* 2011, *50*, C178–C187.
46. Kong, F.; Jin, Y.; Huang, H.; Zhang, H.; Liu, S.; He, H. Laser-induced damage of multilayer dielectric gratings with picosecond laser pulses under vacuum and air. *Opt. Laser Technol.* 2015, *73*, 39–43.
47. Rogov, V.V.; Rublev, N.D.; Krotenko, T.L.; Troyan, A.V. A study of intensity of tribochemical contact interaction between a polishing compound and sapphire in machining. *J. Superhard Mater.* 2008, *30*, 273–275.
48. Larruquert, J.I.; Pérez-Marín, A.P.; García-Cortés, S.; de Marcos, L.R.; Aznárez, J.A.; Méndez, J.A. Self-consistent optical constants of SiC thin films. *J. Opt. Soc. Am. A* 2011, *28*, 2340–2345.
49. Otanicar, T.P.; Phelan, P.E.; Golden, J.S. Optical properties of liquids for direct absorption solar thermal energy systems. *Sol. Energy* 2009, *83*, 969–977.
50. Wang, K.; Wen, W.; Wang, Y.; Wang, K.; He, J.; Wang, J.; Zhai, P.; Yang, Y.; Qiu, P. Order-of-magnitude multiphoton signal enhancement based on characterization of absorption spectra of immersion oils at the 1700-nm window. *Opt. Express* 2017, *25*, 5909–5916.
51. Zawilski, K.T.; Setzler, S.D.; Schunemann, P.G.; Pollak, T.M. Increasing the laser-induced damage threshold of single-crystal $ZnGeP_2$. *J. Opt. Soc. Am. B* 2006, *23*, 2310–2316.
52. Baravian, G.; Chaleix, D.; Choquet, P.; Nauche, P.L.; Puech, V.; Rozoy, M. Oil removal from iron surfaces by atmospheric-pressure barrier discharges. *Surf. Coat. Technol.* 1999, *115*, 66–69.
53. Swain, J.E.; Lowdermilk, W.H.; Milam, D. Raising the surface damage threshold of neutral solution processed BK-7 by pulsed laser irradiation. *Nat. Bur. Stand. Spec. Publ.* 1982, *669*, 292–295.

54. Logothetis, E.M.; Hartman, P.L. Laser-induced electron emission from solids: Many-photon photoelectric effects and thermionic emission. *Phys. Rev.* 1969, *187*, 460–474.

55. Schubert, I. *Wissensspeicher Feinoptik*, 17th ed.; I. Schubert: Apolda, Germany, 2017; pp. 78–90. ISBN 978-3-00-046972-5. (In German)

56. Swain, M.V.; Metras, J.C.; Guillemet, C.G. A deformation and fracture mechanics approach to the scoring and breaking of glass. *J. Non-Cryst. Solids* 1980, *38–39*, 445–450.

57. Smirnov, M.I.; Spiridonov, Y.A.; Karapetyan, A.R. Modern sheet-glass cutting technologies. *Glass. Ceram.*2011, *68*, 6–10.

58. Evans, C.J.; Paul, E.; Dornfeld, D.; Lucca, D.A.; Byrne, G.; Tricard, M.; Klocke, F.; Dambon, O.; Mullany, B.A. Material removal mechanisms in lapping and polishing. *CIRP Ann.-Manuf. Technol.* 2003, *52*, 611–633.

59. Cumbo, M.J.; Jacobs, S.D. Determination of near-surface forces in optical polishing using atomic force microscopy. *Nanotechnology* 1994, *5*, 70–79.

60. Sakata, H. Etude ellipsométrique du mécanisme fondamental de polissage du verre. *Jpn. J. Appl. Phys.*1973, *12*, 173–181.

INDEX

D

E

F

G

H

U

V

W

X

Y